CISM COURSES AND LECTURES

The series presents lecture notes, monographs, edited works and proceedings in the field of Mechanics, Engineering, Computer Science and Applied Mathematics.
Purpose of the series is to make known in the international scientific and technical community results obtained in some of the activities organized by CISM, the International Centre for Mechanical Sciences.

INTERNATIONAL CENTRE FOR MECHANICAL SCIENCES

COURSES AND LECTURES - No. 443

ALUMINIUM STRUCTURAL DESIGN

EDITED BY

FEDERICO M. MAZZOLANI
UNIVERSITÀ DI NAPOLI "FEDERICO II"

Springer-Verlag Wien GmbH

This volume contains 349 illustrations

Originally published by Springer-Verlag Wien New York in 2003
SPIN 10911853

In order to make this volume available as economically and as rapidly as possible the authors' typescripts have been reproduced in their original forms. This method unfortunately has its typographical limitations but it is hoped that they in no way distract the reader.

ISBN 978-3-211-00456-2 ISBN 978-3-7091-2794-0 (eBook)
DOI 10.1007/978-3-7091-2794-0

PREFACE

In the last decades aluminium alloys have been successfully employed in many structural applications. Without counting the aeronautical industry, where there is virtually no metallic material that can replace aluminium, an extensive development in its use is in progress in the transportation industry; in particular in the rail industry (subway coaches, sleeping cars,...), in the automotive industry (motorcars, containers,....) and also in the shipping industry (civil and military hydrofoils, motorboats, sailboats,....).

A parallel trend has been observed more recently in civil engineering structures, where aluminium alloys compete with steel (long-span roofing, bridges, hydraulic structures, off-shore superstructures).

Aluminium alloys represent a large family of materials which covers quite the same range of strength offered by the most commonly used mild steels. To be economical, aluminium structures must be designed to take full advantage of the material properties.

Their corrosion resistance normally makes it unnecessary to protect aluminium structures. The advantage of the lightness of the material cannot always be fully exploited, due to its high deformability, which makes members susceptible to instability. Aluminium is not prone to brittle fracture, but particular attention is required when ductility is a prerequisite.

Thanks to the extrusion fabrication process, the geometrical properties of the cross-section can be enhanced by designing a shape that provides minimum weight together with the highest structural efficiency and functional flexibility.

In particular the lightness of aluminium simplifies erection, allows the transportation of fully prefabricated components, reduces the loads transmitted to foundations, economizes on energy both during production and in operation, and requires less physical labour.

The corrosion resistance of aluminium makes it possible to reduce maintenance costs, and still achieve a superior performance in corrosive environments.

This volume collects the lectures of out-standing international experts, who are all involved in the codification activity of Eurocode 9 on "Aluminium Structural Design". It covers the contents of the Advanced Professional Training course held at CISM-Udine in October 2001 on the design of aluminium structures.

The course aimed to illustrate the basic principles, with particular reference to the fields of transportation and civil engineering – from the properties of the materials and the technological aspects of their application, to the evaluation of the resistance of the structural elements (members and plates) under static, dynamic and fatigue loading conditions.

Particular emphasis has been given to the recent Eurocode 9 (ENV-CEN 1999) as an advanced design tool. Now, the conversion phase of Eurocode 9 already started and it will lead to the EN version within 2004. This incoming activity will take a great profit from the issue of this volume, which can be considered as a background document for the final version of Eurocode 9.

Finally, I would like to express my personal gratitude to Simeone Panico for his precious contribution in the editorial phase.

Federico M. Mazzolani

June 2002

CONTENTS

Chapter 1

DESIGN CRITERIA FOR ALUMINIUM STRUCTURES: TECHNOLOGY, CODIFICATION AND APPLICATIONS

F.M. Mazzolani

Department of Structural Analysis and Design, University of Naples "Federico II", Italy

1. HISTORICAL BACKGROUND AND RANGE OF APPLICATION

1.1. Introduction

The aluminium and its alloys, from a structural engineering point of view, can be considered as a 'new' material (Mazzolani, 1998a). In fact, the first building structures made of aluminium alloy appeared in Europe in the early fifties, when concrete, masonry and steel were the main materials used for civil engineering structures. The aeronautical industry was one of the first fields which benefited of the aluminium alloys physical properties, so much used because there are not other metallic materials which can compete with it. Aluminum alloys are widely and successfully used to build airships, from the early Schwartz and Zeppelin until the modern aircrafts. Nowadays the aluminium alloys are also used in other branches of transportation, such as the rail industry (subway coaches, sleeping cars), the automotive industry (part of cars, containers for trucks, moving cranes) and the shipping industry (civil and military hydrofoils, motorboats). The offshore industry is the largest user of aluminium alloys for structural applications: the helicopter decks and the living quarters of the platforms are an example of structures where aluminium has replaced steel, which in general is the most widely used material for the structural applications. Considering that the first commercial applications of aluminium at the beginning of 20th Century were objects such as mirror frames, house numbers and servings trays, without counting the cooking utensils, which were a major early market, we can recognize the entering into structural engineering for aluminium alloys was not a very simple and easy process. After many difficulties, today it can be recognized that aluminium and its alloys can be considered as a new and competitive material in civil and structural engineering applications (Mazzolani, 1995b, 1999).

1.2. History of the Aluminium

Almost two hundred years are passed since 1807, when an English chemist, Sir Humphry Davy of the Royal Institute of London, foresaw the possibility of isolating the aluminium element from alumina salts, discovered at the end of the eighteenth century by Guyton De Morveau during his studies on the ancient 'allume', which derived from Latin 'allumen', word used in the Egyptian civilization in the sixteenth century B.C. to indicate a material of dubious composition. Twenty years after the Davy's study some traces of the new metal were later isolated by Oersted, while testing alkaline metals discovered by Davy. The first aluminium nugget was obtained with difficulty by the German chemist Whoeler in 1827. Owing to the interest induced by the lightness and ductility of the new material, efforts were therefore made to produce the metal industrially. In 1854 Henry Sainte-Claire developed a process of electrolytic reduction of aluminium from the double chloride of sodium and aluminium. The first aluminium produced was used to coin a medal in homage to Whoeler and to fabricate aluminium eagles to be fixed to the flag lances of the Napoleon III's imperial regiments. In 1866, Paul Louis Touissant Héroult (1863-1914) in France, and Charles Martin Hall (1863-1914) in Ohio, without being in contact, developed the electrolytic reduction process of alumina (Al_2O_3), which all industrial production of aluminium is actually based upon. Héroult and Hall can be considered to be a very impressive case of coincidence of identity: they were born in the same year, studied the same subject, achieved equivalent results, promoted equivalent technological innovations, and died in the same year at the beginning of the First World War, during which aluminium alloys were employed on a large scale for the first time (Mazzolani, 1995a).

1.3. Fields of interest in Civil and Structural Engineering

The physical and mechanical properties of aluminium alloys are the basic reasons of its success as constructional material, able them also to compete with steel in the structural field (Mazzolani 1995a, b, 1998a). The following conclusions can be made:

a. Aluminium alloys represent a wide family of constructional materials, whose mechanical properties cover the range offered by the common mild steels.
b. Corrosion resistance normally makes it unnecessary to protect aluminium structures, even in aggressive environments.
c. The lightness of the material gives advantages in weight reduction, but it is partially offset by the necessity to reduce deformability, which gives a high susceptibility to buckling phenomena.
d. The material itself is not prone to brittle fracture, but particular attention should be paid to those problems in which high ductility is required.
e. The extrusion fabrication process allows to produce individually tailored shapes, according to the requirements which are designed for (Figure 1.1).
f. Large choice for the connection solutions, because the modern technology gives the possibility to use either bolting, riveting and welding, without any difficulties involved.

Figure 1.1: Shapes of extruded profiles

After these preliminary remarks, it is possible to state that aluminium alloys can be economical, and therefore competitive, in those applications in which full advantage is taken of their above prerequisites. In particular, the following three properties must be exploited:

A. **Lightness** makes it possible to:
   - simplify the erection phases;
   - transport fully prefabricated components;
   - reduce the loads transmitted to foundations;
   - economize energy either during erection and/or in service;
   - reduce the physical labour.

B. **Corrosion resistance** makes it possible to:
   - reduce the maintenance expenses;
   - provide good performance in corrosive environments.

C. **Functionally of structural shapes**, due to the extrusion process, makes it possible to:
   - improve the geometrical properties of the cross-section by designing a shape which simultaneously gives the minimum weight and the highest structural efficiency;
   - obtain stiffened shapes without using built-up sections, thus avoiding welding or bolting;
   - simplify connecting system among different components, thus improving joint details;
   - combine different functions of the structural component, thus achieving a more economical and rational profile.

The best fit from the application side can be obtained in some typical cases, which are characterized in getting profit at least of one of the above properties: lightness, corrosion resistance and functionally.

In civil engineering the main cases of structural applications belong to the following groups:

1. Long-span roof system in which live loads are small compared with dead loads. They include reticular schemes of plane and space structures (Figure 1.2).
2. Structures located in inaccessible places far from the fabrication shop, so the transport economy and ease of erection are of extreme importance (Figure 1.3). It is the case of

prefabricated elements such as electrical transmission towers, stair cases, provisional bridges, which can be carried by helicopter or crane completely assembled.

3. Structures situated in corrosive or humid environments. They cover many types, such as swimming pool roofs, river bridges, hydraulic structures and offshore superstructures (Figure 1.4).
4. Structures having moving parts, so that lightness means economy of power under service. They are mainly moving bridges (Figure 1.5a), both for pedestrian and motorcars, as well as the rotating cranebridges on circular pools in the sewage plants (Figure 1.5b).
5. Structures for special purposes, for which maintenance operations are particularly difficult and must be limited, as in case of masts, lighting towers, sign motorway portals, and so on (Figures 6a, b).

Figure 1.2: The space structure in Rio de Janeiro, Brazil

Figure 1.3: A stair-case is erected by crane

More general applications structural applications of the aluminium alloys in structural engineering can be collected in the list of Table 1.1, where they are classified with regard to the requirements of corrosion resistance (C), lightness (L) and functionality of extruded profiles (F), which are determinant in the choice of these material, acting separately or together. The listed applications cover also the above mentioned civil engineering applications (some of them are identified by a star *).

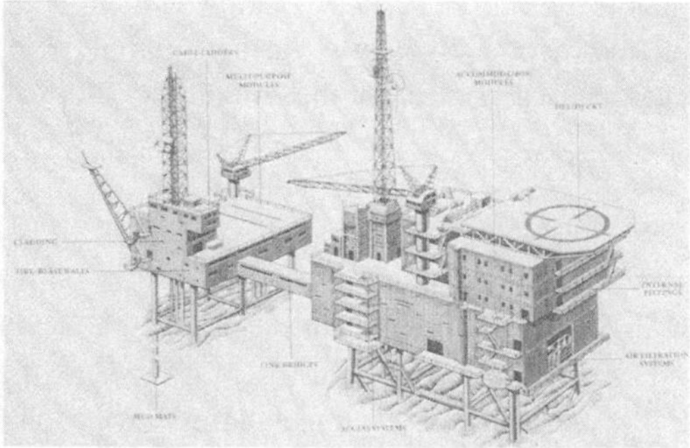

Figure 1.4: Offshore superstructures

a.

b.

Figure 1.5: a. Aberdeen harbour, Scotland (1953); b. Sewage plant in Turin, Italy (1984)

a.

b.

Figure 1.6: The "Information Tower" (a, 1985) and the "ENEL Tower" (b, 1990), both in Naples, Italy

Even if all these kinds of application do not strictly belong to the civil engineering range in the classical sense, it can be noticed that the boundaries of the Building Industry are more and more becoming wider and less traditional; it is sure that the Aluminium Industry will take a good profit from this new scenario.

<table>
<tr>
<td rowspan="2">C
• Storage vessels
• Lamp columns
• Profiled roof and wall cladding
• Support for railway overhead electrification
• Enclosure structures for sewage works
• Sound barriers
• Vehicle restraint systems
• Sewage plant bridges*
• Silos*
• Traffic signal gantries*
• Traffic signal poles*</td>
<td>C+L
• Lighting control loovers
• Flag poles
• Aircraft access bridges
• Transmission towers
• Bridge inspection gantries
• Offshore structures (living quarters, bridges)*
• Tank flotation covers</td>
<td rowspan="2">L
• Crane booms
• Lorry mounted cranes
• Pit props
• Bridges*
• Mobile bridge inspection gantries
• Scaffolding systems
• Ladders
• Cherry pickers
• Telescopic platforms
• Masts for tents</td>
</tr>
<tr>
<td rowspan="2">C+F+L
• Grating planks
• Helidecks*</td>
</tr>
<tr>
<td rowspan="2">C+F
• Domes over sewage tanks*
• Marina landing stages
• Roof access staging
• Dam logs
• Curtain walling
• Overcladding support systems
• Pedestrian parapets
• Chicken house structures
• Wood drying kilns
• Space structures (domes, etc.)*
• Exhibition stands*
• Swimming pool roofs*
• Canopies
• Bus shelters
• Green houses/Glass houses*</td>
<td rowspan="2">F+L
• Access ramps
• Support for shuttering
• Trackways (temporary)
• Elevators for building materials
• Scaffold planks
• Trench supports
• Grave digging supports
• Loading ramps
• Landing mats for aircraft
• Access gangways
• Shuttering support beams
• Military bridges*
• Radio masts
• Shuttering
• Telescopic conveyor belt structures
• Grandstand structures (temporary)
• Building maintenance gantries
• Fabric structure frames</td>
</tr>
<tr>
<td>F
• Prefabricated balconies*
• Conveyor belt structures
• Monorails
• Robot support structures
• Shuttering form work
Tunnel shuttering</td>
</tr>
</table>

Table 1.1: The main structural applications of aluminium alloys in structural engineering

1.4. The main structural applications

In the field of large-span one story buildings, the portal frame scheme has been used since the early fifties for industrial buildings which particularly exploited the functionality of extruded shapes. Some applications of plane schemes with large span (50 to 70 m) are made in the field of aircraft hangars, warehouses, airport building, sport-halls (see Section 4.1.).
The field of space structures for roofing (Figure 1.7) gave rise to a wide number of systems, which particularly utilize the prerequisites of the material technology.

Figure 1.7: The Sport-hall of Quito, Equador

The widest applications in the field of large roofing using reticular space structures (see Section 4.2.) have been done in South America (Figure 1.8).
Many applications of geodetic domes (see Section 4.3.) have been erected in U.S.A. for multipurpose activities with diameters over 100 m (Figure 1.9).

Figure 1.8: The reticular space structure (60x60) Interamerican Center of San Paolo, Brazil

Figure 1.9: The "Spruce Goose" is the world's largest clear-span aluminium dome 415 feet in diameter

All the main typologies used in steel in the field of bridges have been also experienced in aluminium (see Section 5).
It is interesting to observe how the bridge deck can be made of special extruded parts, which are shaped as stiffened plates able to be transversally joined without fasteners, by means of a simple pressure.
A very convenient application is the one of footbridges, they have been successfully built with variable span from 20 to 60 m in many countries (France, Germany, Netherlands, Italy, Canada, Norway).
In case of motor-bridges, the American experience to use concrete-aluminium composite system has been transferred also in Europe (see Section 5.1), pointing out a series of new problems, which deserved a more wide investigation (Mazzolani & Mandara, 1997).
A range of great interest is the one of the moving bridges, due to the energy consuming reduction in service. They have been built both for motorcars and for pedestrians.
A very attractive activity has been developed in the field of suspended bridges. Starting from some first suspended footbridges, this scheme has been successfully revalued in some cases of refurbishment of old suspended bridges of the 19th Century, made of masonry piers and of a combination of iron plus timber as decks (Figure 1.10). They have been upgraded by using a new aluminium alloy deck. This allowed obtaining the following advantages:

- due to the lightness of the new deck, the service load can be increased without reinforcing the existing structures (piers and sometimes cables);
- due to the corrosion resistance, neither any painting protection has been used nor further maintenance is expected.

By means of this system, three bridges have been retrofitted in France (Mazzolani & Mele, 1997, see Section 5.5.). The aluminium deck has been also used in the structural restoration of the "Real Ferdinando" bridge on the Garigliano river, Italy (Mazzolani, 1998c, 2001).
Many towers for electrical transmission lines have been erected in Europe. In particular, the delta scheme with three cables allowed the aluminium solution to be more competitive than steel, due to the special shape of the extruded profiles (L, T, Y), stiffened by lips (see Section 4.4.).

Figure 1.10: The Groslée bridge with 175 m span; the deck is a composite aluminium concrete structure

Lighting towers are frequently used in the urban habitats, specially in France.
Two special towers for parabolic antennas have been recently erected in Naples (Mazzolani, 1989) (Figures 1.6a, b).
Many portal frames for traffic signs on motor-ways have been built in France, Italy, Switzerland and Netherlands.
It is quite obvious that these fields are particularly attractive due to the corrosion resistance.
As hydraulic applications (see Section 4.5.), mention can be made to:

- water pipelines;
- water reservoirs;
- sewage plants.

In the last field, the rotating crane bridges of the cylindrical pools are very often made in aluminium alloys. There are examples in Austria, Germany, Great Britain, Italy. The first application in Italy consisted in 8 bridges of the Turin sewage plant (Mazzolani, 1985) (Figure 1.5b).
Last but not least, the off-shore industry is now utilizing the aluminium alloy structures in many specific applications, which represent a challenge for the development of this new material (see Section 4.6.).
Some specific examples of application will be shown in the fields of buildings and bridges in the following Sections 4 and 5, respectively, where the reasons of the aluminium choice are clearly identified.

2. ALUMINIUM AND ITS ALLOYS AS STRUCTURAL MATERIAL

2.1. Fabrication processes: from mineral to structural elements

Large information on this subject can be found in Sharp (1993).
Pure aluminium oxide, Al_2O_3, is used for aluminium production. The aluminium oxide is obtained from the bauxite (mineral whose components are alumina, iron, oxide, silicon and water of hydration) by the Bayer process. The important part of the refining process then follows: the aluminium compounds are dissolved in NaOH solutions at elevated temperature; the most part of other components of the bauxite do not dissolve and can be removed. The metal is produced by electrolysis in a Hall-Heroult cell of the type illustrated in Figure 2.1. This cell is 9 to 12 m long, 3 to 4 m wide and 1 to a.2 m high, and is made of steel lined with carbon.

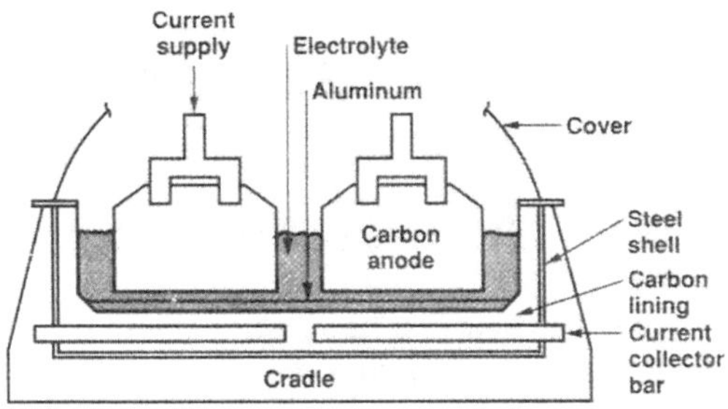

Figure 2.1: Hall-Heroult cell (Sharp, 1993)

The electrolyte is the solution of aluminium oxide in a molten salt solution composed primarily of a mixture of sodium fluoride and aluminium fluoride. The electrical current flows from the carbon anodes through the electrolyte to the carbon liner and the collection bars. The carbon anodes are consumed during the electrolysis. The aluminium collects at the bottom of the cell and is removed. The cells operate at a temperature between 930-980 °C and theoretically 1 kAh of electric current produces 0.336 kg of aluminium. As a consequence, for the production of this metal, considerable electric power must be spent. In fact by means of Hall-Heroult technology, about 25-30 kWh is necessary to produce one kilogram of aluminium. This is the reason why first-fusion aluminium is usually produced in those regions with plentiful cheap electricity. On the other hand, owing to the low melting point of aluminium (660 °C), the fabrication of aluminium alloy does not need a large amount of power.
Plates 5-6 mm thick are produced from slabs by the hot-rolling process. To obtain thinner plates the cold-rolling process is then necessary. This process induces work hardening proportional to the amount and the nature of alloying elements present in the product, and thus

annealing processes are required to increase the ductility. One of the biggest advantages of aluminium alloys is that they can be extruded by the press. This process allows to obtain elements with a complex cross section that is constant over its length. It consists of pushing the metal from an extrusion ingot, or billet, through a die shaped to provide the finished configuration. In particular the extrusion process can be summarised in the following stages (Figure 2.2):

• the billet heated up to the extrusion temperature is put in the press container, whose exit is blocked by the drawplate. The shape of the hole in this drawplate, which is usually made of high-strength tempered steel, gives the shape of the profile to be produced;
• a pressure variable between 250 e 1000 N/mm^2, applied by a pressing bar, pushes the material to pass through the drawplate;
• the profile is then expelled from the drawplate. The length of the extruded product depends on the extrusion ratio, which is the ratio between the cross section of the billet and the cross section of the extruded profile;
• after the extrusion process, a high tension force is applied to the profile on the straightening table.

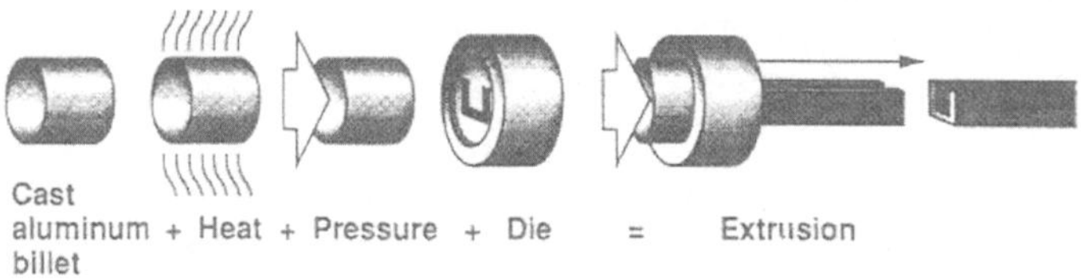

Figure 2.2: The extrusion process (Sharp, 1993)

The process of *straightening for tractioning*, characteristic of aluminium alloys, is applied by means of a straightening table (which can be up to 50 m long) that exerts a tension forces of about 200 t. Even if any cross section shape can be produced by extrusion, its transversal dimensions are limited by the capacity of the press. Nowadays presses of capacities from 1000 to 12500 tonnes are available in Europe. The maximum cross section dimensions can be inscribed either in a circle whose diameter is 500 mm or in a 600 x 200 mm rectangle. Some type of alloys, those so-called "strong" alloys, are less extrudable because of the higher pressure needed to overcome the greater resistance to plastic deformation. The cost of the final product varies with the size of the press and the type of alloy. The cross section of extrusions is constant along the length in most cases. Stepped extrusions can be made (the die needs to be changed during the extrusion), primarily for aircraft industry.
Another part of aluminium manufacturing technology is the *casting process*. This process provides a very attractive way to obtain three-dimensional configuration, like block and rigid shape. The molten metal is simply introduced into a cavity configured essentially as the final part. There are many casting process used nowadays:
• sand casting;
• shell casting;
• die casting.

Differences lie on the type of mold used as well as on the method by which the material is introduced into the mold. In die casting the metal is injected into the reusable dies (steel) at high velocity. Rapid solidification occurs, so the process is capable of high production rate. The components produced by this process are extremely accurate, showing an excellent surface finish. It is used when large productions are requested, so that the initial investment is justified. In shell casting the metal is poured into the cavity of the reusable molds, sometimes with a small vacuum to assist the flow. Generally, the sections are ticker than those in die casting and solidification is slower. In the sand casting process the mold is destroyed each time (cheap molds); solidification and cycle time are slower than for the other processes. However this process is quite economical when the quantity of parts needed is low and the casting is large and complex in geometry.

2.2. Property of additional elements

Aluminium alloys, which are usually called *light alloy*, have been developed in order to increase the strength of the base metal, aluminium, which is very ductile and corrosion resistant. Contrarily to the iron-carbon alloys (steel), aluminium alloys need only a small amount of thermic power for their production. The main elements used in the alloys are magnesium, silicon, zinc, copper and manganese. Nickel, cobalt, chromium, iron, titanium and zirconium are used as additional elements. The main properties of these elements are as follows (Mazzolani, 1995a):

- *Magnesium* reduces the melting point to 451 °C (for this reason is also used as weld metal) and increases work-hardening ability and corrosion resistance to salted water;
- *Silicon* increases the strength and ductility and reduces the melting point. In combination with the magnesium it allows the precipitation hardening;
- *Zinc* drastically increases strength and allows hot or cold precipitation hardening;
- *Copper* gives an even greater strength increase and allows cold precipitation hardening, but reduces corrosion resistance, weldability and ductility;
- *Nickel* can increase the resistance under high-temperature conditions;
- *Titanium* reduces the grain size;
- *Zirconium* is a stabiliser element, which influences the temperability;
- *Chromium* increases the resistance to stress corrosion;
- *Iron* is usually an impurity which can increase the strength of pure aluminium if a low percentage is used.

It seems important to observe that a significant increasing of the mechanical resistance caused by elements like Si, Zn and Cu is a consequence of hardening produced by the precipitation of theirs compounds (Mg_2Si, $CuAl_2$, etc.…), after the alloy has been put in solution.

2.3. Classification of alloy

Aluminium alloys can be classified in several ways.
With respect to the fabrication process, they can be classified as follows:

a) *Casting alloys*, which have to be remelted and then casted. They are delivered under the form of T-bars, which can be very easily remelted;
b) *Wrought alloys,* which have to be hot worked or cold worked without being remelted like happen in hot- and cold-rolling process such as extrusion, forging and drawing. They are delivered under form of slabs, billet or rods, according to the following working process.
The wrought aluminium alloys can be classified according to the treatment as follows:

- *Non-heat-treatable alloys*, also called work-hardenable alloys;
- *Heat-treatable alloys.*

Among the non-heat-treated alloys, there are those such as AlMn, AlMg and AlMgMn in which the strength is increased by cold working. Cold working can be obtained during the rolling process for plates, or via tractioning for structural shapes. After cold working, strength is increased and ductility is decreased. If the material is heated to 350 °C, the alloys return to the initial state. For this reason, the welding process decreases the strength in the so-called *heat-affected-zone.*
The heat-treating process provides strength and can be subdivided as follows:

- Heating up to 450 or 530 °C depending on the alloy;
- Tempering with water or air;
- Ageing at room temperature;
- Ageing at a temperature varying between 120° e 180°C (artificial ageing).

The effects of treating process disappear if the material is heated to between 200 and 350 °C. The heat input from welding process decreases the strength of the alloy to a value which is bounded by the properties of the material in the annealed stage and in the heat-treated stage. AlZnMg alloys recover most of their strength, because of the natural ageing process subsequent to welding.

2.4. Alloy designations

2.4.1. General

In addition to the above classifications according to the fabrication process (casting alloys and wrought alloys) and to the heat treatment (heat-treated alloys and non-heat-treated alloys), aluminium alloys can be also classified with respect to their chemical composition.
This classification allows aluminium alloys to be grouped in different families with similar mechanical and technological behaviour. Unfortunately, although the ISO has attempted to unify the nomenclature for aluminium alloys, there are still some differences, Country by Country. However, the mostly used ways are basically two:

- according to a numerical designation;
- according to an alphanumerical designation

The symbol, which gives the chemical composition, is usually followed by another symbol indicating the manufacturing process.

2.4.2. Numerical designation

This designation is derived from the American nomenclature (Aluminium Association). It is now also being used in Europe in addition to the national nomenclatures. Its alloy identification system employs different nomenclatures from wrought and cast alloys.

Wrought alloys (sheet and plate, extrusion and forging) have a four-digit number (see Tables 2.1 and 2.2); the first digit defines the major alloying element for both wrought and cast alloys. This element is usually 5 percent or less (in weight) in wrought alloys and the same or higher in cast alloys. Most of the alloys contains two to four other elements, but in a much smaller amount than the major alloying one. These alloying elements greatly increase strength in most cases and affect most of the other properties, such as ease of fabrication, corrosion resistance and toughness.

Alloy	Major alloying elements
1xxx	99% minimum aluminium
2xxx	copper
3xxx	manganese
4xxx	silicon
5xxx	magnesium
6xxx	magnesium and silicon
7xxx	zinc
8xxx	other elements
9xxx	unused series

Table 2.1: Wrought Alloy Designation System

Alloy	Major alloying elements
1xxx	99% minimum aluminium
2xxx	copper
3xxx	silicon + (copper and/or magnesium)
4xxx	silicon
5xxx	magnesium
6xxx	unused series
7xxx	zinc
8xxx	tin
9xxx	other elements

Table 2.2: Cast Aluminium Alloy Designation System

In the wrought alloy designation, the last two digits in the 1xxx series give the minimum aluminium percentage above 99.00 percent; the second digit represents the level of impurity. It is equal to zero if the impurities are uncontrolled, and it can vary between 1 and 9 depending upon the level which the impurity content is not to exceed.

In the series 2xxx to 9xxx, the last two digits specify the individual alloys registered for the series. The second digit is equal to zero for the main alloy and varies between 1 and 9 for its modifications.

This type of numerical designation was the first proposed by the Aluminium Association for wrought alloys. Because it is easy to understand, it is now also used in European Countries.

The alloy designation system is similar for casting (see Table 2.2). In this case, the second and the third digits give the minimum aluminium percentage above 99.00 percent for the 1xxx alloys. In the 2xxx to 9xxx groups, the second two numbers are the individual alloys registered in the series. The last number gives the product form: 0 for casting; 1 and 2 (narrower composition limits than 1) for ingot.

In the UK a special numerical designation is used:

(a) One or two digit numbers are used for wrought alloys. The letter N is placed before the two digit numbers for a cold-worked alloy, whereas the letter H is placed before the two digit numbers for heat-treated alloys;
(b) One or two digit numbers preceded by the letter LM are used to represent casting alloys.

Also in Spain and Canada numerical designations different from the American one are used.

2.4.3. Alphanumerical designation

This system is used in the most of the European Countries. It is also used in the ISO normalisation. Each alloy is identified by a group of letter and digit-number divided into two sets. The first group that represent the indication of the base metal, i.e. aluminium, is defined by Al (Italy, Germany, Switzerland, Austria, Sweden, Netherlands, ex-Yugoslavia) or ALP (pure aluminium: Italy) or A (France). In some cases this indication is preceded by another letter (or letters) with the following meanings:

- G casting alloys in rough stage (Italy, Switzerland, Netherlands), sand casted (GS), permanent mould casted (GC), pressure die casted (GP)(Italy);
- GD pressure die casted (Germany);
- GK permanent mould casted (Germany);
- K casting alloys (ex-Yugoslavia)
- P wrought alloys (Italy)

The second group indicates the alloying elements. Sometimes these letters are followed by the percentage of the element. Symbols representing these elements are formed by one letter (France) or two letter (other Countries).
In some Countries, such a Switzerland and Italy, some alloys are represented with a name followed by a number. For example:

Peraluman 460 (corresponding to 5083)
Peraluman 500 (corresponding to 5086)
Estrudal 050 (corresponding to 6060)
Anticorodal 100 (corresponding to 6351)
Unidur 100 (corresponding to 7020)

It should be noted that these last designations are only used in the trade exchange and are not recognised by unifying organisations.

2.5. Fabrication stage designations

Usually, the metallurgic stage of an alloy is identified by a symbol which follows the chemical composition symbol. This symbol can be formed by letters and numbers, which usually differ according to the national designation. According to the Aluminium Association, the fabrication stages are identified by the following symbols:

- *F rough stage fabrication.* This symbol is applied to work-hardened products, in which neither special control over thermal conditions nor strain-hardening is employed. Mechanical properties cannot be defined and therefore guaranteed in these work-hardenable products.
- *O annealed stage.* It is applies to wrought products which are annealed to obtain the lower strength temper and to cast products which are annealed to improve ductility. This is the most ductile stage. Conversely, strength is very low.
- *H work-hardened stage.* This stage represents alloys whose strength is increased by cold working, with or without thermal treatments producing some reduction in strength (soft-annealing). The H is usually followed by two or three letters.
- *W Tempered non-stabilised stage.* An unstable temper applicable only to alloys which spontaneously age at room temperature after solution heat treatment. The symbol is followed by a number indicating the natural ageing period (e.g. W 1/2 h).
- *T Heat-treated stage.* It represents those products which are thermally treated, with or without supplementary strain-hardening. This T is always followed by one or more digits. The heat-treatment can be obtained by a combination of the following process:

  - Solution
  - Tempering
  - Natural ageing
  - Quenching

Sometimes the heat treatment is followed by a cold-working process in order to stabilise the alloy and to eliminate the stresses induced by the tempering process. The symbol H indicate only the strain-hardened alloys. It is applied to products which have their strength increased by strain-hardening, with or without supplementary thermal treatments to produce some reduction in strength. The H is always followed by one or two digits. The possible situations are:

- *H1 cold-worked stage.* This applies to products that are strain-hardened to obtain the desired strength without supplementary thermal treatments. The digit following the H1 indicates the degree of strain hardening.
- *H2 cold-worked and partially annealed stage.* This pertains to products that are strain hardened more than the desired final amount and then reduced in strength to the desired level by partial annealing. The digit following H2 indicates the degree of strain hardening remaining after the product has been partially annealed.
- *H3 cold-worked and stabilised stage.* This applies to products that are strain hardened and whose mechanical properties are stabilised by a low temperature thermal treatment or as a result of heat introduced during fabrication. Stabilisation usually improves the ductility. The digit following the H3 indicates the degree of strain hardening remaining after stabilisation.

The stage T is subdivided as follows:

- *T1: cooled from an elevated temperature shaping process and naturally aged to a substantially stable condition.* Applies to products which are not cold-worked after cooling from an elevated temperature shaping process, or in which the effect of cold-work in flattening or straightening may not be recognised in mechanical property limits.
- *T2: cooled from an elevated temperature shaping process, cold worked, and naturally aged to a substantially stable condition.* Applies to products which are cold-worked to improve strength after cooling from an elevated temperature shaping process, or in which the effect of cold-work in flattening or straightening is recognised in mechanical property limits.
- *T3: solution heat-treated, cold-worked, and naturally aged to a substantially stable condition.* Applies to products which are cold-worked to improve strength after solution heat-treatment, or in which the effect of cold-work in flattening or straightening is recognised in mechanical property limits.
- *T4: solution heat-treated and naturally aged to a substantially stable condition.* Applies to products which are not cold-worked after solution heat-treatment, or in which the effects of cold-work in flattening and straightening may not be recognised in mechanical property limits.
- *T5: cooled from an elevated temperature shaping process and then artificially aged.* Applies to product that are not cold-worked after cooling from an elevated temperature shaping process, or in which the effect of cold-work in flattening or straightening may not be recognised in mechanical property limits.
- *T6: solution heat-treated and then artificially aged.* Applies to products which are not cold-worked after solution heat-treatment, or in which the effect of cold-work in flattening or straightening may not be recognised in mechanical property limits.
- *T7: solutions heat-treated and overaged-stabilised.* Applies to wrought products that are artificially aged after solution heat-treatment to carry them beyond a point of maximum strength to provide control of some significant characteristic. Applies to cast products that are artificially aged after solution heat-treatment to provide dimensional and strength stability.
- *T8: solution heat-treated, cold-worked, and then artificially aged.* Applies to products which are cold-worked to improve strength, or in which the effect of cold-work in flattening or straightening is recognised in mechanical property limits.
- *T9: solution heat-treated, artificially aged, and then cold-worked.* Applies to products which are cold-worked to improve the strength.
- *T10: cooled from an elevated temperature shaping process, cold-worked, and then artificially aged.* Applies to products that are cold-worked to improve the strength, or in which the effect of cold-work in flattening or straightening is recognised in mechanical property limits.
- *T11: cooled after hot-working, work-hardened and naturally aged.* After cooling the material is cold-worked before natural ageing to increase strength.
- *T12: cooled after ho- working, work-hardened and artificially aged.* After cooling and cold -working the material is artificially aged.

The meaning of the Aluminium Association symbols is summarised in Figure 2.3.

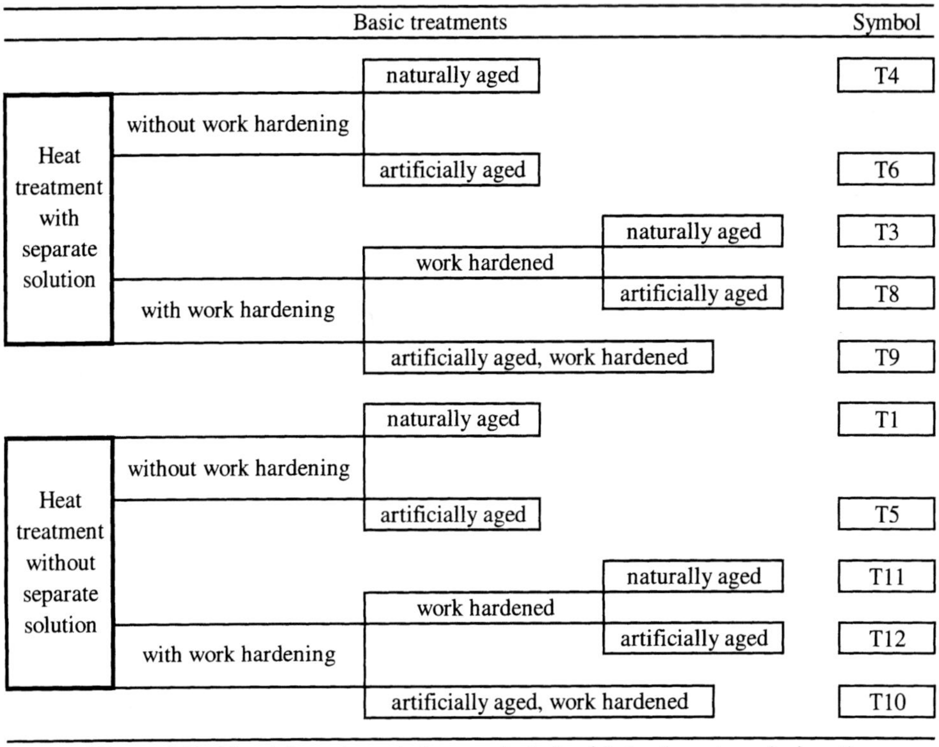

Figure 2.3: Aluminium Association symbols for fabrication stage designation

Special and supplementary treatments are indicated by the following symbols:

- T61: "soft" artificial ageing;
- T66: "hard" artificial ageing;
- TX51: relaxation by tractioning;
- TX52: relaxation by compression;
- TX53: relaxation by heat treatment.

X is the number representing the treatment: 4, 6, 7, etc.

In many Countries nonunified system of representation are used. In UK, for example, particular symbols (M,WP) are used. In Sweden numbers are used (00, 02, 03, etc.).

As an alternative to fabrication designation, it is possible to identify an alloy by mean of parameters indicating the strength. This is the minimum value of ultimate stress (N/mm^2) and it is preceded by a letter varying according to the different Countries:

R	in France
W, F or G	in Germany (for annealed, cold-worked and tempered or cold-worked and partially annealed).

This system of representation is very simple and effective, as it gives the final result of the fabrication process -the strength- without explaining how this result has been achieved.

2.8. Wrought alloys families

In the following paragraph we are going to illustrate the principal mechanical properties and the fields of application of the mostly used wrought alloys.

- *Pure aluminium and alloys of low alloy content (1000 series)*

This series is mainly produced under form of plates with different aluminium percentage (going from 99.0 to 98.8 percent) which can be used in low stressed structures. This unalloyed (pure) composition is primarily used in the electrical and chemical industries. Pure aluminium has a very low strength ($f_{0.2}\cong30$ N/mm^2) but, on the contrary, it is a very ductile material ($\varepsilon_t\cong30$-40 percent). If the material is cold-worked the strength is increased up to $f_{0.2}\cong100$ N/mm^2, whereas the ductility is drastically reduced ($\varepsilon_t\cong3$-4 percent). Pure aluminium is highly corrosion resistant and specific applications, therefore, includes cables, ceilings and tanks.

- *Aluminium-Copper alloys (2000 series)*

This type of alloy can generally be found as profiles, plates and pipes. With heat-treatment, their strength can increase up to $f_{0.2}\cong300$ N/mm^2, with a ductility equal to $\varepsilon_t\cong10$ percent. Since the corrosion resistance of these alloys is not very high, it is necessary to protect them especially when they are used in a corrosive environment (cladding with corrosion resistant alloys, plating or painting processes are used for this purpose). Because of their bad weldability, these alloys are not very used in civil engineering. However, they are used in aeronautical industry in which riveted connections are used.

- *Aluminium-Manganese alloys (3000 series)*

These alloys are not heat-treated; they thus have a slightly higher strength than pure aluminium and they keep a very high ductility which allows very hard cold-forming processes: they are used in the production of cans. On the other hand they are corrosion resistant. Specific applications are panels and roof system (trapezoidal sheeting).

- *Aluminium-Silicon alloys (4000 series)*

The properties of these alloys are similar to those of the preceding series. However, they are not often used, except for welding wires.

- *Aluminium-Magnesium alloys (5000 series)*

Even though these alloys cannot be heat-treated, their mechanical properties could be higher than those corresponding to the 1000, 3000 e 4000 series ($f_{0.2}\cong100$ N/mm^2). The strength can be increased if they are cold worked ($f_{0.2}\cong250$ N/mm^2); the ductility still being high ($\varepsilon_t=10$ percent). The corrosion resistance is also high, especially in marine environment, when the amount of Mg is less than 6 percent. These alloys are often used in welded structures, since their strength is not drastically reduced in the heat-affected zone.

- *Aluminium-Silicon-Magnesium alloys (6000 series)*

Magnesium and silicon are the main alloying elements. They form the intermetallic compound Mg_2Si which readily passes into solution in the aluminium. Hence by heat-treatment the strength of the alloy is increased ($f_{0.2} \cong 250$ N mm^{-2}), with a ductility of $\varepsilon_t = 10$ percent. These alloys are corrosion resistant. They are particularly suitable for extrusion, but also rolled sections as well as pipes can be found. These alloys are used either in welded structures or in bolted and riveted connections.

- *Aluminium-Zinc alloys (7000 series)*

These alloys consist of extruded or rolled heat-treated alloys. They can be subdivided into two subfamilies depending upon the percentage of copper as the third alloying element:

AlZnMgCu alloys are the highest-strength alloys after heat-treatment ($f_{0.2} \cong 500$ N mm^{-2}); conversely, they have low weldability and are not corrosion resistant because of the presence of copper, therefore, requiring protection by plating or painting.
AlZnMg alloys give a strength of $f_{0.2} \cong 200$ N mm^{-2} with a ductility of $\varepsilon_t \cong 10$ percent. They are also corrosion resistant. These alloys are generally used in structural applications; they are particularly useful in welded structures owing to their self-tempering behaviour, which allows to recover the initial strength in the heat-affected zone.

2.6. Aluminium properties

2.6.1. General

The properties of aluminium, which give to metal and its alloys competitive and attractive features for a wide variety of uses, are: appearance, light weight, fabricability, physical properties, mechanical properties and corrosion resistance. In the following, distinction is made between physical and mechanical properties.

2.6.2. Physical properties

The main physical properties of aluminium at room temperature, compared with those of steel and stainless steel, are represented in Table 2.3.
With regard to the most important parameters for structural behaviour, it can be observed that:

		Aluminium	Steel	Stainless Steel
Average weight density	$kg \cdot m^{-3}$	2700	7850	7900
Melting point	°C	658	1450÷1530	1450
Linear thermal expansion coefficient	$°C^{-1}$	$24 \cdot 10^{-6}$	$12 \cdot 10^{-6}$	$17.3 \cdot 10^{-6}$
Specific heat	$cal \cdot g^{-1}$	0.255	0.12	0.12
Thermal conductivity	$Cal \cdot cm^{-1} \cdot s^{-1} \cdot °C^{-1}$	0.52	0.062	0.035
Electrical resistivity	μΩ·cm	2.84	15.5	70
Young's modulus	$N \cdot mm^{-2}$	68500	206000	206000

Table 2.3: Room temperature properties of aluminium, steel and stainless steel

- The density of aluminium is approximately one third of the density of steel (in different alloys it varies from 2600 to 2800 kgm^{-3});
- The thermal expansion coefficient of aluminium is twice that of steel (in different alloys it varies from $19 \cdot 10^{-6}$ to $25 \cdot 10^{-6}$ $°C^{-1}$).

2.6.3. Mechanical properties

The principal mechanical properties can be derived from a tension test. First of all, it can be observed that the Young's modulus is also approximately one third that of steel (in different alloys it varies from 68500 to 74500 Nmm^{-2}). The load-displacement diagram is usually represented by a continuous curve without yielding (tensile stress-strain curves are provided for some alloys in Figure 2.4). The diagram can be divided in three parts:

- A linear elastic portion up the proportional stress f_p (which is usually the stress corresponding to a residual strain of 0.01 percent);
- A non-linear portion up to a "knee";
- The strain-hardening portion after the "knee" (the slope of which depends upon the type of alloy).

Although there is an initial linear elastic portion of the curve, there is not a well defined a yield strength. The yield strength values for all alloys are defined as the elastic limit stress corresponding to a 0.2 percent offset of strain from the initial straight-line portion of the curve ($f_{0.2}$).

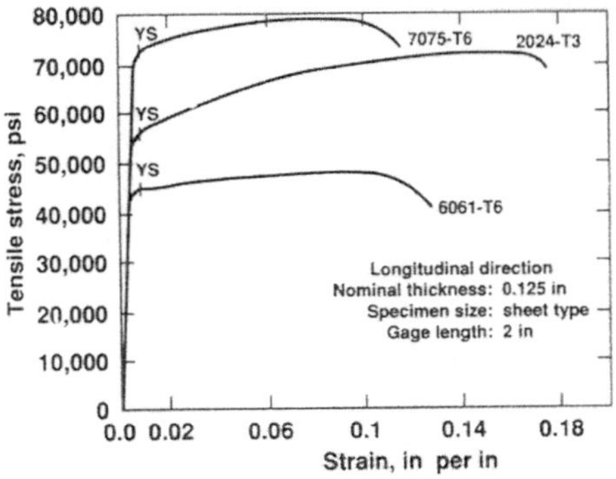

Figure 2.4: Representative stress-strain curves

The stress $f_{0.2}$ is now internationally used in structural computations, with the same meaning as yield stress in steel.

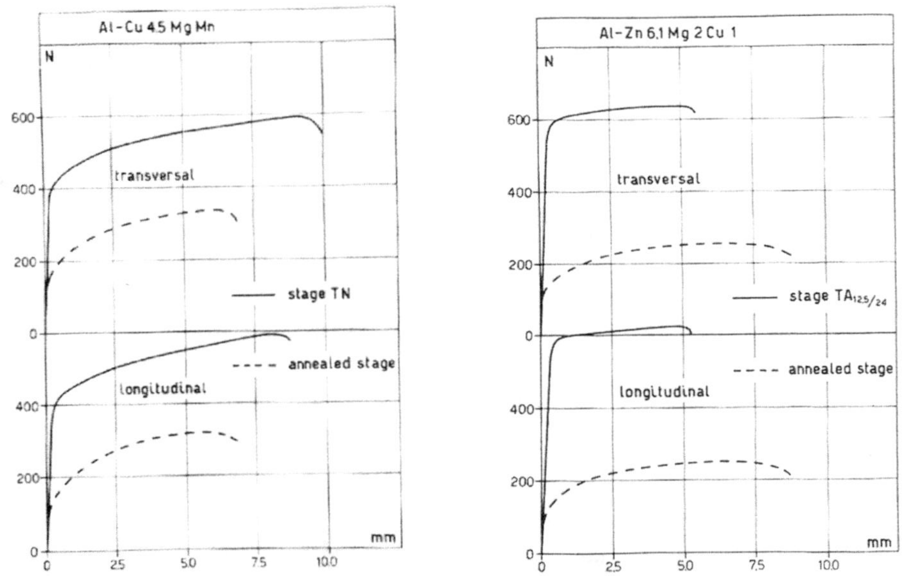

Figure 2.5: The effects of the heat-treatment process

Ultimate stress f_t corresponds to a strain value which can be defined as the limit of "uniform elongation" corresponding to necking of the specimen. At this point stress increases at the neck and large deformation of the specimen cross-section can be observed.

At the same time the testing machine records a load which is decreasing owing to the failure of the specimen, with a value of elongation equal to ε_t. This value is usually used to evaluate the ductility of the material, that is, the capacity of the material to withstand large plastic deformations without experiencing failure. The load-displacement diagrams in Figure 2.5 show how the heat-treatment process drastically changes the behaviour of the material by affecting either strength and ductility.

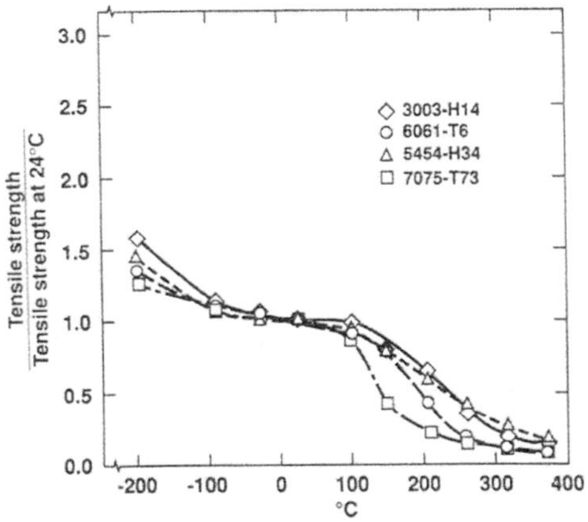

Figure 2.6: Effects of temperature on tensile strength (Sharp, 1993)

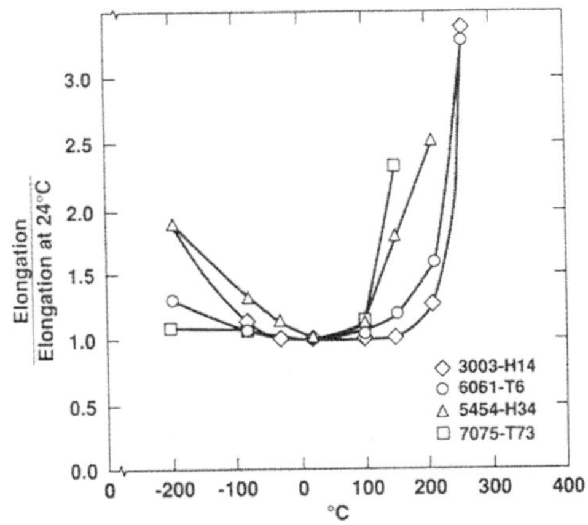

Figure 2.7: Effects of temperature on elongation (Sharp, 1993)

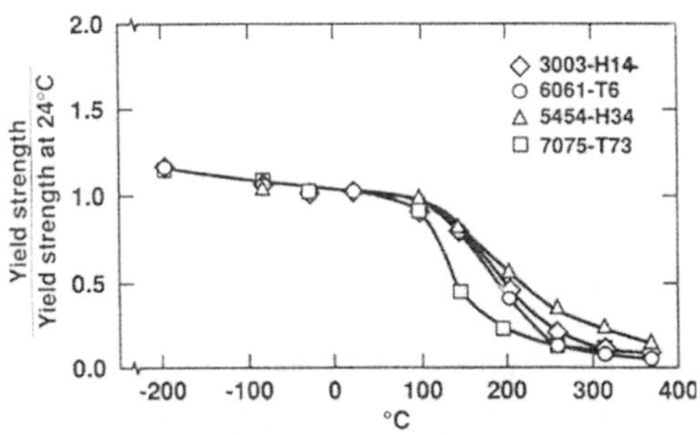

Figure 2.8: Effects of temperature on yield strength (Sharp, 1993)

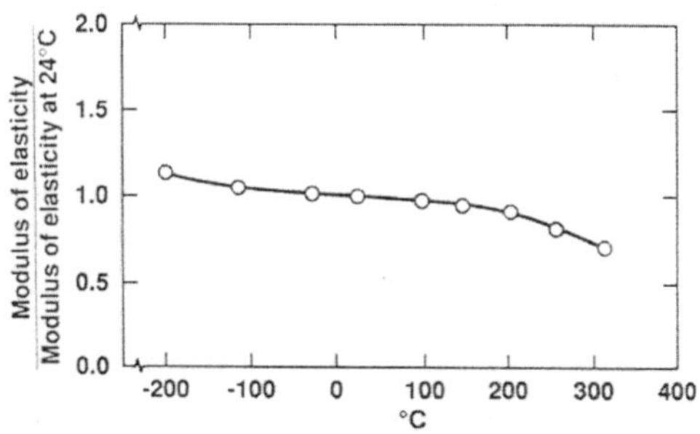

Figure 2.9: Effects of temperature on modulus of elasticity(Sharp, 1993)

Figures 2.6 to 2.9 show examples of the variation of mechanical properties with temperature, referred to the characteristics at room temperature (24 °C) (Sharp, 1993). Noteworthy is that all properties increase with decrease in temperature below room temperature. Aluminium alloys do not show a transition temperature as do most steels, being the transition temperature defined in steel as the temperature below which brittle fracture is most likely to occur. It can be said that aluminium alloys are less susceptible than steel to brittle fracture either at low temperatures or at room temperature and, therefore, they are suitable for applications involving low temperature. Tests to measure fracture toughness are not usually required in specifications. Creep needs to be considered in high-temperature use. Elongation remains high for all temperature, low and high. Finally is very important to say that the mechanical properties of aluminium alloys increase as far as the strain rate increases. Figure 2.10 shows some trends based on tests of various alloys reported in the literature. For most structural applications, it is reasonable and conservative to use test data from quasi-static loading also for problems or applications with higher strain rates.

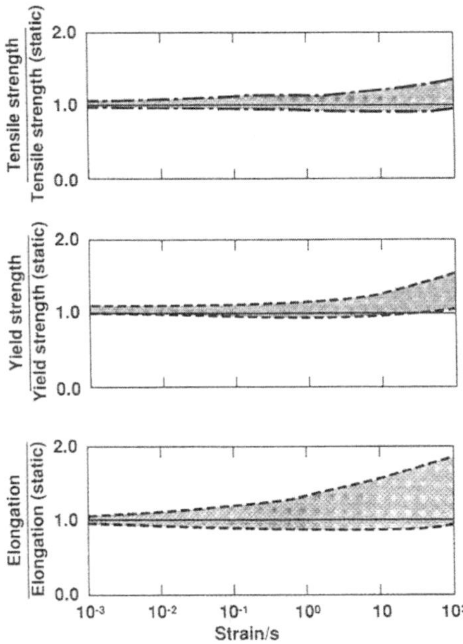

Figure 2.10: Properties under high strain rate (Sharp, 1993)

2.7. Corrosion resistance

An important property of the aluminium and its alloys is the corrosion resistance (Sharp, 1993; Mazzolani, 1995a). Usually aluminium needs no protection against atmospheric or chemical corrosive agents. In contrast to steel, in which the corrosion process in not naturally inhibited (apart from self-protecting steels), corrosion processes in aluminium tend to cease naturally. In general, corrosion can be divided into two classes: when corrosion damages the aesthetics of the structure, and when corrosion damages the safety of structure. Corrosion in aluminium can be considered to belong to the first type. This fact can be explained from an electrochemical point of view. The exposed surface of aluminium combines with oxygen to form an inert aluminium oxide film only a few microns thick (the thickness can be artificially increased to between 20 and 100 microns), which blocks further oxidation. Appropriately alloyed and treated, aluminium can resist corrosion by water, salt and other environmental factors, and by a wide range of other chemical and physical agents. The corrosion resistance of aluminium alloys can be estimated with respect to the alloying elements. The following can be said about additions of the various elements to pure aluminium:

- *Manganese* in small percentages does not significantly vary corrosion resistance;
- *Magnesium* in small percentages (up to 5-6 percent) increases corrosion resistance with respect to some corrosive agents;
- *Silicon* decreases corrosion resistance in some cases, although when combined with magnesium it provides corrosion resistance almost equal to that of pure aluminium;
- *Zinc* together with magnesium provides good corrosion resistance under a proper heat treatment;
- *Nickel, iron and copper,* even if combined with the previous elements, definitely reduce corrosion resistance.

The effects of corrosion depend on chemical composition, fabrication process, heat-treatment and stress field. With respect to common atmospheric corrosion, the following type of corrosion can be defined:

- *Surface corrosion* is the most general type of uniform corrosion on all surfaces. It usually stops immediately after the corrosion of a protective oxide film.
- *Concentrated corrosion* is usually irregularly distributed on different parts of the surface. Unless due to a persistent chemical action, this corrosion ceases naturally. Persistent chemical action can be observed in the case of smoke acting on a ceiling.
- *Intergranular corrosion* is a complex metallurgical process usually present in the 2000 series, which are used in the aeronautical industry. It can be avoided by a specific heat treatment.
- *Lamellar corrosion* produces layers vary similar to rust. It occurs in those products in which the fabrication processes produce a stratified microcrystal structure without a soft-annealing process.
- *Stress corrosion* is the most dangerous type of corrosion. It can be observed when microcracks grow in the tension zones of the stress field owing to the presence of a corrosive agent. Alloys generally used in civil engineering can be considered not susceptible to this type of corrosion, provided they are properly fabricated and erected.
- *Corrosion by contact* (electromechanical). When aluminium is in contact with another metal with a liquid between the two metals, a voltaic pile is established. The material with a lower voltage flows in solution and covers the other metal, unless this has been neutralised. Because the contact voltage of steel with respect to aluminium is 850 mV, it is necessary to avoid direct contact between the two metals and insulating material can be used to separate them. When steel bolts are used to make a connection, the bolts should be galvanized or electroplated. Furthermore this problem is present in a humid and corrosive environment. In this case the contact between aluminium and its alloys and nonmetallic materials such as concrete, mortar, timber and bricks should be avoided, for example using insulating materials (bituminous or epoxy coatings), which can guarantee the absence of water absorption in order to avoid the formation of conductive bridges. Finally it is important to say that for the protection of the surfaces of the metal, the following system can be chosen:

- Painting
- Anodization

- Metal spraying
- Plastic coating
- Surface treatment with chromate solutions.

2.9. Comparing aluminium with steel

The reason for the use of a material in any application are many, but in the final analysis the product must be affordable: the cost of the product must be acceptable to the customer. Generally, aluminium is attractive in many applications, because of a favorable life-cycle cost – the sum of the initial cost of the finished product, the cost of operating or maintaining the product over its life, and the cost of disposing of or recycling it after its useful life. In addition, aluminium has sustained and increased its use in many markets partly because the prince for aluminium relative to that for steel, overall, has decreased gradually over the aluminium industry's 100-year life.
In the following we want to summarise the concepts of the previous sections and to outline the main peculiarities of different alloys (Mazzolani, 1995a).

The main properties of aluminium are:

- *Lightness.* The specific weight is γ=2700 kgm^{-3}, equal to one-third that of steel;
- *Corrosion resistance.* The exposed surface of aluminium combines with oxygen to form a thin inert aluminium oxide film which blocks further oxidation.

Aluminium is also very ductile material ($\varepsilon_t \cong 40$ percent), but its strength is low for structural applications ($f_{0.2} \cong 20$ Nmm^{-2}). In order to increase the material strength, a cold-working process can be used; however this process does not greatly increase strength ($f_{0.2} \cong 100$ Nmm^{-2}) and the ductility is drastically reduced (up to one-tenth of the initial value). Another way of increasing the strength of the material is to alloy aluminium with other elements. After this process the strength can be higher than 100 Nmm^{-2}, with a ductility equal to ε_t=10 percent; however the corrosion resistance decreases.
By means of heat-treatment the strength goes up to $f_{0.2}$=250 Nmm^{-2} in AlSiMg alloys and it can reach $f_{0.2}$=350-400 Nmm^{-2} in AlZn and AlCu alloys.
Figure 2.11 is a summary comparison of some aluminium alloys and the most commonly used mild steels Fe 360 and Fe 510. In this Figure the main mechanical properties are summarised: Elastic limit ($f_{0.2}$), Ultimate strength (f_t) and Young's modulus (E). In general it can be said that aluminium alloys offer a wider range of strength than steel. For this reason the concept of a "fourth dimension" in metal construction seems to be more appropriate to aluminium alloys than to steel. On the other hand, because the Young's modulus is one-third that of steel, more frequent problems of deformation and of instability are present.

	Aluminium Alloys		Steel	
$f_{0.2}$ (f_y)	AlMg4.5Mn	~140	Fe360	~235
[Nmm^{-2}]	AlMgSi1	~260	Fe510	~350
	AlZnMg1	~360		
f_t	AlMg4.5Mn	~280	Fe360	~360
[Nmm^{-2}]	AlMgSi1	~320	Fe510	~510
	AlZnMg1	~410		
E [Nmm^{-2}]	70000		206000	
ε_t (%)	10-25		25-30	
γ [Nm^{-3}]	26500		77000	
α (°C^{-1})	0.00002		0.00001	

Figure 2.11: Comparison of properties between three aluminium alloys and two common mild steels

In the Figure 2.12 the stress-strain curves of AlMgSi alloy and of Fe 360 steel are compared.

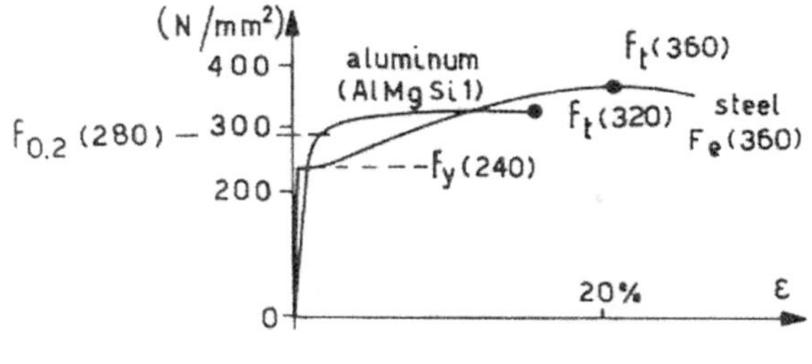

Figure 2.12: Comparison of stress-strain curves of AlMgSi alloy and Fe 360 steel

From this Figure it can be observed that:

- Aluminium alloys have a strain-hardening behaviour which is not preceded by a perfectly plastic branch corresponding to yielding;
- The ultimate deformation is lower than the one of steel;
- The $f_t/f_{0.2}$ ratio is lower that the one of steel.
- The main difference between the two materials is relative to the inelastic range. On the other hand, both materials behave linear elastically up to the elastic limit, which basically represents the working range of structures.

Finally an other important parameter is represented by the ratio between strength f_t or $f_{0.2}$ and the specific weight γ. The figure 2.13 shows that this ratio is extremely positive for aluminium alloys. However it is not always possible to take advantage of this benefit offered by aluminium alloys, especially when the material is under compression, because of the small value of the Young's modulus the instability phenomena is more likely to occur than in steel structures.

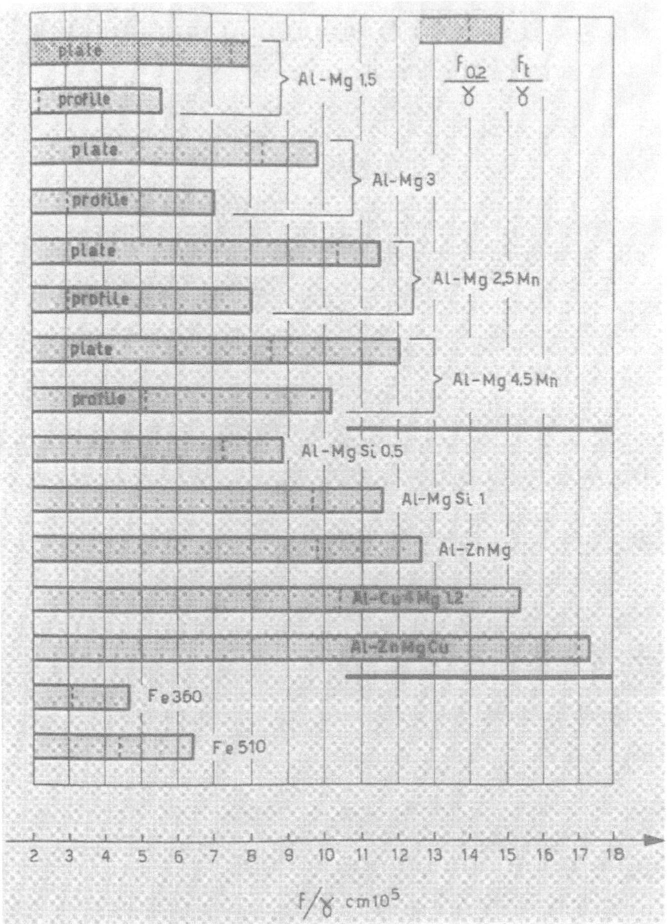

Figure 2.13: The "strength rating"

Further observations regarding aluminium alloy structures have to be pointed out:

- The structures in aluminium alloys are more sensitive to thermal variations because the coefficients of thermal expansion of this metal is twice that of steel. This fact has to be taken into account when designing support apparatus;
- Residual stresses produced by constraining thermal deformations are about 30 percent lower than those in steel structures, because they are proportional to the product αE;
- Aluminium alloy structural components can be manufactured by rolling, extrusion, casting and drawing processes. The extrusion process is of particular interest as it allows fabrication of profile of any shapes (useful for those shapes not obtainable by a rolling process).

2.10. Characterisation of the σ-ε law

2.10.1. General

The first difficulty that is usually encountered in the theoretical analysis of static and stability problems for aluminium alloy structures is related to the idealisation of the tensile material properties. It was shown in the previous paragraph that aluminium alloys commonly used in structures possess mechanical properties considerably different from each other, also in relation with the fabrication process and heat-treatment. Furthermore, the continuous behaviour of the σ-ε curve, obtainable in a tension test, cannot be simplified to an elastic/perfectly plastic behaviour as is done for mild steel. In this case is necessary to consider more complex models. Among them, the Ramberg-Osgood continuous model represents the most commonly used in research activity, giving a very good agreement with experimental results (Mazzolani, 1995a).

2.10.2. The Ramberg-Osgood law

A generalised law $\varepsilon = \varepsilon(\sigma)$ has been proposed by Ramberg and Osgood for aluminium alloys:

$$\varepsilon = \frac{\sigma}{E} + \left(\frac{\sigma}{B}\right)^n \tag{2.1}$$

where E is the Young's modulus at the origin, and B and n have to be determined by experiments (Figure 2.15).

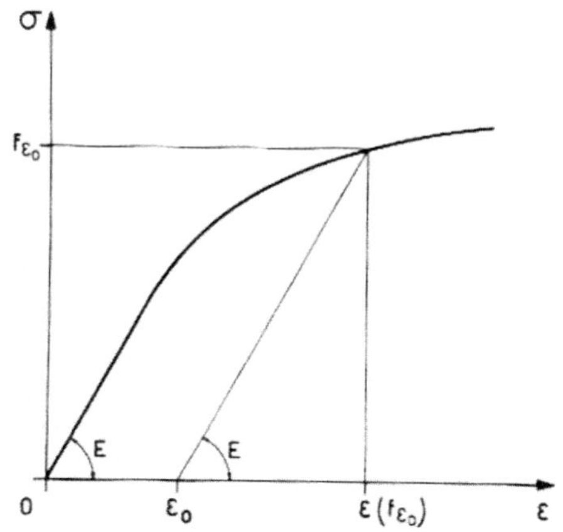

Figure 2.14: Continuous model of the form $\varepsilon = \varepsilon(\sigma)$

If an elastic limit stress is defined such as the one producing a residual strain when the specimen is unloaded, the above-mentioned equation assumes the following form (Figure 2.14):

$$\varepsilon = \frac{f_{\varepsilon o}}{E} + \left(\frac{f_{\varepsilon o}}{B}\right)^n \tag{2.2}$$

By substituting

$$\frac{f_{\varepsilon o}}{E} = \varepsilon - \varepsilon_o \tag{2.3}$$

in (2.2), it results:

$$\varepsilon_o = \left(\frac{f_{\varepsilon o}}{B}\right)^n \tag{2.4}$$

Therefore

$$f_{\varepsilon_o} = B \cdot \sqrt[n]{\varepsilon_o} \tag{2.5}$$

By assuming $\varepsilon_o = 0.002$ or $\varepsilon_o = 0.001$, then alternatively we obtain:

$$f_{0.2} = B\sqrt[n]{0.002} \quad ; \quad f_{0.1} = B\sqrt[n]{0.001} \tag{2.6}$$

whose ratio gives:

$$\frac{f_{0.2}}{f_{0.1}} = \sqrt[n]{2} \tag{2.7}$$

This equation relates the exponent n to the strain-hardening parameter $f_{0.2}/f_{0.1}$ and hence, according to the Sutter's classification, to the heat-treatment of the material. The exponent n of the Ramberg-Osgood law is therefore characteristic of the strain-hardening rate of the inelastic portion of the σ-ε diagram (Figure 2.15). It can be expressed by means of following expression:

$$n = \frac{\ln 2}{\ln\left(\frac{f_{0.2}}{f_{0.1}}\right)} \tag{2.8}$$

When the ratio $f_{0.2}/f_{0.1}$ tends to 1, the value of *n* tends to infinity, and Eq. 2 then represents the behaviour of mild steel.

		$f_{0.2}/f_{0.1}$	n
Linear law			
		2	1
Sutter	I class		
		1.060	11.89
	II class		
		1.045	15.75
	III class		
		1.030	23.45
IV class			
		1	∞
mild steel			

(a)

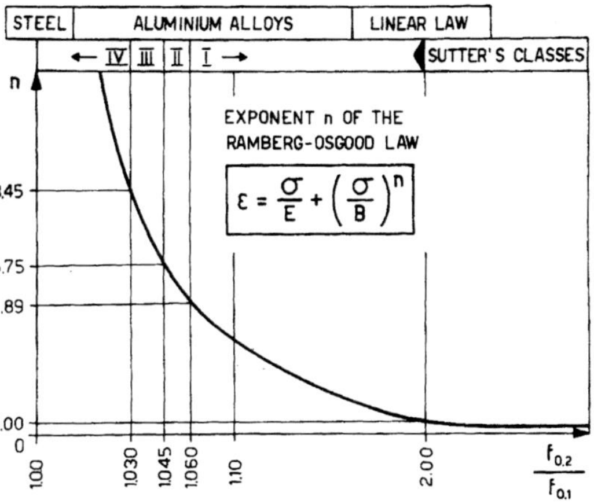

(b)

Figure 2.15: Comparison between the exponent n of the Ramberg-Osgood law and the values of the ratio $f_{0.2}/f_{0.1}$ (b), which are related to the Sutter's classes (a)

By substituting the first of the equations (2.6), the equation (2.1) assumes the classical form:

$$\varepsilon = \frac{\sigma}{E} + 0.002 \cdot \left(\frac{\sigma}{f_{0.2}} \right)^n \tag{2.9}$$

The Ramberg-Osgood law is widely used, because it is able to predict the actual behaviour of aluminium alloys. It has been used also in the computations carried out by the ECCS Committee on Aluminium Structures when developing the European Recommendations for Aluminium Alloy Structures (1978). It is also mentioned into an Annex of Eurocode 9. The equation (2.9) can be rewritten in the following more general form:

$$\varepsilon = \frac{\sigma}{E} + \varepsilon_o \cdot \left(\frac{\sigma}{f_{\varepsilon_o}} \right)^n \tag{2.10}$$

which can be successfully applied to other structural metallic and non-metallic materials as well as to aluminium. The meaning of the new parameters is:

f_{ε_o}: conventional elastic limit;
ε_o:residual strain corresponding to the elastic limit;
n: the exponent equal to:

$$n = \frac{\ln \alpha}{\ln(f_{\varepsilon_o} / f_{\varepsilon_o / \alpha})} \tag{2.11}$$

where $\alpha > 1$.

In particular, ε_o=0.002 and α=2 corresponds to the classical formulation. The perfectly plastic behaviour for ε_o=0 corresponds to the stress-strain law of brittle materials such as plastic and composite materials. Some σ–ε curves obtained from the equation (2.10) for various values of n and for ε_o=0.001 and 0.002 are given in Figure 2.16. It has already been observed that the behaviour of mild steel is a particular case of this formulation, and with $n \rightarrow \infty$ the σ–ε curve is independent of the value of ε_o. In practice, values of n greater than 100 are enough to provide an elastic-perfectly plastic curve.

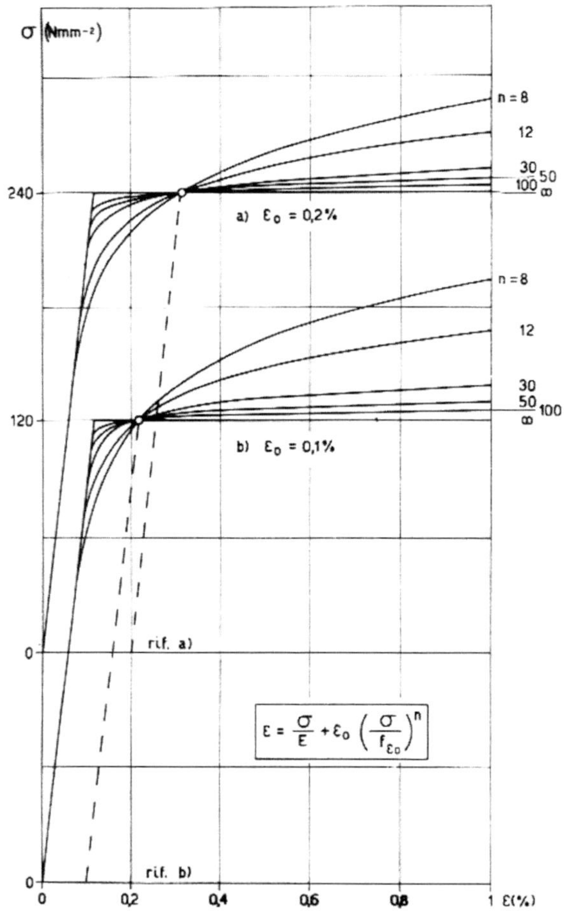

Figure 2.16: Some σ–ε curves obtained from the equation (2.10) for various values of n and for ε_o = 0.001 and 0.002

However, the Ramberg-Osgood law gives its best results when used for aluminium alloys and high-strength steel, where the constitutive law is usually characterised by the conventional values of $f_{0.1}$ and $f_{0.2}$.

2.11. The Structural imperfections

2.11.1. Definition of industrial bar

The definition of "industrial bar", used for the first time for the steel structures during the 1960s, details the unavoidable imperfections produced during the fabrication process of structural components. The random imperfections cannot be ignored since they affect, sometimes in a significant way, the load-bearing capacity of the members (Mazzolani, 1995a). The "industrial bar" has replaced the "ideal bar", which is usually assumed as an isotropic, homogeneous, perfectly linear bar without residual stress. Usually the imperfection can be divided in two types:

- *Geometrical imperfections* characterize the differences between the nominal and the actual shape of the structural element, either for the longitudinal dimensions or for the transverse dimensions of the cross-section;
- *Mechanical imperfections* can be grouped into the following types:

  - *Residual stresses* are self-equilibrated internal stresses in metal member, closely dependent upon the industrial manufacturing process, which arise in a member when it undergoes non-uniform plastic deformation. The inhomogeneous deformation field which generates residual stresses in aluminium alloy profiles is caused by thermal processes, such as cooling after extrusion and welding, and by mechanical processes, such as cold-rolling and straightening by tractioning;
  - *Inhomogeneous distribution of mechanical properties* is characterised by a scattered deviation between the measured values and the nominal value at different points in the cross-section of a profile;
  - *The Bauschinger effect* gives rise to a different resistance in compression and in tension, which is produced by the tractioning process used for straightening aluminium extruded profiles.

2.11.2. Measurement of the geometrical and mechanical imperfections

During the seventies, geometrical and mechanical imperfections have been measured for the first time on the aluminium alloy bars made of extruded and welded shapes. The experimental activity has been developed by means of the following tests:

a Measurement of geometrical imperfections;
b Measurement of the distribution of longitudinal residual stresses;
c Measurement of the distribution of the $\sigma - \varepsilon$ relationship in tension and in compression;
d Stub column test;
e Buckling test.

The corresponding results allowed to characterise the aluminium alloy bar as "industrial bar" with their geometrical and mechanical imperfections. The first issue has been the interpretation of these

imperfections by means of mathematical models, which have been used in the simulation calculation in order to extrapolate the evaluation of the load carrying capacity obtained by tests on a given number of specimens to all possible combinations of material and shapes.
From the result of imperfection measurements, the following can be summarized (Mazzolani, 1995a; 1995c):

- Out-of-straightness of aluminium extruded bars is usually less sever than in steel, being limited by the value of about L/2000; in case of welded bars this limit increases to L/1300.
- Variations of dimension are present in extruded tubes where the scatter in the thickness reaches 9 percent (Figure 2.17); in the case of welded double T profiles the eccentricity of the webs in the weak axis direction reaches the value of L/1600, but when added to the initial out-of-straightness never overcomes the value of L/1000, which can be considered as an upper bound to be used in calculations.

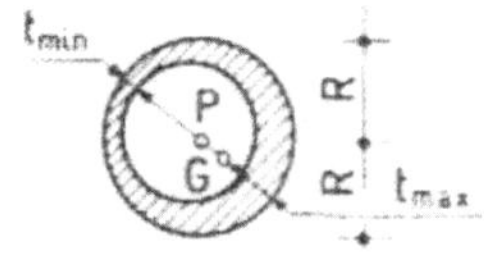

$$t = (t_{max} + t_{min})/2$$

$$\frac{\Delta t}{t} = \frac{t_{max} - t}{t}$$

Figure 2.17: Thickness variation in hollow sections

- Residual stresses in the extruded profiles of any shapes, whatever the heat-treatment, have very low values, so they play a very small effect on the load-bearing capacity of members, so for practical purposes they can be neglected. On the contrary, they are not negligible in welded profiles, where the elastic limit of the welding metal is reached near the welds. So, their influence must be taken into account in checking stability, even if in average the residual stress distribution in aluminium alloy is less severe than in steel (Figure 2.18);
- Distribution of mechanical properties can be considered highly homogeneous in extruded profiles and therefore ignored. On the contrary, in case of welded shapes the distribution is strongly influenced by the technological treatment, giving rise to different lowering effect near the weld. In case of work-hardened alloys the decrease of strength is about 10 percent only, but in case of heat-treated alloys it reaches 40-50 percent. The result of an investigation on welded profiles show that three different types of stress-strain curves can be identified in the three regions near the weld (Figure2.19):

A. Unaffected parent material;
B. Partially affected parent material;
C. Heat-treated zones around the weld metal

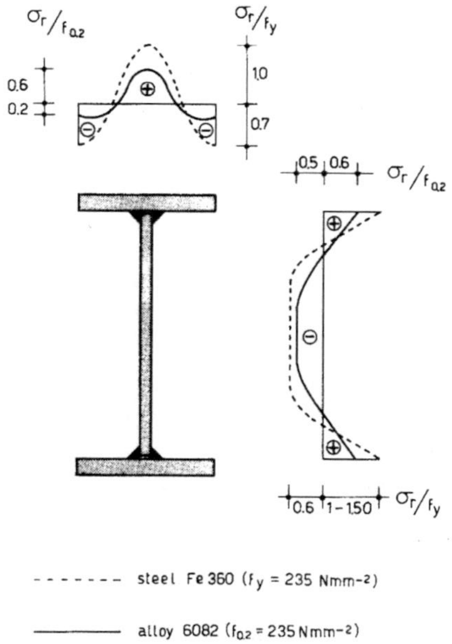

σ_w^+ highest tensile residual stress in the web
σ_w^- highest compressive residual stress in the web

Figure 2.18: Comparison between residual stress distribution in steel and aluminium

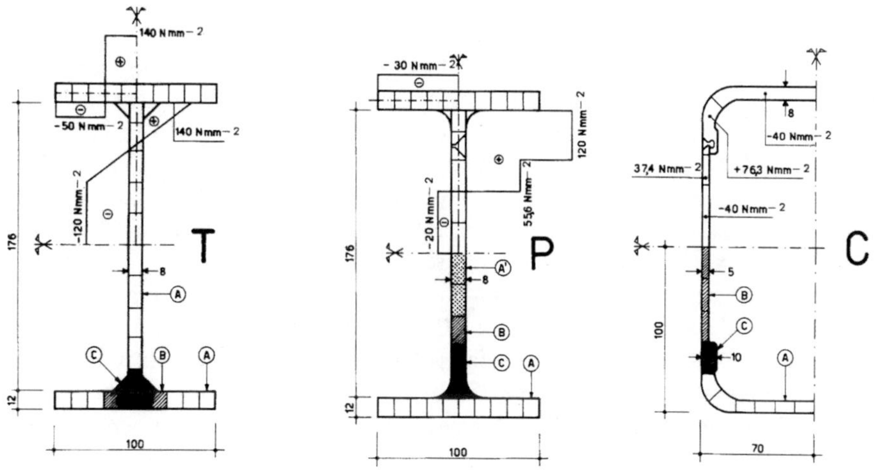

Figure 2.19: Variation of mechanical properties around the weld

For heat-treated alloy welded sections the actual distribution of the proof stress must be, therefore, considered in the evaluation of load-bearing capacity, by means of appropriate models.

3. DEVELOPMENT OF CODIFICATION

3.1. Development of the International codification

Owing to the increasing use of aluminium alloys in constructions, several Countries have published specifications for the design of aluminium structures. It is due to the efforts of the ECCS Committee for Aluminium Structures and of its working groups that the first edition of the European Recommendations for Aluminium Alloy Structures became available in 1978 (European Convention for Constructional Steelworks, 1978). These Recommendations represent the first international attempt to unify computational methods for the design of aluminium alloy constructions in civil engineering and in other structural applications, by using a semi-probabilistic limit state methodology (Mazzolani, 1995a).

Since 1970 the ECCS Committee on Aluminium Alloy Structures has carried out extensive studies and research, in order to investigate the mechanical properties of materials, their imperfections and their influence on the instability of members. On the basis of these data, for the first time, the aluminium alloy members have been characterised as "industrial bars", in accordance with the current trends of the safety principles in metallic structures (Mazzolani, 1995c).

Among the research programs in this field, undertaken with the co-operation and support of several European Countries, buckling tests on extruded and welded built-up members were carried out at the University of Liège, in co-operation with the University of Naples and the Experimental Institute for Light Metals of Novara, Italy.

During the seventies, the use of "ad hoc" simulation methods which allow all the geometrical and mechanical properties, together with their imperfections, to be taken into account, has led to satisfactory results in the study of the instability phenomena of columns and beam-columns. The analysis of these experimental and numerical results demonstrated the major differences between the behaviour of steel and aluminium. In particular the buckling curves, valid for extruded and welded bars with different cross-sections and different alloys, have been defined and they have been used in many national and international Codes, including the Eurocode 9 many years after. During the eighties the UK (BS 8118), Italian (UNI 8634), Swedish (SVR), French (DTU), German (DIN 4113) and Austrian (ON) specifications have been published or revised, taking into account in many cases the ECCS Recommendations.

In the last decade the research reached satisfactory levels also in other fields, such as the local buckling of thin plates and its interaction with the global behaviour of the bar, the instability of two-dimensional elements (plates, stiffened plates, web panels) and the post-buckling problems of cylindrical shells.

The time being is characterised by the activity in progress for the preparation of the EN version of the Eurocode for Aluminium Alloy Structures (EC9), within the Committee CEN-TC 250/SC9 (Chairman Mazzolani).

3.2. The contents of EC 9

The ENV version of Eurocode 9 in basically composed by three documents, which were approved in 1997 (CEN, 1998).

The complete work has been done in the period 1993-1997 by four Project Teams:

PT1 : "*Members*" (convener Ph. Bulson);
PT2 : "*Connections*" (convener F. Soetens);
PT3 : "*Fatigue*" (convener R. Jaccard);
PT4 : "*Fire*" (convener L. Twilt).

Three document of Eurocode 9 are:

- *Part 1.1: General rules (pr ENV 1999-1.1);*
- *Part 1.2: Structural fire design (pr ENV 1999-1.2);*
- *Part 2: Structures susceptible to fatigue (pr ENV 1999-2).*

They have been published as European Standards in 1999 to be submitted to public inquiry during the so-called conversion period from ENV to EN.
The contents of Part 1.1 "***General Rules***" follow, as close as possible, the format of all Eurocodes, by including these Chapters:

Chapter 1: General;
Chapter 2: Basis of design;
Chapter 3: Materials;
Chapter 4: Serviceability limit states;
Chapter 5: Ultimate limit states (members);
Chapter 6: Connections subjected to static loading;
Chapter 7: Fabrication and execution;
Chapter 8: Design assisted by testing.

Some detailed topics are treated in the following Annexes to Part 1.1:

A. Test of slip factor;
B. Materials selection;
C. Calculation of internal forces and moments;
D. Methods of global analysis;
E. Analytical models for stress strain relationship;
F. Frame stability;
G. Behaviour of cross sections beyond elastic limit;
H. Lateral torsional buckling;
I. Torsional properties of thin-walled sections.

Part 1.2 "***Structural fire design***" is made of four Chapters:

Chapter 1:General;
Chapter 2: Basis principles and rules;
Chapter 3: Material properties;
Chapter 4: Structural fire design.

Completed by three Annexes:

A. Properties of aluminium alloys (not listed in Part 1.1);
B. External aluminium alloy structures;
C. Configuration factor.

Part 2 "***Structures susceptible to fatigue***" contains six Chapters:

Chapter 1:General;
Chapter 2: Basis of design;
Chapter 3: Loading;
Chapter 4: Stress analysis;
Chapter 5: Fatigue strength;
Chapter 6: Quality requirements.

and five Annexes:

A. Stress analysis;
B. Guidance on assessment by fracture mechanics;
C. Testing for fatigue design;
D. Inspection and workmanship acceptance levels;
E. Fatigue strength improvement of welds.

3.3. The main features of EC9

The unavoidable complexity of a code on Aluminium Structures is essentially due to both the nature of the material itself (much more "critical" and less known than steel), which involves the solution of difficult problems and demands careful analysis. In this case the need for the code to be educational as well as informative and not only normative has been particularly determinant (Mazzolani, 1998b).
The present edition of the Eurocode 9 is based on the most recent results which has been achieved in the field of aluminium alloy structures, without ignoring the previous activities developed within ECCS (Mazzolani, 1995c) and in the revision of outstanding codes, like BS 8118 (Bulson, 1992).
The innovative contents of Eurocode 9 Part 1.1 "General Rules" is basically given by the introduction, for the first time in a structural aluminium code, of the analysis of the inelastic behaviour starting from the cross-section up to the structure as a whole. The classification of cross-section has been done on the basis of experimental results, which come from an "ad hoc" research project supported by the main representatives of the European Aluminium Industry, which provided the material for specimens.
The output has been the assessment of behavioural classes based on the *b/t* slenderness ratio, according to an approach qualitatively similar to the one used for steel, but with different extension of behavioural ranges, which have been based on the experimental evidence (Mazzolani *et al.*, 1996b) and confirmed by numerical simulation (Mazzolani *et al.*, 1997).
For members of class 4 (slender sections), the check of local buckling effect is done by means of a new calculation method which is based on the effective thickness concept. Three new buckling curves for slender sections has been set up by considering both heat-treated and work-

hardened alloys, together with welded and non-welded shapes (Landolfo and Mazzolani, 1995; Landolfo and Mazzolani, 1997).

The problem of the evaluation of internal actions has been faced by considering several models for the material constitutive law from the simplest to the most sophisticated, which give rise to different degrees of approximation. The global analysis of structural systems in inelastic range (plastic, strain hardening) has been based on a simple method which is similar to the well known method of plastic hinge, but it considers the typical parameters of aluminium alloys, like absence of yielding plateau, continuous strain-hardening behaviour, limited ductility of some alloys (Mandara and Mazzolani, 1995).

The importance of ductility on local and global behaviour of aluminium structures has been emphasized, due to the sometime poor values of ultimate elongation, and a new "ad hoc" method for the evaluation of rotation capacity for members in bending has been set up (Mazzolani and Piluso, 1995).

For the behaviour of connections, a new classification system has been proposed according to strength, stiffness and ductility (Mazzolani et al.,1996a). This approach is now under numerical check (De Matteis *et al.*, 1998) and an experimental research project on T-stub connections is going to be completed (De Matteis *et al.*, 2001a, 2001b).

Fire Design is a transversal subject for all Eurocodes dealing with structural materials. For Aluminium Structures it has been codified for the first time according to the general rules which assess the fire resistance on the bases of the three criteria: Resistance (*R*), Insulation (*I*) and Integrity (*E*).

As it is well known, aluminium alloys are generally less resistant to high temperatures than steel and reinforced concrete. Nevertheless, by introducing rational risk assessment methods, the analysis of a fire scenario may in some cases results in a more beneficial time-temperature relationship and thus makes aluminium more competitive and the thermal properties of aluminium alloys may have a beneficial effect on the temperature development in the structural component (Forsén, in Mazzolani, 1995).

The knowledge on the fatigue behaviour of aluminium joints has been consolidated during the last 30 years (Mazzolani, 1995a). In 1992 the ECCS Recommendations on Fatigue Design of Aluminium Alloy Structures have been published (Kosteas, 1992), representing a fundamental bases for the development of Eurocode 9. It was decided to characterise Part 2 of EC9 in a general way, giving general rules applicable to all kind of structures under fatigue loading conditions with respect to the limit state of fatigue induced fracture. It has been done contrary to steel, for which Part 2 is dealing with bridges only. Three design methods has been introduced:

- Safe life design;
- Damage tolerant design;
- Design by testing.

Five basic groups of detail categories have been considered:

- non-welded details in wrought and cast alloys;
- welded details on surface of loaded members;
- welded details at end connections;
- mechanically fastened joints;
- adhesively bonded joints.

The use of finite elements and the guidance on assessment by fracture mechanism have been suggested for stress analysis. The importance of quality control on welding has been particularly emphasised.

3.4. Classification of cross-sections

The behaviour of members strictly depends on the shape of the cross-section and, therefore, the model to be used in structural analysis must be related to the capability of members to reach a given limit state, such as:

a) **elastic buckling limit state**, characterised by the onset of local instability phenomena in the compressed parts of the section;
b) **elastic limit state**, corresponding to the attainment of the proof stress in the most stressed fibres of the cross-section;
c) **plastic limit state**, corresponding to the complete yielding of the cross-section under the hypothesis of elastic-perfectly plastic material;
d) **collapse limit state**, corresponding to the actual strength of the cross-section considering hardening effects.

In Eurocode 9 cross-sections are classified with reference to the above limit states in four classes (Figure 3.1):

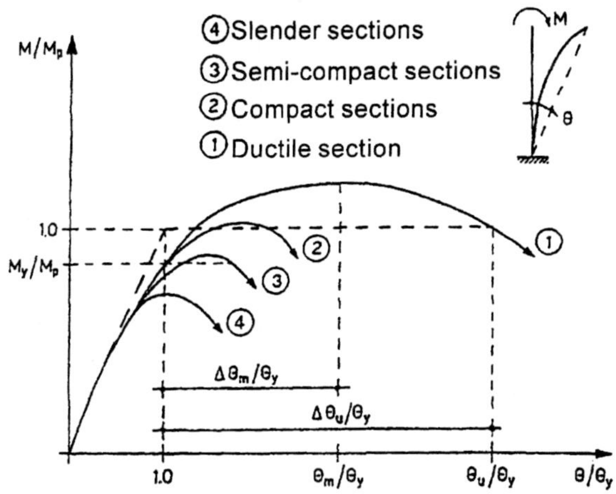

Figure 3.1: Classification of cross-sections

Class 1: ductile sections, which develop all the collapse resistance without any problem of local buckling and with the full exploitation of the hardening properties of material until the ultimate value of deformation depending on the type of alloy (limit state d).
Class 2: compact sections, which are capable to develop the plastic ultimate resistance without full exploitation of the hardening properties of material, which is prevented by the onset of

plastic instability phenomena (limit state c).

Class 3: semi-compact sections, which are capable to develop the elastic limit resistance only without getting into inelastic range owing to instability phenomena, which prevent the development of important plastic deformations, giving rise to a substantially brittle behaviour (limit state b).

Class 4: slender sections, whose behaviour is governed by the occurring of local buckling phenomena, which produce a reduction of the effective resistant section without plastic deformations, giving rise to a remarkably brittle behaviour (limit state a).

The following basic types of elements are identified in the classification process:

a) flat outstand element;
b) flat internal element;
c) curved internal element.

These elements can be unrinforced, or renforced by longitudinal stiffening ribs or lips or bulbs (Figure 3.2).

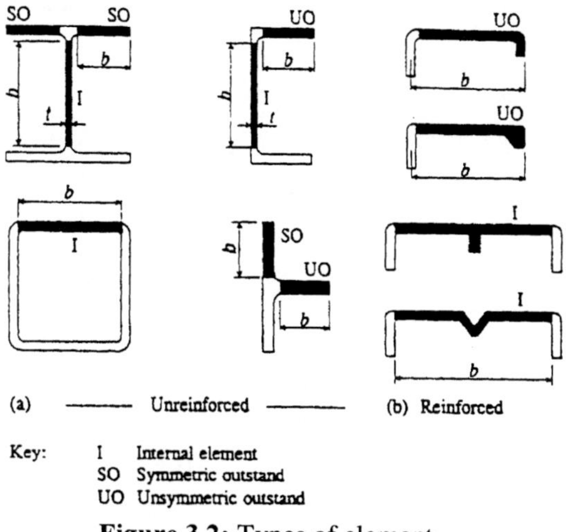

Figure 3.2: Types of element

The basic parameter to decide which class a given cross-section belongs to is the *b/t* ratio between the width and thickness of each element. A research project has been expressly developed on aluminium sections (square and rectangular hollow sections, channels), in order to identify the influence of the width-to-thickness ratios on the local buckling phenomenon. On the basis of stub-column tests on heat-treated unwelded hollow sections, the following limit values for the *b/t* ratio have been set up (Mazzolani *et al.*, 1996b):

- class 1: $b/t \leq 11$;
- class 2: $11 < b/t \leq 16$;
- class 3: $16 < b/t \leq 21$;
- class 4: $b/t > 21$.

Eurocode 9 relates the classification of elements in a cross-section to the value of the

slenderness parameter β, which is defined according to the type of elements as a function of the b/t ratio.
In case of plane unstiffened elements, β is related to the stress gradient (Figure 3.3):

$$\beta = g\, b/t \quad or \quad \beta = g\, d/t \tag{3.1}$$

where:

b: is the width of an element;
d: is the depth of web element;
t: is the element thickness;
g: is the stress gradient coefficient, given by the expressions
$g = 0.70 + 0.30\psi$, $-1 < \psi < 1$
$g = 0.80 / (1+\psi)$, $\psi \leq -1$
where ψ is the ratio of the stresses at the edges of the plate under consideration related to the maximum compressive stress.

In the case of plane-stiffened elements, more complex formulations are provided in order to take into account three possible buckling modes:

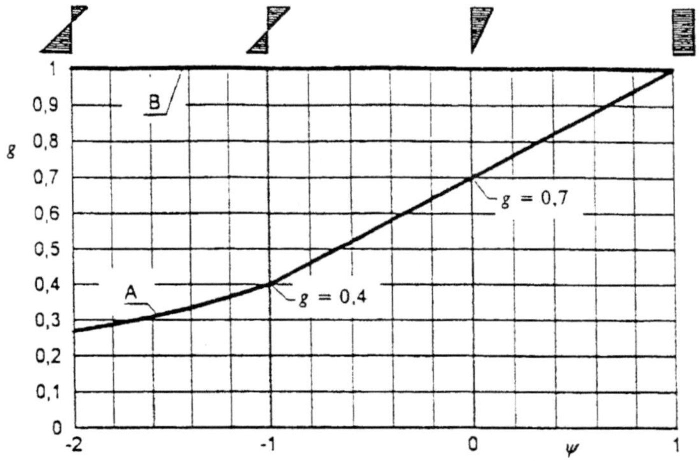

Figure 3.3: Values of g

1. the stiffened element buckles as a unit, so that the stiffener buckles with the same curvature as the element (Figure 3.4a);
2. the sub-elements and the stiffener buckle as individual elements with the junction between them remaining straight (Figure 3.4b);
3. this is a combination of Modes 1 and 2, in which sub-element buckles are superimposed on the whole element (Figure 3.4c).

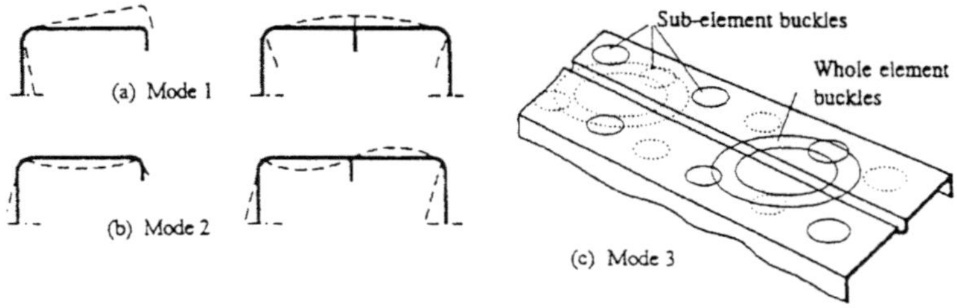

Figure 3.4: Buckling mode for flat reinforced elements

Table 3.1 reports the element classification, where the value of the limit parameters β_1, β_2 and β_3 are given in Table 3.2, being $\varepsilon = \sqrt{250/f_{0.2}}$ and $f_{0.2}$ the 0.2% proof strength in MPa.

Elements in beams		Elements in struts	
$\beta \leq \beta_1$	Class 1	$\beta \leq \beta_2$	Class 1 or 2
$\beta_1 < \beta \leq \beta_2$	Class 2	$\beta_2 < \beta \leq \beta_3$	Class 3
$\beta_2 < \beta \leq \beta_3$	Class 3	$\beta_3 < \beta$	Class 4
$\beta_3 < \beta$	Class 4		

Table 3.1: Element classification

Elements	Heat-treated unwelded			Heat-treated welded or non-heat-treated unwelded			Non-heat-treated unwelded		
	β_1	β_2	β_3	β_1	β_2	β_3	β_1	β_2	β_3
Outstand	2ε	4.5ε	6ε	2.5ε	4ε	5ε	2ε	3ε	4ε
Internal	11ε	16ε	22ε	9ε	13ε	18ε	7ε	11ε	15ε

Table 3.2: Values of β_1, β_2 and β_3

3.5. Resistance of cross-sections

3.5.1. Evaluation of ultimate axial load

The load-bearing capacity of cross-sections under axial load, excluding overall buckling phenomena of the member, can be evaluated with reference to the above mentioned limit states and the corresponding behavioural classes (Mazzolani and Piluso, 1995).
The value of axial load for a given limit state can be expressed by the generalised formula:

$$N_{Rd} = \alpha_{Nj}\, A\, f_d \tag{3.2}$$

being:

f_d the design value of strength, which can be assumed equal to the conventional elastic limit ($f_{0.2}$);

A the net cross-sectional area;

α_{Nj} a correction factor, given in Table 3.3, depending on the assumed limit state, where:

A_{eff} is the effective cross-sectional area, evaluated accounting for local buckling phenomena, in case of class 4.

When welded sections are involved, a reduced value A_{red} of the net cross sectional area shall be used, evaluated by accounting for HAZ (heat-affected zones).

Axial load	Limit state	Section class	Correction factor
N_u	Collapse	Class 1	$\alpha_{N,1} = f_t/f_d$
N_{pl}	Plastic	Class 2	$\alpha_{N,2} = 1$
N_{el}	Elastic	Class 3	$\alpha_{N,3} = 1$
N_{red}	El. Buckling	Class 4	$\alpha_{N,4} = A_{eff}/A$

Table 3.3: Ultimate Axial Load

Bending moment	Limit state	Section class	Correction factor
M_u	Collapse	Class 1	$\alpha_{M,1} = \begin{cases} \alpha_5 = 5 - (3.89 + 0.00190n) / \alpha_0^{[0.27+0.0014n]} \\ \alpha_{10} = \alpha_0^{[0,21\log(1000n)]} \cdot 10^{\left[7.96 \cdot 10^{-2} - 8.09 \cdot 10^{-2} \log(n/10)\right]} \end{cases}$ (depending on the alloy)
M_{pl}	Plastic	Class 2	$\alpha_{M,2} = \alpha_o = Z/W$
M_{el}	Elastic	Class 3	$\alpha_{M,3} = 1$
M_{red}	Elastic buckling	Class 4	$\alpha_{M,4} = W_{eff}/W$

Table 3.4: Ultimate Bending Moment

3.5.2. Evaluation of ultimate bending moment

The load-bearing capacity of cross-sections under bending moment can be evaluated with reference to the above mentioned limit states and the corresponding behavioural classes (Mazzolani and Piluso, 1995)

The value of bending moment for a given limit state can be expressed by the generalised formula:

$$M_{Rd} = \alpha_{Mj}\, Mj\, W f_d \tag{3.3}$$

being:

f_d the design value of strength;

W the section modulus;

α_{Mj} a correction factor, given in Table 3.4, depending on the assumed limit state, where:

$n = f_{0.2}$ (in daNmm^{-2}) is the exponent of Ramberg - Osgood law representing the material behaviour;

α_5 and α_{10} are the generalised section shape factors corresponding to ultimate curvature values χ_u equal to $5\chi_{el}$ and $10\chi_{el}$, respectively, being χ_{el} the elastic limit curvature;

α_0 is the geometrical shape factor;

Z is the plastic section modulus;

W_{eff} is the section resistance modulus evaluated accounting for local buckling phenomena, in case of class 4.

When welded sections are involved, reduced value W_{red} and Z_{red} of section resistance and plastic modulus shall be used, evaluated by accounting for HAZ (heat-affected zones).

5.5.3. Combined axial force and bending

Cross-sections subject to combined axial force N_{Sd} and bending moments $M_{y,Sd}$ and $M_{z,Sd}$ shall satisfy the following equations:

$$\left(\frac{N_{Sd}}{\alpha_{Nj}\, A\, f_d}\right)^{\xi_o} + \frac{M_{y,Sd}}{\alpha_{Mj,y}\, W_y\, f_d} \leq 1 \tag{3.4}$$

and

$$\left(\frac{N_{Sd}}{\alpha_{Nj}\, A\, f_d}\right)^{\eta_o} + \left(\frac{M_{y,Sd}}{\alpha_{Mj,y}\, W_y\, f_d}\right)^{\gamma_o} + \left(\frac{M_{z,Sd}}{\alpha_{Mj,z}\, W_z\, f_d}\right)^{\xi_o} \leq 1 \tag{3.5}$$

where:

f_d is the design value of strength;

A is the net cross-sectional area;

W_x, W_z are the section moduli for maximum stress, if subjected only to moment about the y-y and z-z axes, respectively;

α_{Nj} is a correction factor, given in Table 3.3, depending on the assumed limit state;

$\alpha_{Mj.y}$, $\alpha_{Mj.z}$ are the correction factors for bending about the y-y and z-z axes, given in Table 3.4, depending on the assumed limit state;

η_o, γ_o, ξ_o are exponents given by the following relationships:

$\eta_o = 1$ or, alternatively, $\eta_o = \alpha_{Mj.y}^{2}\ \alpha_{Mj.z}^{2}$ but $1 \leqslant \eta_o \leqslant 2$

$\gamma_o = 1$ or, alternatively, $\gamma_o = \alpha_{Mj.z}^{2}$ but $1 \leqslant \gamma_o \leqslant 1.56$

$\xi_o = 1$ or, alternatively, $\xi_o = \alpha_{Mj.y}^{2}$ but $\xi_o \geqslant 1$

The abovementioned interaction equations are provided by Part 1.1 of Eurocode 9 for I-sections without localized welds or holes. Slightly different expressions are proposed for solid cross-sections, hollow sections and other cross-sections.

3.6. Stability of members

3.6.1. Column buckling

The buckling curves proposed by ECCS and included in EC9 have been based upon the available experimental and theoretical results, with the following assumptions (Mazzolani & Frey 1983):

a) The geometrical imperfection is usually assumed to be represented by an initial out-of-straightness with a midspan deflection $v_0 = L/1000$.
b) The influence of the shape of the cross - section and of its variation of thickness is negligible in case of open symmetrical sections, but must be considered in case of open asymmetrical sections (T, U etc.) or box - sections.
c) The σ - ε law plays an important role in the shape of buckling curves and the Ramberg - Osgood law can be used with satisfactory approximation, by assuming the values $E = 70000$ Nmm^{-2} and $10n = f_{0.2}$ ($f_{0.2}$ in Nmm^{-2}), being n its exponent.
d) Nevertheless, from test and computation, it can be assumed that the behaviour of the alloys commonly used can be divided into two classes: the first comprises the heat-treated alloys (with $f_{0.2}$ from 200 to 300 Nmm^{-2}) and the second the non-heat-treated alloys (with $f_{0.2}$ from 100 Nmm^{-2} to 200 Nmm^{-2}).
e) Residual stresses and variations of the elastic limit along the cross-section are negligible imperfections in extruded members, but not in welded.
f) Since experimental results were in very close agreement with the numerical analysis, simulation computations can be systematically used to elaborate the basic data.

Starting from these statements, the main problem was how to select buckling curves. At first glance, it seemed that the principal aspects of the problem meriting consideration were the cross-section shape (box or open section, symmetrical or asymmetrical) and the material treatment (heat-treated or not-heat-treated alloy).
If all these variables were taken into account, too many curves would have resulted. A reduction can be obtained, by considering:

- the necessity to cover just the alloys commonly used for structural applications;
- the advantage of having as few buckling curves as possible;
- the scatter of the buckling curves in the $\overline{N} - \overline{\lambda}$ plane from the worst to the most favourable case;
- the practical deviations of the main properties such as σ - ε law, yield stress, shape of the profile, etc.

As a result of this study, it was decided to adopt three nondimensional buckling curves (a), (b) and (c), which cover the extruded aluminium bars with a guaranteed elastic limit greater than or equal to 100 Nmm^{-2} (Figure 3.5) These three curves have been computed with the following data (Figure 3.6):

a) double T cross-section in the minor axis bending with n=20;
b) tubolar section with an eccentricity of 10 percent and n = 15;
c) triangular box section in the minor axis bending with an eccentricity equal to 10 percent and n = 10.

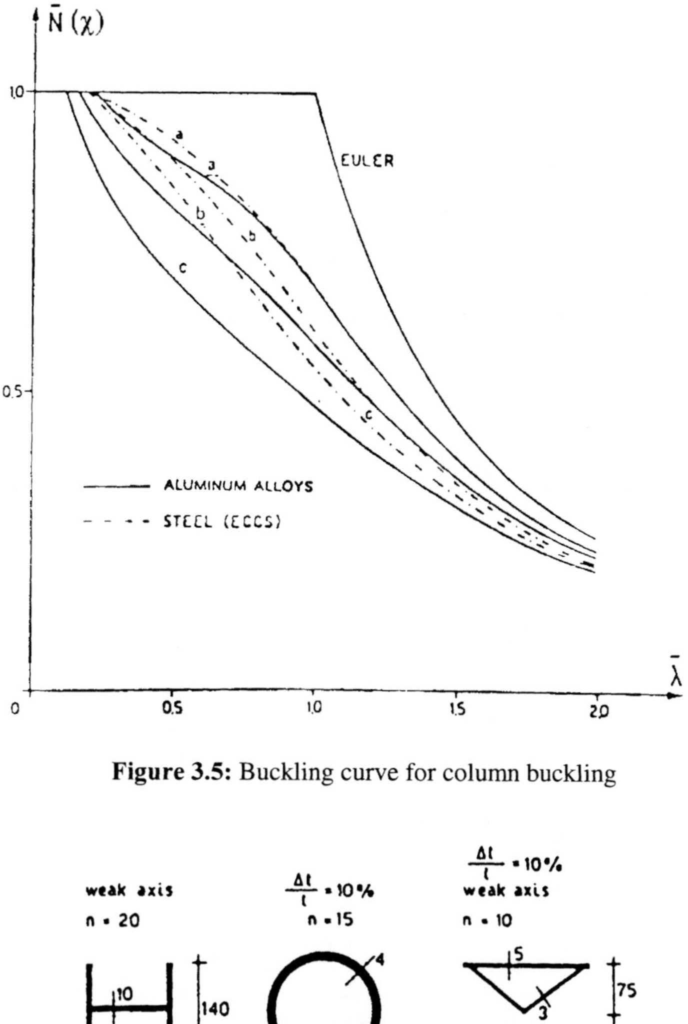

Figure 3.5: Buckling curve for column buckling

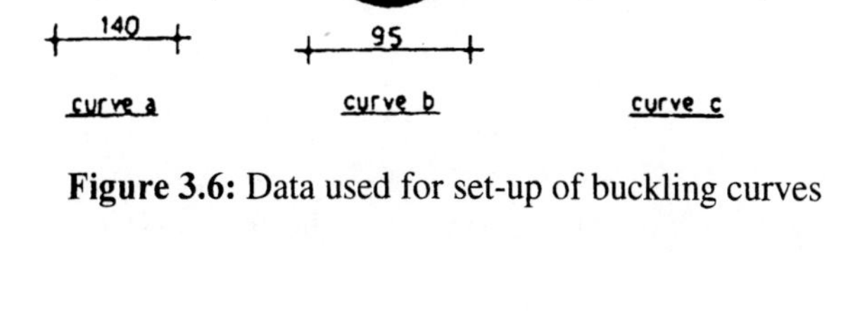

Figure 3.6: Data used for set-up of buckling curves

Axial load	Limit state	Section class	
Shape of cross-section	Cross-section with respect to an axis perpendicular to bending plane	Aluminium alloy with elastic limit $f_{0.2} \geq 100\text{Nmm}^{-2}$	
		heat-trated	not- heat-trated
Open Solid	symmetrical asymmetrical	a b	b c
Hollow	Elastic	b c	c c

Table 3.5: Choice of the buckling curve for aluminium alloy columns

The cross-sections corresponding to curves (a) and (b) have been used in the experimental research. The cross-section of curve (c) combines both of the unfavourable effects due to asymmetry and eccentricity. The three buckling curves must be used according the data of Table 3.5.
As shown in Figure 3.5, the range of the $\overline{N} - \overline{\lambda}$ plan covered by these curves is wider than the one of the corresponding a, b, c curves of steel, particularly due to the lowering shape of curve c.
These curves have been validated by wide comparisons with test and simulation results. The comparison (Figure 3.7) with the main codes (Mazzolani & Valtinat, 1987) have shown that substantially the ECCS method provides in average the most economical results, even if curve (c) seemed to be too conservative (Mazzolani, 1981).
With codification purpose only, the ECCS Committee decided to simplify the method, by limiting the use of curves (a) and (b) for symmetrical profiles of heat-treated and not-heat-treated alloys, respectively. The curve (c), which takes into account all possible imperfections including asymmetry, has been only referred to as a lower bound.

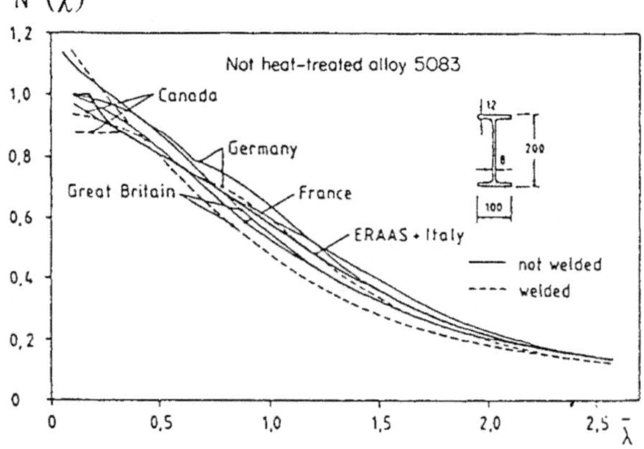

Figure 3.7: Comparison among different codes

Appropriate reduction factors have been introduced in the provisions, in order to consider both the effect of asymmetry and the lowering due to residual stresses and heat-affected zones in welded profiles.
The buckling load can be evaluated by means of the following formula proposed by ECCS and also used in EC9 (the notation used is derived from Eurocode 9):

$$N_{b,Rd} = \chi \cdot \eta \cdot k_1 \cdot k_2 \cdot f_d \cdot A \tag{3.6}$$

being:

f_d the design value of strength;
A the cross-sectional area of the bar;
χ the reduction factor, equal to $\overline{N}$ for ECCS, given by buckling curves (Figure 3.5) as a function of the slenderness ratio $\overline{\lambda}$;
η the factor to allow for any loss of effectiveness due to the presence of class 4 elements, equal to 1 for class 1, 2 or 3 and equal to A_{eff} / A for class 4;
k_1 a reduction coefficient which takes into account the asymmetry of the cross-section given in Table 3.6;
k_2 a reduction coefficient taking account of welding effects given in Table 3.6;

<table>
<tr><th colspan="2"></th><th>heat-trated alloys</th><th>not- heat-trated alloys</th></tr>
<tr><td rowspan="3">k_1</td><td>symmetric cross section</td><td>$k_1 = 1$</td><td>$k_1 = 1$</td></tr>
<tr><td rowspan="2">asymmetric cross section</td><td>$k_1 = 1 - 2.4\psi^2 \dfrac{\overline{\lambda}^2}{(1+\overline{\lambda}^2)(1+\overline{\lambda})^2}$</td><td>$k_1 = 1 - 3.2\psi^2 \dfrac{\overline{\lambda}^2}{(1+\overline{\lambda}^2)(1+\overline{\lambda})^2}$</td></tr>
<tr><td colspan="2">$\psi = (\gamma_{max} - \gamma_{min})/h$ where γ_{max} and γ_{min} are the distances between the section edges and the centroid in the plane of buckling and h is the cross-section depth</td></tr>
<tr><td rowspan="2">k_2</td><td>longitudinal welds</td><td>$k_2 = 1 - \left(1 - \dfrac{A_1}{A}\right)10^{-\overline{\lambda}} - \left(0.05 + 0.1\dfrac{A_1}{A}\right)\overline{\lambda}^{1.3(1-\overline{\lambda})}$
with $A_1 = A - A_{haz}(1-\rho_{haz})$
in which A_{haz} is the area of HAZ and ρ_{haz} is the HAZ softening factor</td><td>$k_2 = 1 + 0.04(4\overline{\lambda})^{(0.5-\overline{\lambda})} - 0.22\overline{\lambda}^{1.4(1-\overline{\lambda})}$</td></tr>
<tr><td>transversal welds</td><td>$k_2 = \rho_{haz}$</td><td>$k_2 = \rho_{haz}$</td></tr>
</table>

Table 3.6: Choice of the buckling curve for aluminium alloy columns

3.6.2. Flexural-torsional buckling

The method proposed by ECCS is based on the equation:

$$M_D = M_{pl}\left(\frac{1}{1+\overline{\lambda}_M^{2n}}\right)^{\frac{1}{n}} \tag{3.7}$$

with n=2.

The range of applicability of this formula is based on the following conditions:

- The cross-section has double symmetry (double T sections with equal flanges).
- Loads act in the plane of the web.
- Distortion of the cross-section and local buckling phenomena are not taken into account.
- Supports do not allow lateral displacements and rotations outside the bending plane.

It has been checked that:

$$M_{max} \leq M_D \qquad \text{and} \qquad M_{max} \leq W f_d \tag{3.8}$$

being:

M_{max} the maximum bending moment caused by design loads;
M_D the ultimate flexural-torsional buckling moment;
W the section modulus;
f_d the design value of strength.
The M_D value is given by the relationship

$$M_D = \frac{M_{pl}}{\sqrt{\left(1+\overline{\lambda}_M^4\right)}} \tag{3.9}$$

where:

$\overline{\lambda}_M = \sqrt{M_{pl} / M_{cr,D}}$ is the slenderness ratio;

$$M_{cr,D} = \psi_1 \frac{\pi}{L_{c,h}}\sqrt{EI_y GI_T}\sqrt{1+\frac{\pi^2}{k^2}}$$

$$k = L_{c,h}\sqrt{GI_T / EI_\varpi}$$

with the well known meaning of symbols.

According to EC9, the buckling resistance moment is given by the following formula:

$$M_{b,Rd} = \chi_{LT} \cdot \alpha_{Mj} \cdot W \cdot f_d \tag{3.10}$$

being:

f_d the design value of strength;

W the section modulus;

α_{Mj} a correction factor, given in Table 3.4, subject to the limitation $\alpha \leqslant Z / W$;

χ_{LT} the reduction factor given by buckling curves (Fig. 3.8) as a function of the slenderness ratio $\overline{\lambda}_{LT}$.

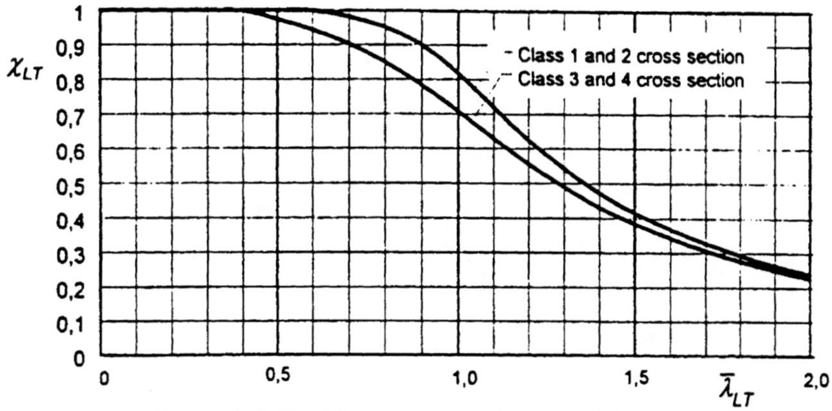

Figure 3.8: Buckling curve for lateral-torsional buckling

3.6.3. Beam-columns

On the basis of the experimental results, the ECCS Committee decided to adopt the most conservative formulation, which was the one proposed by Clark-Massonnet, for both plane buckling and flexural-torsional buckling (Mazzolani, 1995a).

The method was also extended to biaxial bending.

For plane bending, it has to be checked that:

$$\frac{N}{N_c} + \frac{\mu}{\mu - 1}\frac{M_{eq}}{M_u} \leq 1 \qquad \text{and} \qquad \frac{N}{A} + \frac{M_{\max}}{W} \leq f_d \tag{3.11}$$

It should be noted that the ultimate moment M_u is equal to:

M_D in the case of flexural-torsional buckling;

M_{pl} in the case of plane buckling.

However the equivalent moment M_{eq} is calculated as follows:

M_a and M_b are the end moments of the bar, with $|M_b| \geq |M_a|$; $M_1 = \beta |M_b|$, with $\beta = 0.6 + 0.4(M_a / M_b) \geq 0.4$; M_2 is the absolute value of the maximum bending moment on the axis of the bar with transverse loads applied (when the moment diagram has no peaks between the ends, $M_2 = |M_b|$. If there are no transverse loads, it is assumed that $M_2 = 0$.

The value to be used as M_{eq} is the greater of M_1 and M_2 defined in this way. The value of M_{max} is the greater of M_b and M_2.

For biaxial bending, it has to be checked that:

$$\frac{N}{N_c}+\frac{\mu_y}{\mu_y-1}\frac{M_{eq,y}}{M_{u,y}}+\frac{\mu_z}{\mu_z-1}\frac{M_{eq,z}}{W_z\cdot f_d}\leq 1 \qquad \text{and} \qquad \frac{N}{A}+\frac{M_y}{W_y}+\frac{M_z}{W_z}\leq f_d \tag{3.12}$$

where the subscript y signifies major axis properties and the subscript z signifies minor axis properties.
The $\mu = \sigma_{cr} A / N$ ratio, which characterizes the amplifying coefficient, can be calculated by using the slenderness ratio corresponding to the bending plane.
The EC9, for beam-columns subject to combined axial force N_{Sd} and bending moments $M_{y,Sd}$ and $M_{z,Sd}$ with I shaped and similar cross-section without localized welds or holes, proposed the following criterion:

$$\left(\frac{N_{Sd}}{\chi_z\,\alpha_{Nj}\,A\,f_d}\right)^{\eta_c}+\left(\frac{M_{y,Sd}}{\chi_{LT}\,\alpha_{Mj,y}\,\mathrm{W_y}\,f_d}\right)^{\gamma_c}+\left(\frac{M_{z,Sd}}{\alpha_{Mj,z}\,\mathrm{W_z}\,f_d}\right)^{\xi_{zc}}\leq 1 \tag{3.13}$$

where:

f_d is the design value of strength;
A is the net cross sectional area;
W_x, W_z are the section moduli for maximum stress, if subjected only to moment about the y-y and z-z axes;
α_{Nj} is a correction factor, given in Table 3.3, subject to the limitation $\alpha \leqslant 1$;
$\alpha_{Mj.y}$, $\alpha_{Mj.z}$ are the correction factors for bending about the y-y and z-z axes, given in Table 3.4, depending on the assumed limit state;
η_c, γ_c, ξ_{cz} are exponents given by the following relationships:
$\eta_c = 0.8$ or, alternatively, $\eta_c = \chi_z\,\eta_o$ but $\eta_c \geqslant 0.8$
$\gamma_c = \gamma_o$
$\xi_{cz} = 0.8$ or, alternatively, $\xi_c = \chi_z\,\xi_o$ but $\xi_{cz} \geqslant 0.8$
χ_z is the reduction factor for buckling when one or both flanges deflect laterally (buckling in the y-z plane or lateral-torsional buckling);
χ_{LT} is the reduction factor for lateral-torsional buckling given by buckling curves (Fig. 3.8).

3.7. The approach for slender sections

The effect of local buckling in slender members is allowed for by replacing the true section by an effective one, which is obtained by using a local buckling coefficient ρ_c to factor down the thickness of any slender element which is wholly or partly in compression. The coefficient ρ_c is provided through curves as a function of the factor β/ε, being β a slenderness parameter which depends on the b/t ratio, the stress gradient for the element concerned and $\varepsilon = (250/f_{0.2})^{0.5}$.
Three design curves have been proposed (Landolfo and Mazzolani, 1997) and introduced in EC9 (Figure 3.9), referring to the following cases:
Curve A: unwelded elements in heat-treated alloy ($n > 10$)
Curve B: welded elements in heat-treated alloy and unwelded plates in non heat-treated alloy

$(n \leq 10)$
Curve C: welded elements in non heat-treated alloy.
According to EC9, a preliminary distinction between internal and outstand elements give rise to the design curves expressed in the general form:

$$\rho_c = \frac{\delta_1}{\beta/\varepsilon} - \frac{\delta_2}{(\beta/\varepsilon)^2} \tag{3.14}$$

being δ_1 and δ_2 two coefficients, whose values are approximately given in the Table 3.7.

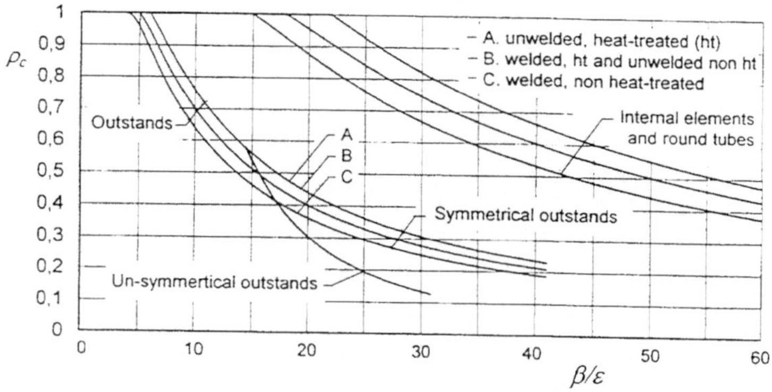

Figure 3.9: Relationship between ρ_c and β/ε

	internal			outstand		
Curve	δ_1	δ_2	$(\beta/\varepsilon)_0$	δ_1	δ_2	$(\beta/\varepsilon)_0$
A	32	220	22	10	24	6
B	29	198	18	9	20	5
C	25	150	15	8	16	4

Table 3.7: Values of δ_1, δ_2 and $(\beta/\varepsilon)_o$

3.8. Evaluation of internal actions

Some indications regarding the methods of global analysis both elastic and plastic, to be used in the calculation of structures, are provided in EC9, Part 1.1 (Annex C) (CEN, 1998).
Elastic global analysis relies on the assumption that the stress-strain relationship of the material is linear up to failure, independently of the stress level (Figure 3.10 A1). This assumption may be kept for both first order and second order analysis, even when the resistance of the cross sections is evaluated according to its ultimate load bearing capacity in post-elastic range. In order to take into account the plastic moment redistribution within the structure, the peak elastic moment can be increased or decreased by up to 15%, provided the new internal forces and moments remain in equilibrium and the cross-section have sufficient ductility to allow for the plastic redistribution. For this reason all members, where the moments are reduced, must have Class 1 or at least Class 2 cross-section. When such conditions are fulfilled, then elastic global

analysis can be used in all cases.
For more refined calculation, elastic global analysis may be also applied by assuming that the stress-strain relationship of the material is not linear (Figure 3.10 A2), the value of the instantaneous tangent modulus depending on the stress level. In order for allowing the exploitation of the inelastic feature of material, all the members existing in the structure must have Class 1 or Class 2 cross-sections. The characterisation of the law of the material must take into account the actual strain-hardening behaviour of the alloy.
Plastic global analysis may be used only when member cross-sections satisfy requirements specified for Class 1 cross-sections and provided that the aluminium alloy has sufficient ductility. Cross-section of Class 2, 3 and 4 are not allowed. For Class 1 cross-sections the check of the deformation capacity is required in relation to the ductility demand of the structural scheme.
Plastic analysis may be carried out by assuming for the material the following behavioural models:

- Rigid-Perfectly plastic (Figure 3.10 B1);
- Elastic-Perfectly plastic (Figure 3.10 B2);
- Inelastic-Perfectly plastic (Figure 3.10 B3).

The three model differs from each other by the assumption made on the material behaviour in the elastic range, which can be either rigid, elastic or inelastic. When it is assumed to be rigid (Figure 3.10 B1), then elastic deformations of cross-sections and members may be neglected and all plastic deformations are assumed to be concentrated at plastic hinge location. It can be used when the elastic pre-collapse deformations are comparatively small and their magnitude is not so relevant as to involve second order effects.
In the elastic-perfectly plastic analysis, the behaviour of material is assumed to be linear up to a limit level of stress corresponding to the yielding (Figure 3.10 B2). As a consequence of this, the behaviour of the cross-sections can be generally assumed to be elastic-plastic as well, at least for small values of the geometrical shape factor ($\alpha < 1.2$). As a consequence of this, it is possible to assume that the plastic deformations are concentrated in correspondence of the plastic hinge location. The transition from elastic to plastic range will be more or less gradual depending on both load condition and section shape.
A more generality can be achieved if the elastic branch of the material law is assumed to be non linear, before reaching the horizontal plateau, as in the third of the above cases (Figure 3.10 B3). Accordingly, the non linear behaviour of sections is considered in the evaluation of the deformation occurring in a given member before the formation of the plastic hinge. For this method, a discretised F.E.M. approach is recommended, in order to closely represent the non linear behaviour of the structure.
EC9 also allows for the effect of strain-hardening to be taken into account. For this purpose, the material modulus used for the plastic analysis can be modified by substituting the horizontal plastic branch, with an increasing one, evaluated according to the hardening feature of the alloy. The following options are covered:

- Rigid-Hardening (Figure 3.10 C1);
- Elastic-Hardening (Figure 3.10 C2);
- Generically inelastic (Figure 3.10 C3).

Rigid- and elastic-hardening analyses are quite similar to the corresponding rigid- and elastic-plastic ones, in the sense that they are based on a concentrated plasticity model relying on the concept of plastic hinge (Figure 3.10 C1 and C2). The main difference stands in the evaluation of the post-elastic response, which depends on the hardening feature of the alloy, as well as on its available ductility. For this reason, the analysis is assumed to be concluded when a given limit value of deformation is reached in the material.

In the most general case of structural analysis, called "Generically inelastic", both material and sections are idealised according to their actual stress-strain and generalised force-displacement relationship, respectively (Figure 3.10 C3).

The transition from the elastic to the plastic range is gradual and the achievement of the ultimate limit state is defined by the attainment of a given limit values of strength or deformation. Contrary to the all previous cases, the "Generically inelastic" approach cannot adopt the simple concentrated plasticity idealisation, based on the concept of plastic hinge, but should use refined discretised approaches, e.g. F.E.M. simulation, to display the whole of its accuracy in the prediction of structural inelastic behaviour.

Different analytical models for interpreting the stress-strain relationship, according to various degree of accuracy, are given in EC9, Part 1.1 (Annex E).

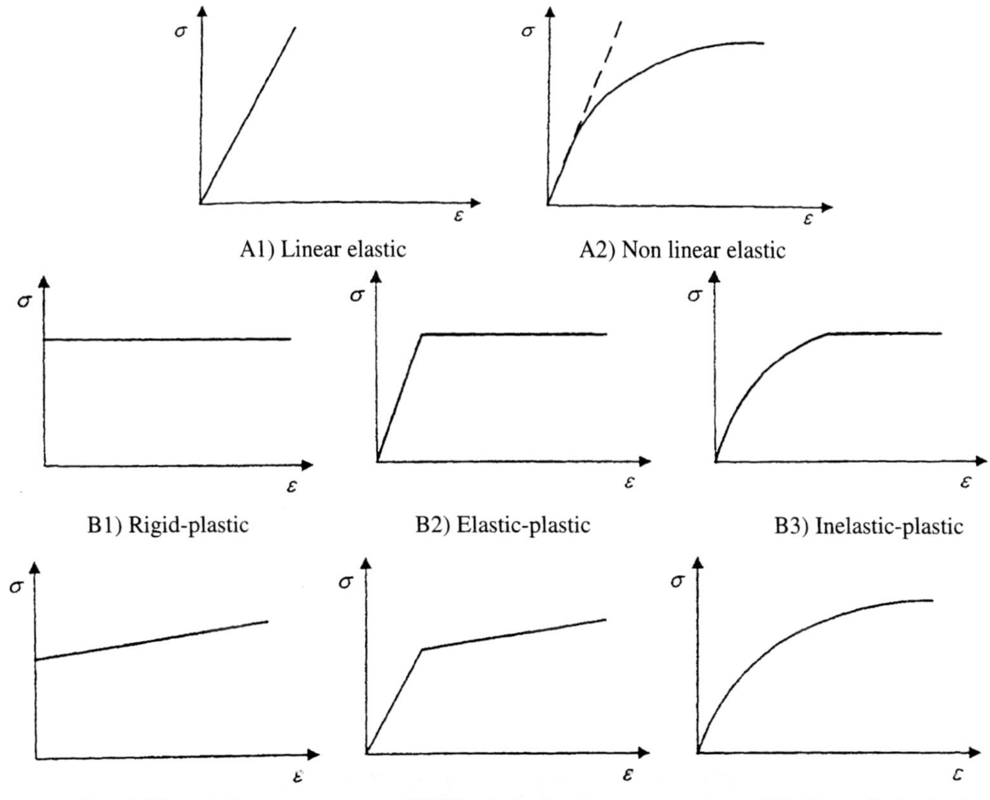

Figure 3.10: Material behavioural models

3.9. Evaluation of ductility demand

The ductility demand of a given structure under the design loads can be evaluated in several ways, depending on how the external actions are applied to the structures. A rigorous definition of the ductility demand is only possible if the load is applied to the structure through a system of impressed displacements. In this case, regardless of the structure strength capability, the ductility demand can be defined as the maximum value of a deformation parameter which the structure is able to reach in a load process in which a generical displacement parameter is assumed as independent variable. However, in most cases, the structure is loaded by means of a force system increasing up to collapse. In such conditions, the ductility demand would be nominally infinite, because when the ultimate value of load is attained, the structure has no longer possibility to resist the external loads and the plastic flow into the collapsed sections would be unlimited. Thus, the ductility demand can be defined only in a conventional way. For generical truss- or beam-made structures, EC9 indicates the following ways to evaluate the ductility demand:

1. The ductility demand is defined as the required rotation in the most developed plastic hinge when the plastic mechanism is attained. The structure is solved by means of a concentrated plasticity approach based on the concept of plastic hinge. The maximum required strain can be evaluated provided that as convenient length for the plastic hinge is assumed.
2. The ductility demand is defined as the required rotation in the most developed plastic hinge evaluated when the plastic hinge idealisation provides the same load bearing capacity as predicted by a more accurate, inelastic method of analysis based on a discretised model. In this case the structure should be solved by means of both methods, in order to compare the obtained results.
3. The ductility demand is not evaluated on the base of the structural scheme, but it is defined "a priori" as a function of the maximum elastic strain of the alloy. The corresponding load bearing capacity can be evaluated in a simpler way by applying the plastic hinge method, in which a modified value of the conventional yield stress is adopted, in order to take into account the actual behaviour of the alloy in terms of both available ductility and strain hardening.

Among the three methods suggested by EC9, the first one is purely conventional, since is based on a concentrated plasticity idealisation, which hardly corresponds to the actual structural behaviour at collapse. The second one requires for the structure to be calculated two times, with a concentrated plasticity approach, as well as with a F.E.M. numerical simulation. For this reason, it would result in a higher computation cost, which is not suitable for practical application. Eventually, the third one, could represent a good method for an accurate inelastic analysis of aluminium alloy structures, without disregarding the actual mechanical features of the material. Furthermore, because of its inherent simplicity, it can be profitably used as a design method for structures at the ultimate limit state. In fact, from the point of view of application, it is quite similar to the classical plastic hinge method applied for steel structures. EC9 provides some indications about the use of such method for the plastic analysis of structures, by considering the actual mechanical properties of the different aluminium alloys. Additional rules for the evaluation of rotation capacity are supplied in Annex G.

3.10. The plastic hinge method

The problem of how to apply the concentrated plasticity approach, namely the "plastic hinge method", is faced in Annex D of EC9, Part 1.1 (CEN, 1998). This part of EC9 is concerned with Class 1 cross-sections only, i.e. with structures for which collapse occurs due to the attainment of ultimate deformation in a certain number of sections. When cross-section belonging to Class 2, 3 and 4 are used, local buckling phenomena must be taken into account for the evaluation of both load bearing capacity and available ductility (Mandara & Mazzolani, 1995)
For rigid-perfectly plastic (Figure 3.10 B1) or elastic-perfectly plastic (Figure 3.10 B2) behavioural curves, the ultimate moment M_u in a plastic hinge is given by:

$$M_u = \alpha_o \, \eta f_{0.2} \, W \tag{3.15}$$

being:

- $f_{0.2}$ the 0.2% proof strength;
- W the section modulus;
- α_o the geometrical shape factor;
- η a numerical parameter depending on α_o and on the conventional available ductility of the material.

For rigid-hardening (Figure 3.10 C1) or elastic-hardening (Figure 3.10 C2) behavioural courves, the conventional yield moment M_y and the ultimate moment M_u in a plastic hinge are given respectively by:

$$M_y = \alpha_o \, \eta f_{0.2} \, W \tag{3.16}$$

$$M_u = \alpha_\xi \, \eta f_{0.2} \, W \tag{3.17}$$

being:

- $f_{0.2}$ the 0.2% proof strength;
- W the section modulus;
- α_o the geometrical shape factor;
- α_ξ equal to α_5 or α_{10} depending on the ductility feature of the alloy (for the definition of α_5 and α_{10} see Table 3.4);
- η a numerical parameter depending on α_o and on the conventional available ductility of the material.

The correction coefficient η of the previous equations 3.15, 3.16 and 3.17 is provided in Figure 3.11, where, the conventional ultimate curvature χ_u defined through the conventional curvature at elastic limit state χ_e, shall be chosen according to the ductility properties of the alloy:

$\chi_u = 5\,\chi_e$ for brittle alloys;

$\chi_u = 10\,\chi_e$ for ductile alloys.

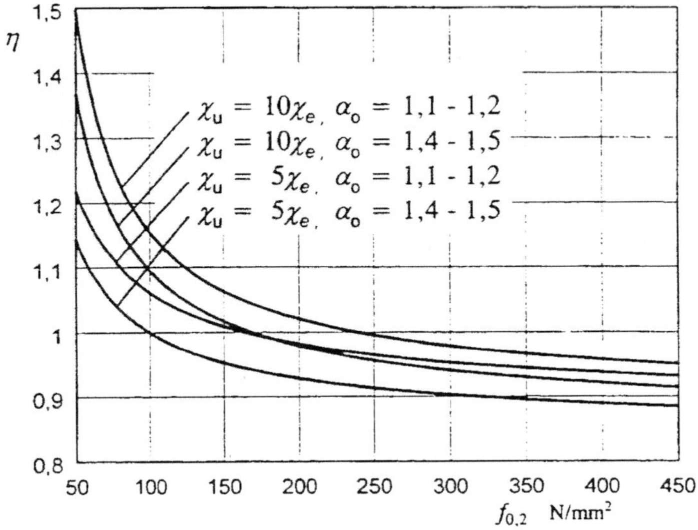

Figure 3.11: Value of the correction coefficient η

3.11. Classification of connections

According to Eurocode 9 (CEN, 1998), connections may be classified by considering their capability to restore the behavioural properties (elasticity, rigidity, strength and ductility) of the connected member. With respect to the global behaviour of the connected member, two main classes are defined:

- *Fully restoring connections*, where the behavioural properties are always higher than those of the connected member.
- *Partially restoring connections*, where the behavioural properties do not reach those of the connected member.

With respect to the single behavioural property of the connected member, the following classifications can be assumed (Figure 3.12):

Classification according to rigidity

- Rigidity restoring connections (rigid);
- Rigidity non-restoring connections (semi-rigid);

depending on whether the initial stiffness of the connected member is restored or not, regardless of strength and ductility.

Classification according to strength

- Strength restoring connections (full strength);
- Strength non-restoring connections (partial strength);

depending on whether the ultimate strength of the connected member is restored or not, regardless of rigidity and ductility.

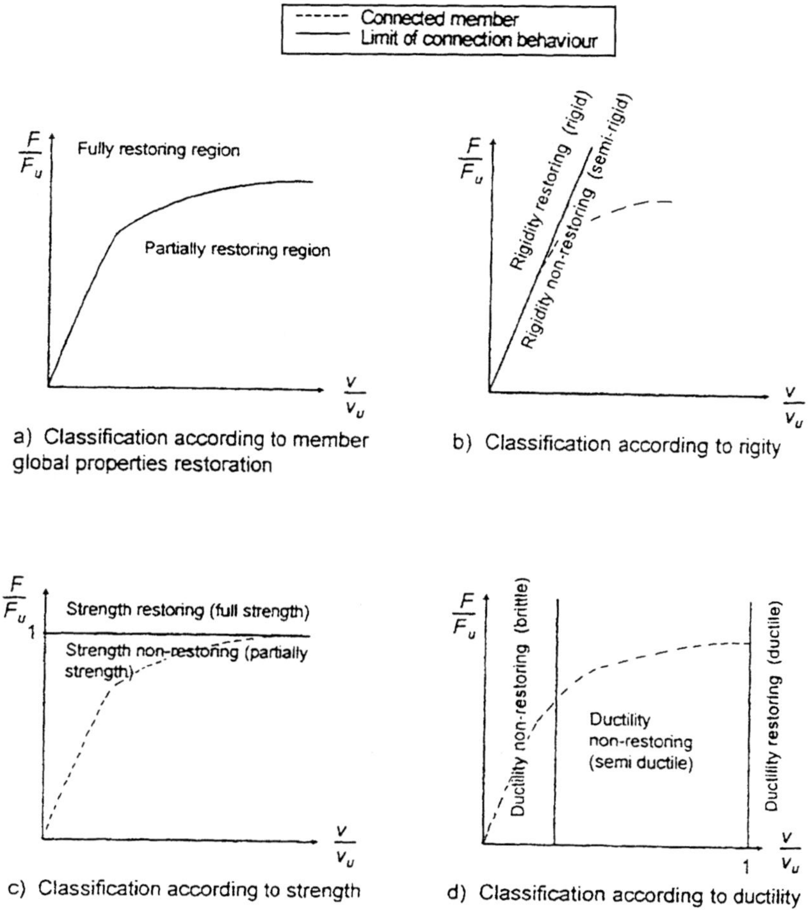

Figure 3.12: Classification of connections

Classification according to ductility

- Ductility restoring connections (Ductile);
- Ductility non-restoring connections (Semi-ductile or Brittle with regard to the level of deformation);

depending on whether the ductility of the connected member is higher or lower than that of the connected member, regardless of strength and rigidity.

In addition, the Eurocode 9 establishes a strict relationship between the type of connection according to the above classification and the calculation method used for the global analysis of the structure, as it is shown in the Table 3.8.

Method of global analysis	**Type of connections which must be accounted for**	**Type of connection which may be ignored**
ELASTIC (Linear, Non linear)	Semi-rigid connections (full or partial strength, ductile or non-ductile with or without restoring of member elastic strength) Partial strength connections (rigid or semi-rigid, ductile or non-ductile) without restoring of member elastic strength	Fully restoring connections Rigid connections (full or partial strength, ductile or non-ductile) with restoring of member elastic strength Partial strength connections (rigid, ductile or non-ductile) with restoring of member elastic strength
PLASTIC (Rigid-Plastic Elastic-Plastic Inelastic-Plastic)	Partial strength connections (rigid or semi-rigid, ductile or non-ductile) without restoring of member elastic strength	Fully restoring connections Partial strength, ductile connections (rigid or semi-rigid) with restoring of member elastic strength Full strength connections
HARDENING (Rigid-Hardening Elastic-Hardening Generically Inelastic)	Partially restoring connections	Fully restoring connections

Table 3.8: General design requirements

4. ALUMINIUM STRUCTURES IN BUILDINGS AND SPECIAL STRUCTURES

4.1. Framed structures

The first aluminium structural system used in building construction has been the portal frame scheme for industrial applications (CIDA, 1972). Contrary multi-storey schemes are not used in practice. The portal frame allows to exploit the functionally of extruded shapes and to use prefabricated systems. Typical examples are: the upper chords made of a double T section where the flange is ondulated (in the rolling mill roof in Kreuzlingen, CH, 1953, Figures 4.1e,f); the purlines made of double T section where the flanges are sloped (in the prefabricated rural building, I, 1967, Figures 4.1c,d). Among several types of constructions using plane framed structures, the following can be mentioned:

- prefabricated systems for one-story buildings (Figures 4.1a,b);
- hangars (Figures 4.2a,b);
- roofing (Figures 4.2c,d);
- warehouse (Figures 4.2e,f);
- aerostation (Figures 4.2g,h).

4.2. Reticular space structures

The necessity to cover a large surface without internal supports has addressed the civil engineer to the use of light structural materials as the aluminium alloys. In these cases, where the incidence of the structural weight is fundamental, the utilization of the aluminium can represent a valid alternative to steel. On the other hand, if we considere the complete absence of maintenance, the advantages increase above all for those structures situaded in humid environments, such as the swimming pool roofs. Neverthless, the growth of the aluminium in this structural field was initially hindered due both to great computational complexity, owing to the high number of degree of freedom, and to the difficulty of fabrication of joints able to connect many bars in several directions. While the "calculation problem" has already been solved thanks to the modern computer programs, for the second problem there are many proposed solutions (Mazzolani & Piluso, 1988), which utilize the wrought properties of the material. Among them, the following systems can be mentioned: GYROTRON, M-DEC, MERO, NODO-NON-NODO, SPHEROBAT, SPHAR, TRIODETIC and SPACE-DAL, many of them use both steel and aluminium.

Several applications of reticular space structures can be found in South America (Figures 4.3a,b,c,d,e,f,g,h,i,l). A very spectacular space structure has been erected for the Interamerican Exibition Center of San Paulo in Brazil (1969). This structure covers an area of about 67600 m^2 with a mesh 60x60 m. The depth of the reticular layer is 2,36 m. It was interelly site-bolted on the ground and after lifted at the final level of 14 m by means of 25 cranes located in the corners of the mesh, in the position of the actual supports. The weight of the reticular structure was 16

kg/m^2; the number of bars was 56820 and their total length one after another was 300 km. The erection time was extraordinary quick (27 hours!), by using a number of 550000 bolts in 13724 nodes. The materials were: aluminium alloys of 6063 and 6351series T6 for cylindrical bars, Al 99,5 for trapezoidal sheeting and galvanized steel bolts for connections (Figures 4.3a,b,c).
After this very challenging example, many other important buildings have been erected in Brazil (Figures 4.3d,e,f,g,h,i) by using aluminium reticular space systems. Among them, very similar is the case of the International Congress Center in Rio de Janeiro, where the same mesh 60x60 m has been used, covering in total 33000 m^2 (Figures 4.3d,e,f). More recently, an aluminium reticular system has been used for erecting the Memorial Pyramid in La Baie (Quebec, Canada), built by the Remac Industry, in order to remember the damage due to a flood arised during the 80's.
Particularly interesting is the solution proposed for roofing the tribunes of the Olimpic Stadium in Rome (Figures 4.4a,b,c,d) for the Mundial Games of 1990 in Italy (Mazzolani & Piluso, 1988). Such joint is made of a vertical extrusion cross-shaped element bolted to an upper horizontal square plate, this allowing for the exploitation of the advantages of the extrusion process. The system involves the use of aluminium bars with stamped ends, which are connected to nodes with bolts.
A more recent application in Italy has been the roof of the Congress Center in Alghero, a square area of 60x60 m covered by the Space-Dal system of the Ponteggi Dalmine in aluminium version (Figures 4.4e,f) (Ballio *et al.*, 1993).
A new italian system, so-called Geo-system (Aprile & Benedetti, 1998), for reticular space structures has been used for many applications in Italy and abroad (Figure 4.5). It is based on cylindrical bars with conic ends and hollow spherical nodes.

4.3. Domes

This type of structure represents the most challenging application of the aluminium alloys in the structural field. The lightness property of the material and the exploitation of the particular configuration allow the realisation of structures of considerable dimensions. Usually, these structures are employed for covering sporting houses, exhibition centres, congress halls, etc. In many cases these applications are very interesting for the rapidity of erection, the connection systems and the remarkable dimensions. In the Figures 4.6 and 4.7 some applications of the reticular and geodetic domes in aluminium alloys are shown.
The first reticular dome made of aluminium was the so-called "Dome of Discovery" erected in London for the South Bank Exhibition during the Festival of Britain (1951), composed by three directional reticulated arches, with a diameter of 110 m and 24 kg/m^2 weight (Figures 4.6a,b). At that time, it was the largest dome and a remarkable technological achievement, but unfortunately its life was very short, because "tragically, it was demolished by vengeful politicians" some years later (Cruickshank, 1995).
After few years (1959), the first geodetic aluminium dome was erected for covering the "Palasport" in Paris, by using the Kaiser Aluminium system with 61 m diameter and 20 m height (Figure 4.6c), which is very similar to the dome built later on in Guayaquil (Equador) (Figure 4.6d).

More recently, interesting structural systems for geodetic domes made of aluminium have been set-up in U.S.A.. The Tem-cor system, very easy to erect (Figures 4.6e,f,g), is used both for roofing industrial plants with ecological purposes (Figure 4.6h) and for nice public buildings (Figure 4.6i). The Conservatek system is used for large roofing, like auditoriums, congress centers and sport halls (Figure 4.7). It is famous for the “Spruce Goose” dome, which is the largest in the world with a diameter of 415 feets (see some erection phases in Figures 4.7e,f,g,h,i,j).

4.4. Special structures

This category basically covers the vertical structures whose function is to support fixed elements which, for theirs particular use, must be located at a given distance from the ground. The prevalent dimensions can be horizontal, as in case of portal frames for traffic sign (Figure 4.8), or vertical, as in case of antennas, electrical transmission towers and lighting towers (Figure 4.9). The extrusion fabrication process can improve the geometrical properties of cross-sections in such a way to obtain the minimum weight and the highest structural efficiency. In addition, the lightweight of aluminium allows prefabricated systems, very easy for transportation and erection, giving rise to competitive solutions in comparison with other materials (Figures 4.8, 4.9 and 4.10).

Many towers for electrical transmission lines have been erected in Europe (Boulogne - sur - Mer, F, 1956; Seitevara - Ligga, S, 1965; Nantes - Vannes, F, 1969; Marseille, F, 1969). Some special stand-by towers for quick repair of electrical transmission lines have been produced in Italy (1967).

Two important aluminium towers have been erected in Naples (see Figure 4.10). The first is the tower for parabolic antennas of the Electrical Department of Naples erected in 1986. This design received the international award “Hundred Years of Aluminium” (Mazzolani, 1989). The reason of the aluminium choice is basically due to lightness (the tower has been erected on the lop of an existing r.c. stair-case) and corrosion resistance (no problems of maintening). Its heigth is 35 m from the top of the stair-case (in total 50 m about). It is composed by a cylinder 1800 mm internal diameter and 20 mm thickness. The fabrication was shop-welded, by dividing the total heigth in three parts. Than, it was field-bolted during the erection (see some fabrication and erection phases in Figures 4.10a,b,c,d,e,f).

The second example is the “Information Tower” near the football stadium in Naples, which has been equiped by antennas and screens in order to follow from out-side the stadium games (Figures 4.10g,h,i).

4.5. Hydraulic structures

The most important property of aluminium used in hydraulic applications is the corrosion resistance. For this reason, the utilization of the aluminium as structural material represents the only alternative to the realization of structures which require low maintenance costs (Figure 4.11). Further, the lightness is undoubtedly an additional advantage, which is essential when there

are moving parts, as in the case of the rotating crane bridges for large settling circular tanks in water sewage treatment plants (Figures 4.11c,d).
In this range the aluminium alloys can have favourable possibilities to be more convenient than steel, provided that a suitable design is able to exploit their main physical properties. In particular, the corrosion resistance allows to eliminate any protection also in presence of corrosive environment, while the lightness corresponds to energy saving during the operating phases of the plant (Mazzolani, 1985).
The first applications of aluminium alloys in sewage plants appeared in U.K. during the sixties. Afterwards the Ranshofen-Berndorf produced in Austria a series of crane bridges for circular tanks in aluminium-magnesium alloys for spans varying from 7 to 15 m.
A further application in 7020 (T6) alloy was presented by Koser (1982). The cross-section of the girder is a channel composed by two corrugated webs, which transmit the wind loads and railing loads down to the bottom plate.
Pipelines represent another hydraulic application, where the solutions in steel are often disregarded because of high maintenance cost. In the Figure 4.11 some examples of hydraulic structures are shown.

4.6. Offshore structures

It seems important to underline that nowadays the offshore applications can be considered the main future trend for aluminium alloys (Figure 4.12). In fact, they offer to this industry enormous benefits under form of cost savings, ease of fabrication and proven performance in hostile environments. Stair towers, mezzanine flooring, access platforms, walkways, gangways, bridges, towers and cable ladder systems can all be constructed in pre-fabricated units for simple assembly offshore or at the fabrication yard (Figure 4.12a).
Mobility and ease of installation are maintained even for large structural elements, such as link bridges and telescopic bridges (Figure 4.12b).
Helidecks have been made by using aluminium alloy since the early seventies, so they have now a fully tried experience (Figure 4.12c). Moreover, they are designed to be modular and bolted connections, allowing quick erection, easy shipping and handling. In addition, they offer weight reduction of up to 70% over steel, meeting the highest safety standards and proving up to 12% cost saving.
Complete crew quarters and utilities modules, from large purpose-built modules to flexible prefabricated units, have been recently developed. The modules may be used singly or assembled in group to form multi-story complexes, linked by central transverse corridors and stair towers (Figures 4.12d,e,f).
In many cases, the off-shore structures are unique and "world-firsts", requiring extensive applied research in material technology, joint configurations, fatigue design and fastening methods. Therefore, the usual drivers for encouraging a switch from steel to aluminium are (ALCAN OFFSHORE):

- Lower maintenance cost: with aluminium alloys, corrosion is reduced to insignificant levels. Also, certain structures like offshore bridges are difficult to maintain and require special safety considerations during maintenance.

- Lower weight: aluminum's exceptional strength-to-weight ratio means weight savings of 50-70% over equivalent steel structures. It is estimated that a one-tonne weight saving in the superstructure of a platform can give up to a 6-tonnes saving in the overall structure.
- Lower cost: long term cost reduction through maintenance cost savings are well known. That is also achieved from initial cost savings at installation owing to lower lift costs, easier fabrication and faster installation.
- Longer life: the durability of the alloy means that it usually outlasts the structure it is placed on and can be reused on new platforms.
- Recycling: aluminium has high residual value and is genuinely and economically recyclable.

a.

b.

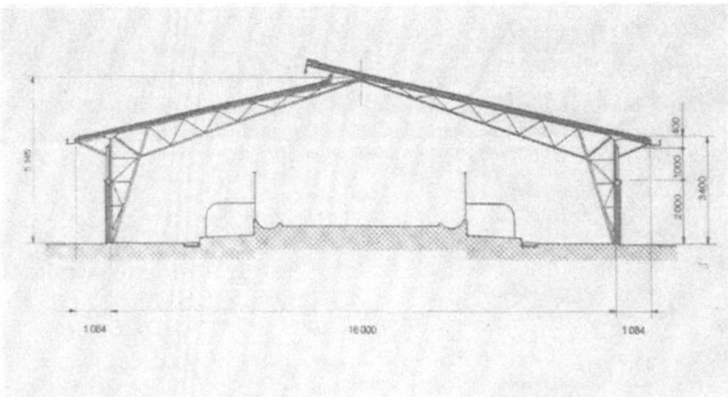

c.

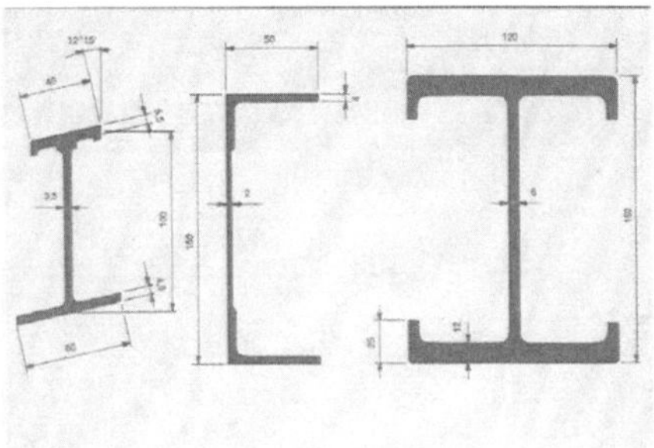

d.

e.

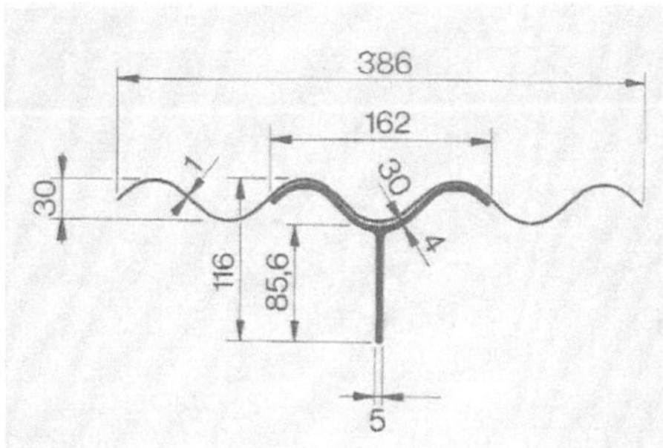

f.

Figure 4.1: Prefabricated Structures (CIDA, 1972)

a	"Trelement" building system (Germany)
b	Prefabricated club-house (France)
c,d	Prefabricated rural building (Italy)
e,f	Rolling mill roof (Krenzlingen, Switzerland)

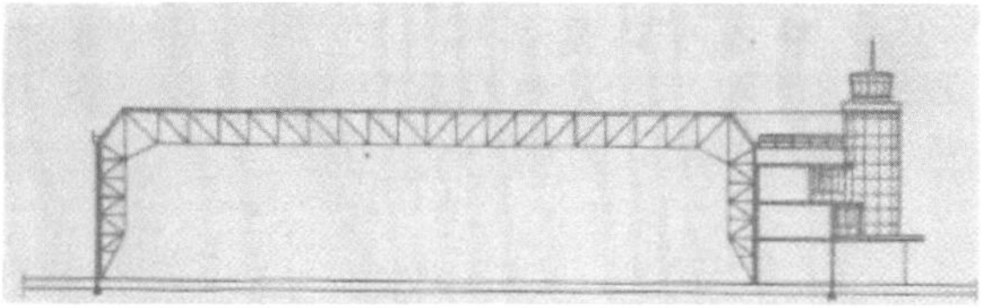

a.

b.

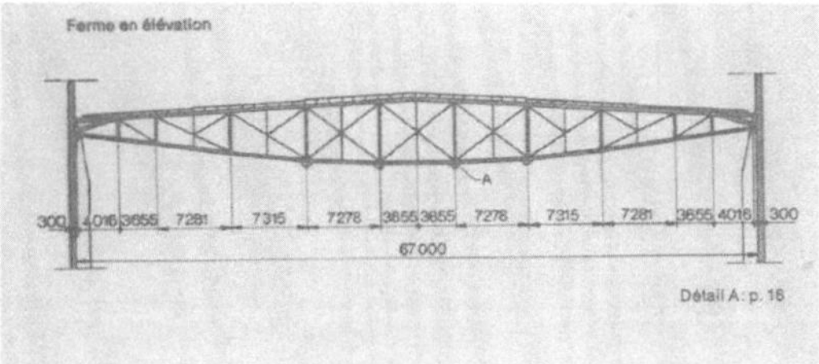

c.

d.

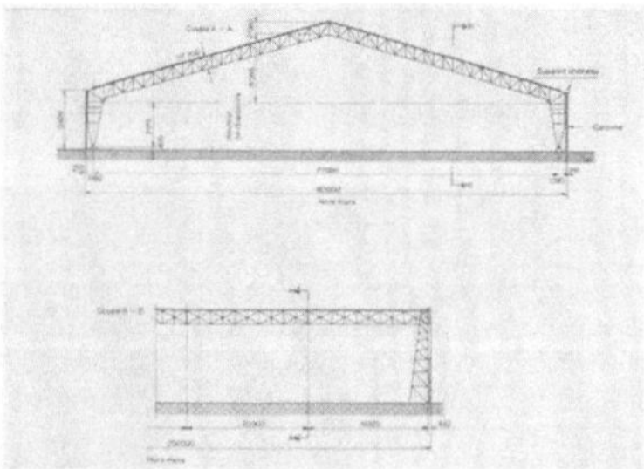

e.

f.

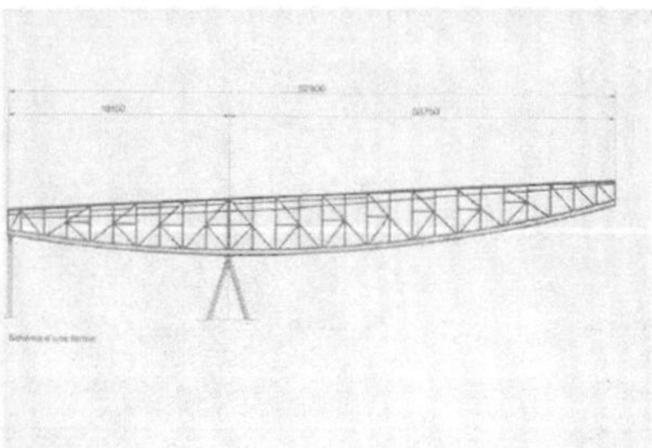

g.

h.

Figure 4.2: Plane Structures (CIDA, 1972)

a,b Hangar (Hatfield, England)
c,d Sporthall (Gand, Belgium)
e,f Warehouse (Antwerp, Belgium)
g,h Melsbroek airport (Brussels, Belgium)

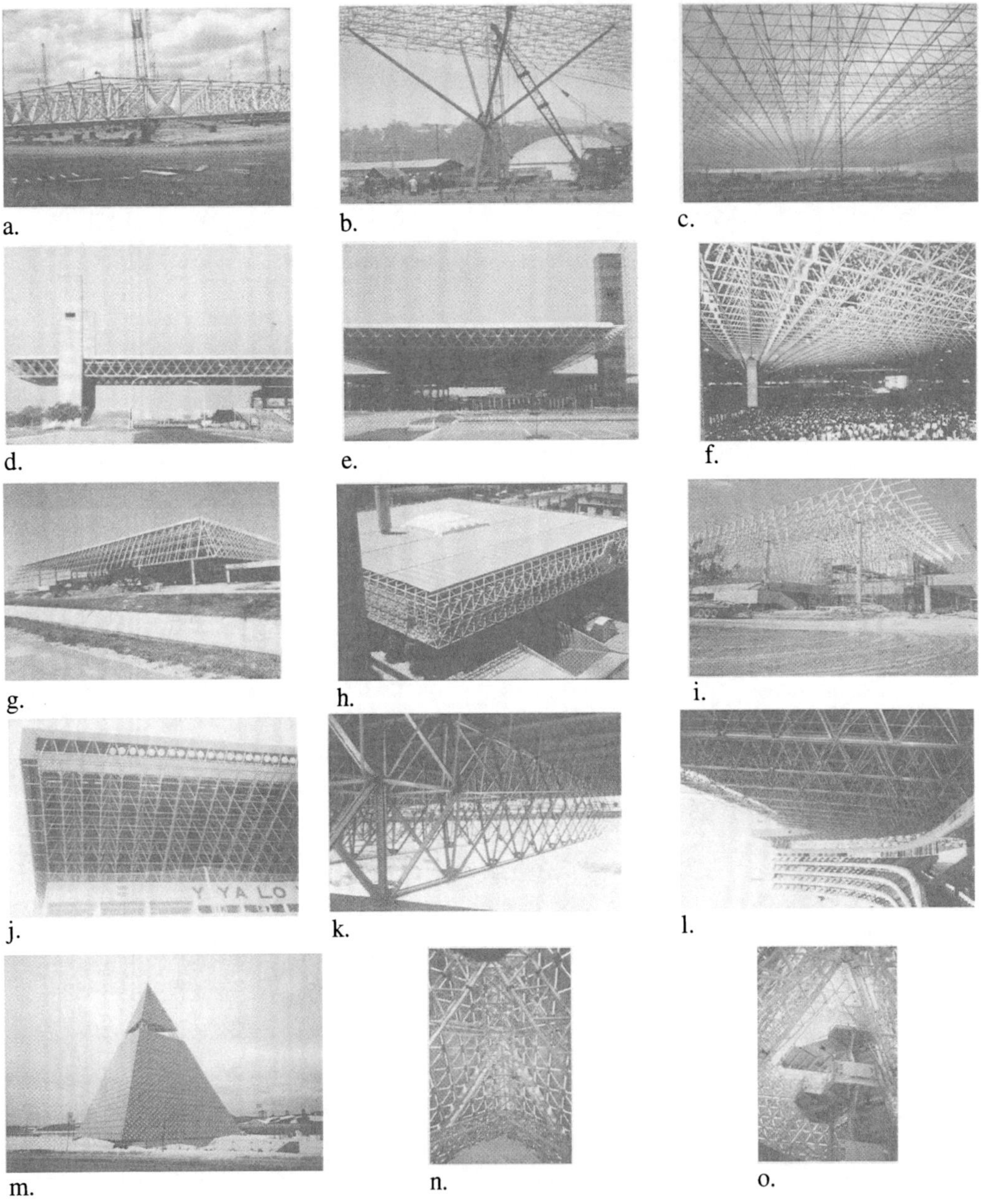

Figure 4.3: Reticular space structures

a,b,c Erection phases of the Interamerican Exhibition Center of San Paulo (Brazil)
d,e,f The International Congress Center of Rio de Janeiro (Brazil)
g,h,i Industrial buildings (Brazil)
j,k,l Roof of the tribune of the footbal stadium in Guayaquil (Equador)
m,n,o The Memorial Pyramid in La Baie (Quebec, Canada)

a.

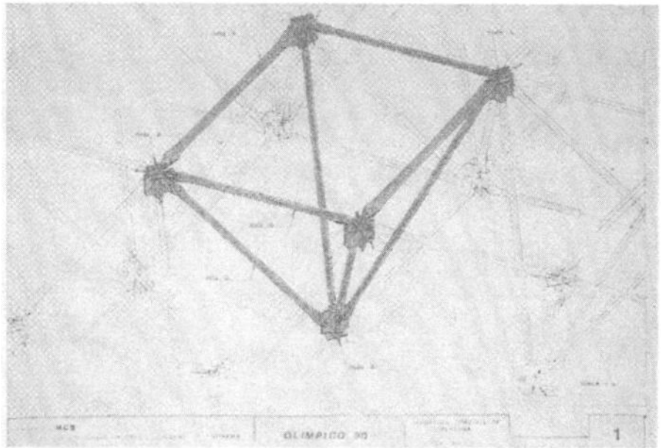
b.

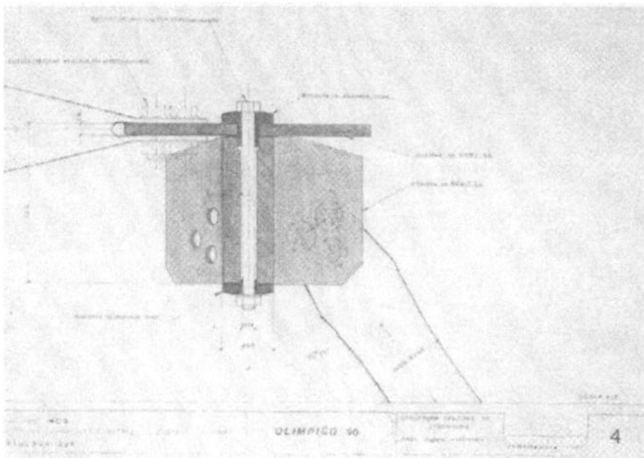
c.

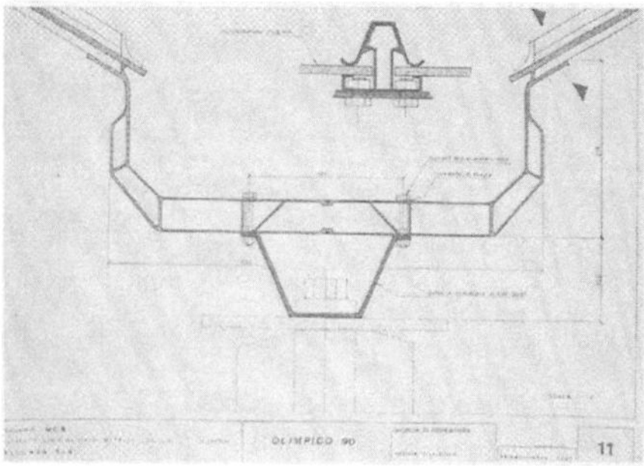
d.

e.

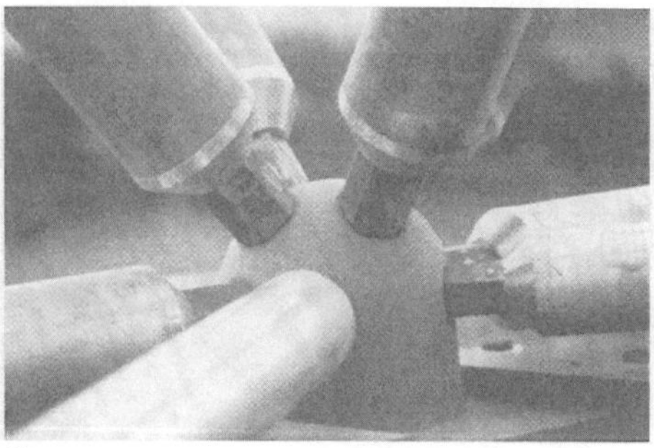
f.

Figure 4.4: Reticular space structures in Italy
a,b,c,d A proposal for the roof of the Olimpic Stadium in Rome
e,f The structure of the Congress Center of Alghero

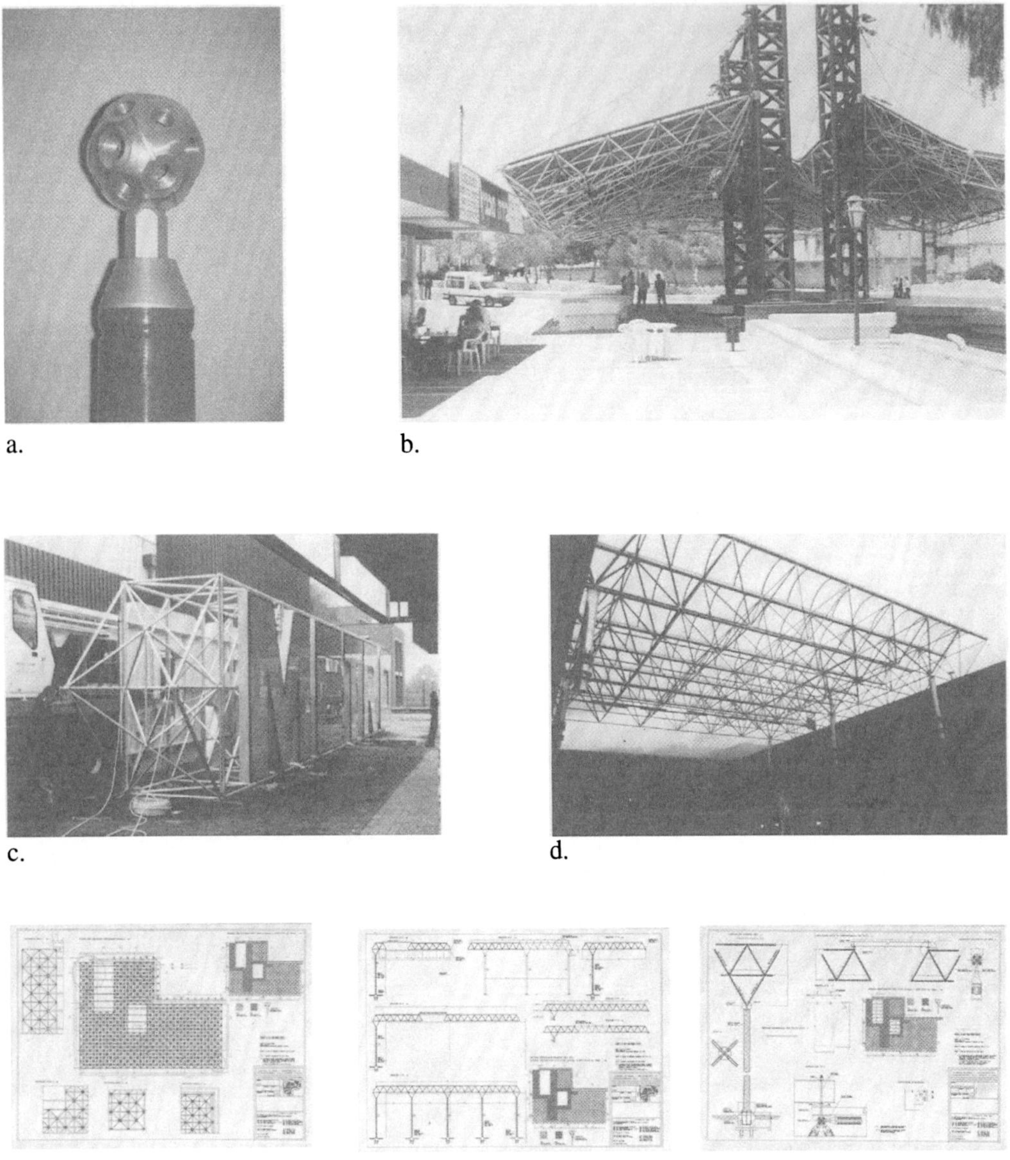

Figure 4.5: The Geo-system (Italy) and some applications

Figure 4.6: Domes

a,b	Dome of Discovery built in London in 1951
c	The Palasport of Paris (1959)
d	The geodetic dome of Guayaquil (Equador)
e,f,g	The Tem-cor system (U.S.A.) : erection phases
h,i	The Tem-cor system (U.S.A.) : applications

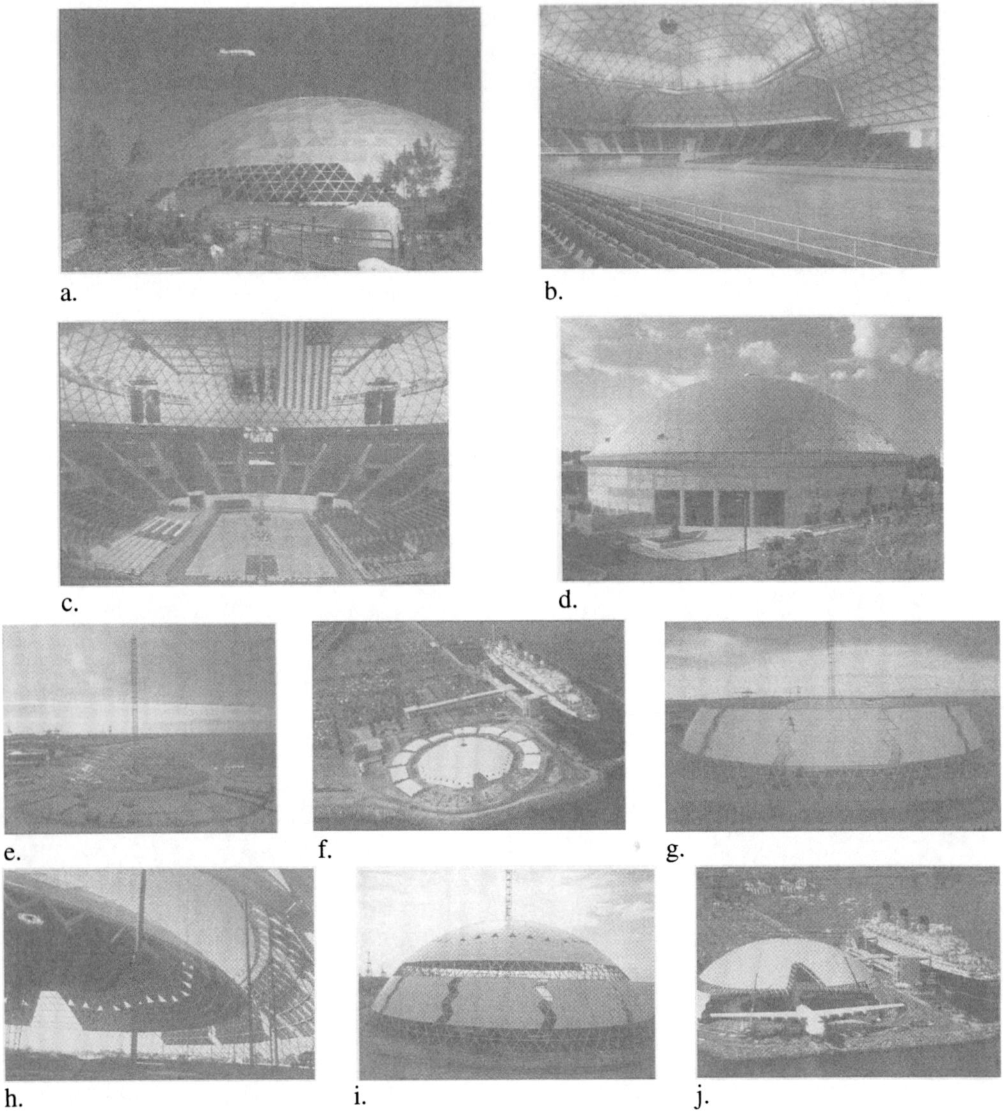

Figure 4.7: Domes

a	Epcot Center (Florida)
b	Bell County Arena (Temple, Texas)
c	Baylor University Ferrell Events Center (Waco, Texas)
d	University of Connecticut
e,f,g,h,i,j	Spruce Goose Dome: erection phases (Long Beach, California)

a. b.

Figure 4.8: Motorway sign supports

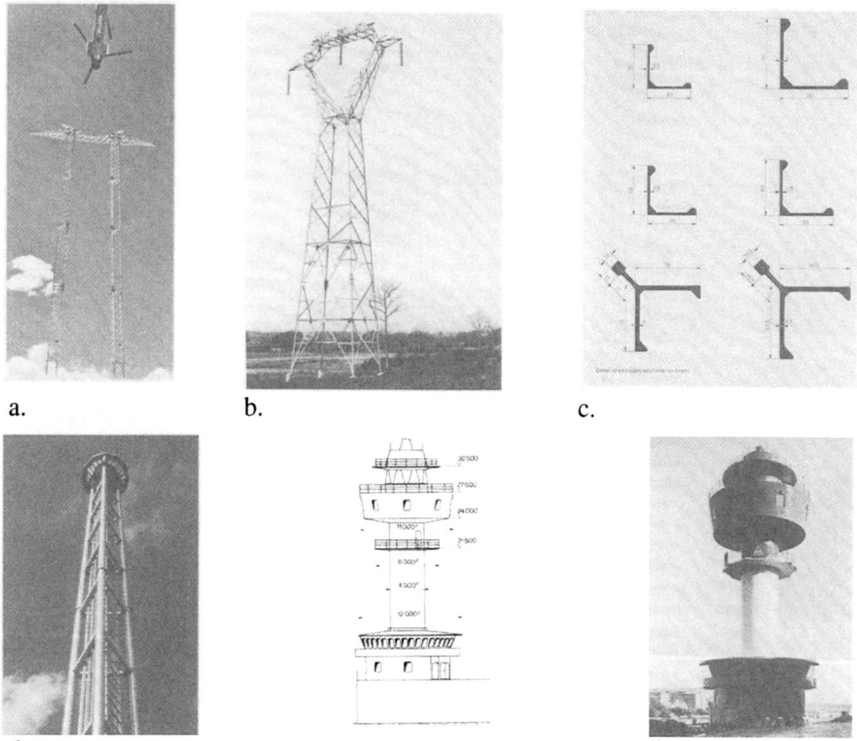

a. b. c.

d. e. f.

Figure 4.9: Towers

a,b,c Electrical transmission towers and typical extruded cross-sections

d Lighting tower

e,f Lighthouse in Kiel (Germany)

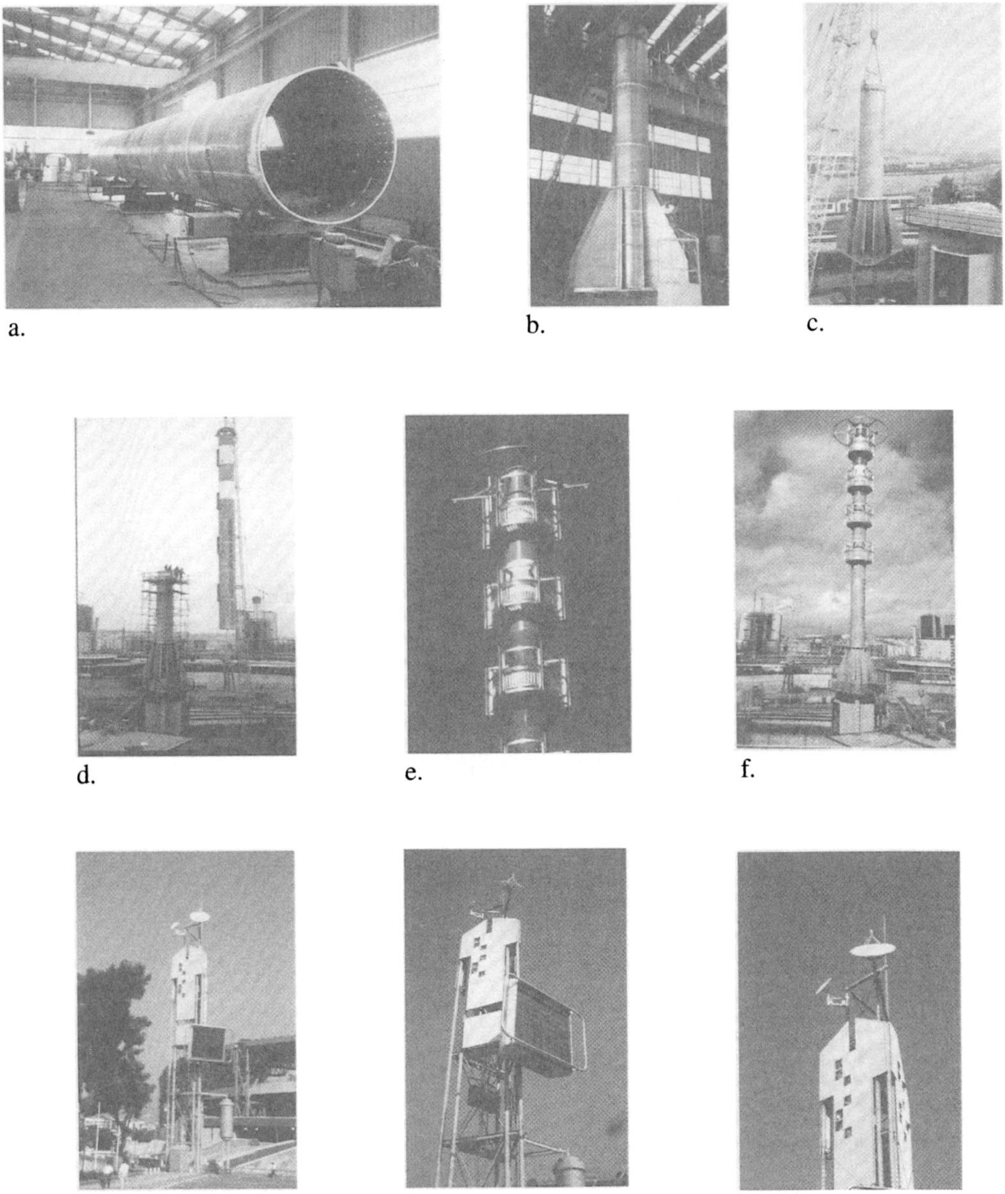

Figure 4.10: Towers in Italy

a,b,c,d,e,f The Enel Tower: fabrication and erection phases (Naples)

g,h,i The Information Tower (Naples)

a.

b.

c.

d.

Figure 4.11: Hydraulic structures
a Reservoir: erection phases
b Pipeline
c,d Sewage plant (Turin)

a.

b.

c.

d.

e.

f.

Figure 4.12: Offshore structures
a General view
b Bridge
c Helydeck
d,e,f Housing: phases of fabrication

5. ALUMINIUM STRUCTURES IN BRIDGES

5.1. Motorway bridges

The properties of lightness and corrosion resistance of the aluminium alloys have allowed to extend their use in the field of bridges, covering all the main structural types in which the steel normally is used.

The Arvida Bridge (Figures 5.1a,b,c) in Quebec, Canada (1950), was one of first motorway bridge made of aluminium alloy. It built according to the Maillart's scheme with a total span of 150 m, an arch of 87 m of span and total weight of 200000 kg. Such scheme cannot be considered as an optimum solution for aluminium alloys, but this choice is justified, considering that the bridge has been built in territory of Canadian Industry of Aluminium. So, it can be considered as an experiment, which represented in due course a kind of challenging symbol for future structural developments of this new material.

The technology based on the use of composite structures made of aluminium beams and concrete decks has been also experienced (Figures 5.1d,e,f). Concrete-aluminium composite systems have been used in some bridges built since the sixties in USA and later in France.

Figure 5.1d shows the cross-sections of three composite aluminium and concrete bridges built in USA. The first type, having triangular cross-section beams in 6061-T6 alloy, was widely tested at the Lehigh University, and showed a good structural performance. It was erected in Amityville, Long Island (span = 65 m), in Sykesville, Maryland (span = 90 m) and on the Appomattox river, Virginia (span = 30 m). The second type is characterized by arch-like thin-walled (1/8 inch) aluminium beams. The cross-section is stiffened by extruded profiles placed at the ends of the lower triangles. The electrochemical corrosion of the aluminium components in contact with the concrete slab was prevented by means of a double chromium-zinc film. Some experimental checks have shown the economic feasibility of using aluminium against steel for spans greater than 23 m. The last section in Figure 5.1d shows details of the bridge built at Des Moines, Iowa, whose girders are made of double T Al-Mg (5456 and 5083) aluminium alloys.

In spite of these relevant examples, several behavioral aspects of composite aluminium-concrete systems are still not completely known and further investigations, both theoretical and experimental, are needed for a thorough understanding of the response of aluminium-concrete to different working conditions (Figures 5.1e,f).

In fact, a large variety of problems are posed by the peculiar features of aluminum alloys when considered as a component of a mixed system. They include (Bruzzese *et al.*, 1989, 1991):

- The possibility of applying the reinforced concrete conventional theory, which has already proved to be valid for steel and concrete composite sections;
- The influence of the low value of the elasticity modulus ratio of the two components, as compared with steel;
- The verification of the assumption that cross-sections are also plane in the inelastic range and remain plane up to failure;
- The experimental check of the flexural behavior;
- The effects of concrete viscosity;

- The behavior of the mechanical connection system between the aluminium beam and the concrete slab and its influence on the bending and shear response of the whole structure;
- The behavior of particular connection systems, for example adhesive bonded systems, whose performance remains somewhat unexplored in the case of aluminium;
- The chemical interaction between the two materials, resulting in a possible attack on the alloy by the concrete component, as well as in its corrosion due to interaction, through the concrete cover, with steel reinforcements in the slab;
- The influence on the structural behavior of the different thermal expansion coefficients of concrete and aluminium alloys.

5.2. Military bridges

A new important field of application is the one of military bridges in which lightness and corrosion resistance play a fundamental role (Figure 5.2).
Modern military support bridges are commonly lightweight aluminium structures with a high coefficient of material utilization and, compared to civil bridges, very low crossing numbers but high crossing loads. For high mobility off-road and on-road, these bridges are modular systems consisting of one or more bridge segments dimensioned with the goal of minimum structural weight. The main problem of these bridge structures is the fatigue. In fact, because of stringent weight limitation, they are subjected to low cycle fatigue, with high stress ranges applied less than 10^4 times. At present, it is possible to reach 40 meter of span with prefabricated elements easy to transport and to erect. The main examples have been developed in Great Britain (Figures 5.2a,b) and Germany (Figures 5.2c,d,e). A new military bridge of 60 m span has been proposed in Italy during the 80s, but the design process stopped for political reasons.

5.3. Foot bridges

This is a structural typology where the aluminium alloys are successfully employed, owing to low live load (Figure 5.3). Another benefit derives from the use of orthotropic plates of aluminium which can be simply obtained through extrusion shapes joined together by means of easy and effective connections (Figures 5.4a,b). These elements are studied to be resistant and lightweight. Last property allowed the transport fully prefabricated structure. Many structural schemes can be used, like simply supported girders with box section (Figures 5.3a,b), continuous beams (Figure 5.3c), cable stayed (Figures 5.3d,e,f,i,j,k). Additional advantages due to lightness are evident in case of moving bridges (Figures 5.3g,h). Examples of aluminium foot bridges can be found in France, Germany, The Netherlands, Italy, Canada.

5.4. Deck reparation

Road bridges are long-term investments with an expected lifetime of at least 50 years. In many cases bridges are in such a poor condition after a much shorter time that they need to be repaired or replaced (Figure 5.4). A severe climate, the use of road salt and increasing traffic loads are three major factors causing deterioration of bridge structures. Deterioration of concrete decks is mainly due to the first two factors. In areas with poor soil conditions, increasing traffic loads have caused damage to foundations and supporting structures. In case of severe damage to concrete decks, the usual course of action is to replace the deck with a new one. There are several methods for undertaking this. If the foundations are damaged, or if the bridge has to carry increased loads, then reinforcement of selected area is usually necessary, often at great cost.

Bridges are structures in which a considerable part of the load-bearing capacity is allocated to carrying its own weight. For a steel girder composite bridge most of the dead load lies in the concrete deck. By replacing the concrete deck with a lighter one it is possible to increase the allowable live load without reinforcing the main structure or the foundations.

A lightweight system for replacing damaged concrete bridge decks has been developed and used in Sweden (TALAT, 1995). The basic concept is an orthotropic plate of aluminium (Figure 5.4b). Weight reduction has made it possible to use the existing foundations, supports and structural systems for increased allowable live loads on bridges that otherwise would have to be replaced.

This orthotropic plate is built up of hollow aluminium extrusions. The extrusions are fitted together by means of a tongue and a groove in the upper flange. This type of connection transfers shear force from one extrusion to the other. At the same time it allows each extrusion to rotate independently of the neighbouring ones. The hollow section is used to create a high degree of torsional stiffness in the extrusion. Experimental tests have been done to validate this solution (Figures 5.4c,d).

In many cases the aluminium bridge deck can be a very competitive technical and economic alternative to the conventional concrete bridge deck. Even though the cost for the aluminium deck is approximately the same as for a corresponding concrete deck, the aluminium deck has numerous advantages (Figures 5.4e,f). They are (TALAT, 1995):

- Very short time for construction works, which greatly reduces the total investment costs in many cases. For example, the replacement of a concrete bridge deck by an aluminium deck was achieved in just three days in order to disturb the traffic as little as possible. In addition, costs for a temporary bypass road could be avoided.
- In many cases the reduction of the dead load can make it possible to use the existing bridge foundations and main girders. This is maybe the greatest advantage and should be studied carefully for each particular case.
- After completion, the costs for maintenance are also kept at a minimum, because the corrosion resistance will also be very good.

Examples of bridge refurbishment by means of aluminium alloys are frequent in U.S.A. (Figures 5.4g,h)

5.5. Structural restoration

The aluminium alloys can be considered as a family of materials that exhibit a wide range of mechanical properties depending on the type of alloy and the technological treatments. Therefore it is possible to identify alloys that have strength comparable to the common structural steel and, at the same time, a weight that is one third of the steel one. Due to these specific combined characteristics, a high structural effectiveness can be obtained through the use of aluminium alloys. In addition, in the special case of retrofitting suspension bridges, the lightness of the material adopted for the deck allows for reducing the stress state both in the steel cables and in the old masonry piers, thus giving the possibility of keeping some existing structural elements as they are, in some cases without any intervention. As further consequence of the reduced structural weight, i.e. of the dead loads, an increase of the live loads, i.e. of the maximum weight of vehicles that the bridge must sustain, can be obtained. Finally the corrosion resistance of the aluminium leads to avoid any surface protection, even in the case of bridges crossing rivers, thus reducing both the initial and final maintenance costs.

For these reasons, the retrofitting of suspension bridges can take profit of the specific aluminium alloy characteristics for obtaining the maximum structural effectiveness, particularly if compared to classical solutions based on the use of steel.

By the way, during the seventies a rehabilitation program of ancient suspension bridges, built between the end of the 19th century and the beginning of the 20th century, has been developed in France. The old structures were made of wooden deck, masonry piers, steel girders and steel suspension chains. The adoption of aluminium alloy floor girders has been done in the retrofit project of three bridges: the Montmerle (Figures 5.5a,b,c) and the Trevoux (Figures 5.5d,e,f) bridges on the Saone river, the Groslée bridge (Figures 5.5g,h,i,j) on the Rône river. It allowed for conserving as much as possible some of the old structural elements, and brought to very effective solutions, both from the cost and structural performance points of view (Landolfo *et al.*, 1989; Mazzolani & Mele, 1997).

In the Montmerle bridge (two 80 m bays), the use of aluminium, both for the two truss beams with bolted connections and for the deck slab (see extruded shapes in Figure 5.5c), led to the possibility of increasing the weight of the road vehicles, while keeping both the existing cables and piers without significant strengthening. In the retrofit of the Groslée bridge (a single 174 m long bay) the deck structure is made of three longitudinal aluminium truss girders, connected to a light reinforced concrete slab. In this scheme a remarkable example of co-operation among different structural materials, each of them utilized in an optimum working condition, can be observed: the old masonry piers, the high strength steel suspension cables, the stainless steel suspension ties, the aluminium alloy reticular girders with high strength steel bolted joints and the light reinforced concrete slab floor. Therefore, the rehabilitated structure appears as a peculiar composite construction.

More recently (1998), a structural retrofit project of the oldest Italian suspension bridge, the "Real Ferdinando" bridge on the Garigliano river (Mazzolani, 1998c) (Figures 5.5j,k,l,m,n), based on the use of aluminium alloy girders, has been realized in the context of a wider rehabilitation program of the zone. The original bridge had a single 85 m bay scheme and the suspension system consisted of two pairs of steel chains, connected to steel ties that sustain the two steel longitudinal truss girder and the wooden transverse beams and deck (Mazzolani, 1988, 1990).

Restoration solution, designed in order to satisfy the requirements of: (a) historic preservation, (b) stiffening of the deck structures, and (c) adoption of innovative technologies and materials, is based on the use of aluminium alloy girders, allowing for the conservation of the original geometrical configuration and appearance. The comparison between the results of structural analyses showed the following advantages of the aluminium with respect to an alternative steel solution: significant reduction of the dead loads; reduction of the required section area of the suspension cables; elimination of the costs necessary for corrosion protection treatments; reduction of repair and strengthening measures required for the masonry piers; easier and less expensive transporting and erection operations due to the lightness of the structural elements.
The "Real Ferdinando" bridge was the first suspension bridge in Italy (1832) and now it is the first suspension bridge made of aluminium (Mazzolani, 2001).

a.

b.

c.

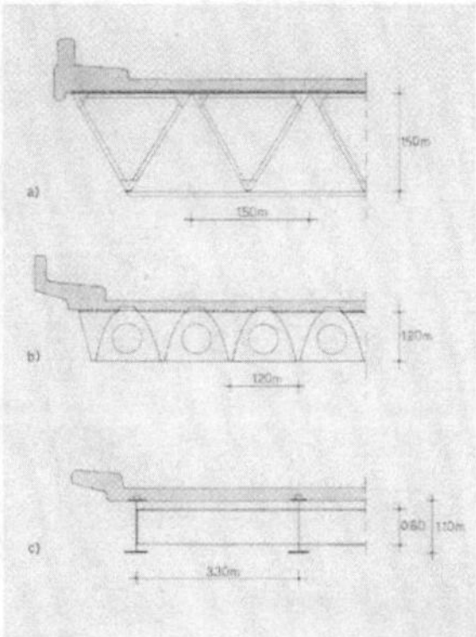
d.

e.

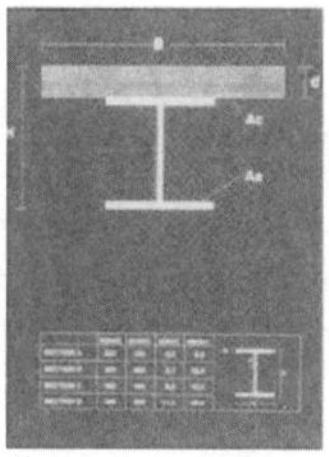
f.

Figure 5.1: Motorway bridges
a,b,c Arvida bridge in Quebec (Canada)
d,e,f Composed aluminium-concrete bridges: sections, test and theory

a.

b.

c.

d.

e.

Figure 5.2: Military bridges
a,b U.K. bridges
c,d,e German bridge (Dornier): erection phases

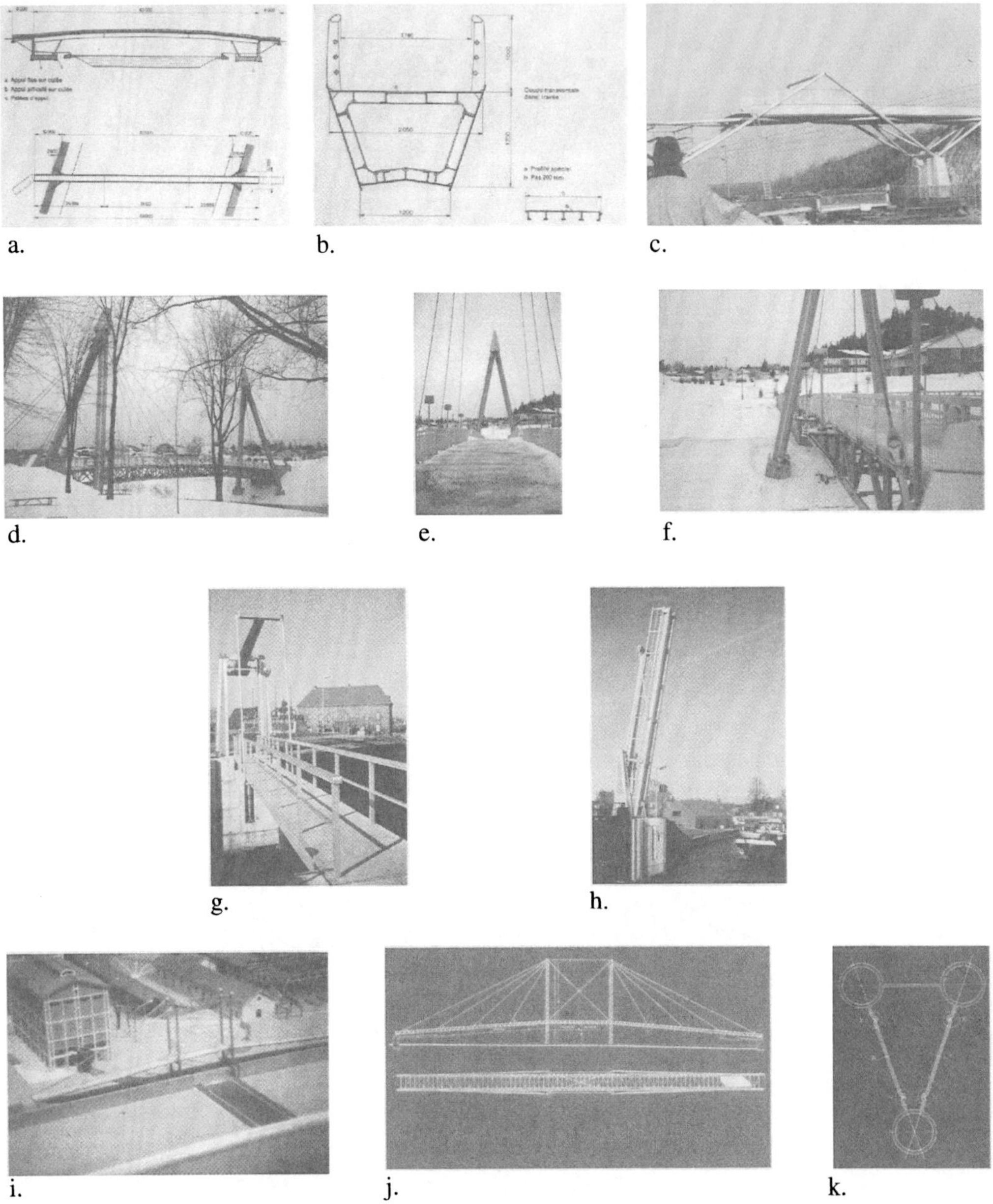

Figure 5.3: Foot bridges

a,b	Foot bridge in Hem-Lenglet (France)
c	Foot bridge in Villepinte (France)
d,e,f	Foot bridge in Jonquière (Quebec, Canada)
g,h	Moving foot bridge in Oldersum (Germany)
i,j,k	A cable-stayed foot bridge, designed for the City of Science in Naples (Italy)

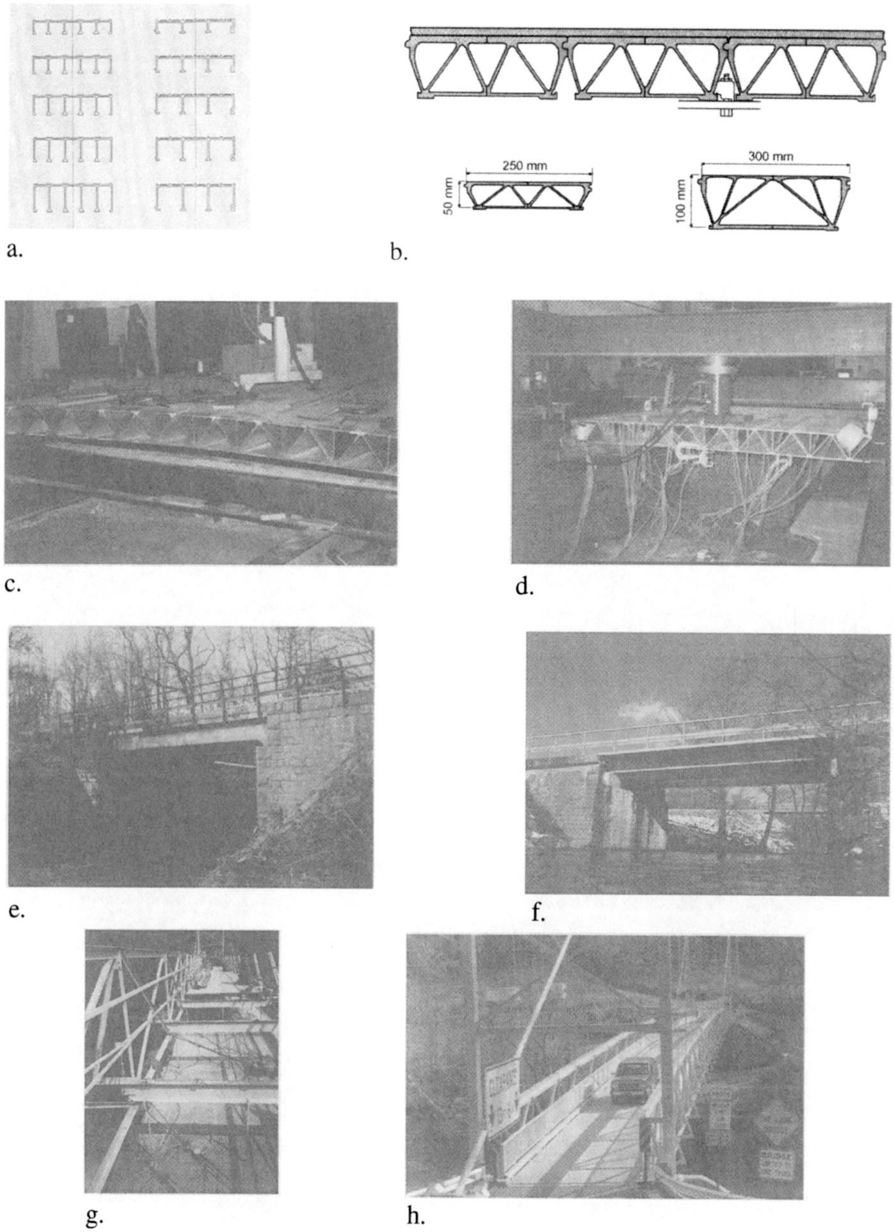

Figure 5.4: Deck reparation

a,b Extruded decks

c,d Test on extruded decks

e,f Substitution of r.c. deck with aluminium deck

g,h Examples of bridge refurbishment in U.S.A.

Figure 5.5: Old suspension bridges restoration

a,b,c	Montmerle (France)
d,e,f	Trevoux (France)
g,h,i,j	Groslée (France)
k,l,m,n,o,p,q	Real Ferdinando (Italy)

Acknowledgements

The author is sincerely grateful to the three PhD students Luigi Fiorino and Simeone Panico of the University of Naples "Federico II" and Maurizio Manganiello of the University of Chieti "G. D'Annunzio", who attended the CISM course on Aluminium Structural Design and gave their help in organizing the lecture notes and transforming them into a chapter format.

References

ALCAN OFFSHORE. *The design guide on the use of aluminium in offshore structures.*

Aprile, A., Benedetti, A. (1998). On the optional design of aluminium space frame structures. *Costruzioni Metalliche,* **n.1**.

Ballio, G., Bonauguro, M., Buzzi, R., Vintani, A. (1993). The aluminium alloy covering of the Alghero (SS) Convention Center. *XIV C.T.A. Congress, Viareggio, October,* 1993.

Bruzzese, E., Cappelli, M., Mazzolani, F.M. (1989). Experimental investigation on aluminium-concrete beams. *Costruzioni Metalliche,* **n.5**.

Bruzzese, E., De Martino, A., Mandara, A., Mazzolani, F.M. (1991). Aluminium-concrete systems: behavioural parameters. *International Conference on Steel & Aluminium Structures,* Singapore, May.

Bulson, P.S. (1992). The New British Design Code for Aluminium BS 8118. In *Proceeding of the 5th International Conference on Aluminium Weldments*, INALCO, Munich.

CEN-TC250/SC9 (1998). Eurocode n. 9 "Design of aluminium structures". ENV 1999 - 1.1; 1999 - 1.2; 1999 - 2.

CIDA (1972). Structures in aluminium. Aluminium - Verlag, Dusseldorf.

Cruickshank, D. (1995). The Dome of Discovery. *Architectural Review*, January, 1995.

De Matteis, G, Landolfo, R. and Mazzolani, F.M. (2001a). Experimental analysis of aluminium T-stubs: framing of the research activity. In *Proceeding of the 8th International Conference on Joints in Aluminium*, INALCO, Munich.

De Matteis, G, Mandara, A. and Mazzolani, F.M. (1998). Numerical Analysis for T-Stub Aluminium Joints. In *Proceeding of the 1rst International Conference Engineering Computational Technology*, Edinburgh, Scotland.

De Matteis, G, Mandara, A. and Mazzolani, F.M. (2001b). Calculation method for aluminium T-stubs: a revision of EC3 annex J. In *Proceeding of the 8th International Conference on Joints in Aluminium*, INALCO, Munich.

European Convention for Constructional Steelworks (1978). European Recommendation for Aluminium Alloy Structures. ECCS doc.

Forsén, N.E. (1995). Fire Resistance, Chapter 10. In Mazzolani F.M. (1995). *Aluminium Alloy Structures (second edition)*, E & FN SPON, an imprint of Chapman & Hall, London.

Koser, J. (1982). Optmization of a welded aluminium girder. *Second International Conference on Aluminium Weldments.* Munich, May 1982.

Kosteas, D. (1992). European Recommendation for Fatigue Design of Aluminium Structures, In *Proceeding of the 5th International Conference on Aluminium Weldments*, INALCO, Munich.

Landolfo, R., Mazzolani, F.M. (1995). Different approaches in the design of slender aluminium alloy sections. In *Proceedings of ICSAS '95*, Istanbul.

Landolfo, R., Mazzolani, F.M. (1997). The Background of EC9 design curves for slender sections. In in *Volume in honor of Prof. J. Lindner*.

Landolfo, R., Mazzolani, F.M., Mele, E. (1989). Riesame dei ponti sospesi di Montmerle e Groslée: confronto fra acciaio e alluminio ("Re-exam of the Montmerle and Groslée suspended bridges: comparison between steel and aluminium"), *XII C.T.A. Congress, Capri, Italy,* October.

Mandara, A., Mazzolani, F.M. (1995). Behavioural aspects and ductility demand of aluminium alloy structures. In *Proceedings of ICSAS '95*, Istanbul.

Mazzolani, F.M. (1981). European recommendations for aluminium alloy. Structures and their comparison with national standards. In *7th International light metal congress*, Vienna.

Mazzolani, F.M. (1985). A new aluminium crane bridge for sewage treatment plants. In *Proceedings of the 3rd International Conference on Aluminium Weldments.* Munich.

Mazzolani, F.M. (1988). L'impiego delle leghe di alluminio nel restauro dei ponti sospesi ("The use of the aluminium alloys in the refurbishment of suspended bridges"). *3rd National Congress ASS.I.R.CO.* Catania, Italy, November.

Mazzolani, F.M. (1989). Torre in Lega di alluminio per antenne paraboliche (Tower in aluminium alloy for parabolic antennas). *Alluminio e Leghe,* **n. 2**.

Mazzolani, F.M. (1990). Il restauro strutturale del ponte Real Ferdinando sul Garigliano ("The structural restauration of the Real Ferdinando bridge on the Garigliano river"). *Costruzioni Metalliche,* **n.2**.

Mazzolani, F.M. (1995a). *Aluminium Alloy Structures* (second edition), E & FN SPON, an imprint of Chapman & Hall, London.

Mazzolani, F.M. (1995b). Globaal overzicht van constructieve aluminium toepassingen in Europa. *Aluminium in Beweging,* Utrecht.

Mazzolani, F.M. (1995c). Stability problems of aluminium alloy members: the ECCS methodology. In *Structural Stability and Design* (eds. Kitipornchai S., Hancock G.J. and Bradford M.A.), Balkema, Rotterdam.

Mazzolani, F.M. (1998a). New developments in the design of aluminium structures. In *Proceedings of the 3rd National Conference on Steel Structures.* Thessaloniki. Greece.

Mazzolani, F.M. (1998b). Design of aluminium structures accordino to EC9. *Proceeding of the Nordic Steel Conference,* Norway, September.

Mazzolani, F.M. (1998c). L'alluminio ed il restauro strutturale dei ponti sospesi: il "Real Ferdinando" sul Garigliano. *Restauro* **n.146**/98, October-December, 1998.

Mazzolani, F.M. (1999). The structural use of aluminium: design and application. *Int. Conf. ICSAS, Espoo,* Helsinki, Finland.

Mazzolani, F.M. (2001). The use of aluminium in the restoration of the "Real Ferdinando" bridge on the Garigliano river. In *Festschrift Ehren Von Prof. Dr. Ing. Günther Valtinat* (J. Priebe & U. Eberwien eds). Anzeiger, Rhauderfehn, September 2001.

Mazzolani, F.M., Frey, F. (1977). Buckling behaviour of aluminium alloy extruded members. *Proceedings of the Second International Colloquium on Stability of Steel Structures.* Liège, April 1977.

Mazzolani, F.M., Mandara, A. (1997). Plastic Design of Aluminium - Concrete Composite Sections: a Simplified Method. *Proceedings of the International Conference on composite Construction - Conventional and Innovative.* Innsbruck, Austria.

Mazzolani, F.M., Mele, E. (1997). Use of Aluminium Alloys in Retrofitting Ancient Suspension Bridges. *Proceedings of the International Conference on composite Construction - Conventional and Innovative.* Innsbruck, Austria.

Mazzolani, F.M., Piluso, V. (1988). Le possibilità d'impiego delle leghe d'alluminio nel campo delle strutture reticolari spaziali ("The possibilities of use of aluminium alloys in the field of reticular space structures"). *L'edilizia e l'industrializzazione,* **n.6.**

Mazzolani, F.M., Piluso, V. (1995). Prediction of rotation capacity of aluminium alloy beams. In *Proceedings of ICSAS '95*, Istanbul.

Mazzolani, F.M., Valtinat G. (1987). ECCS activity in the field of buckling of aluminium alloy members. In *1st International conference on steel and aluminium structures,* Cardiff, (eds. Narayaman R.), Elsevier applied science.

Mazzolani, F.M., De Matteis, G., Mandara, A. (1996a). Classification system for aluminium alloy connections. In *Proceedings of IABSE Colloquium*, Istanbul.

Mazzolani, F.M., Piluso, V., Rizzano, G.V. (1997). Numerical simulation of aluminium stocky hollow members under uniform compression. In *Proceeding of 5th International Colloquium on Stability and Ductility of Steel Structures*, SDSS '97, Nagoya.

Mazzolani, F.M., Faella, C., Piluso, V., Rizzano, G. (1996b). Experimental analysis of aluminium alloy SHS-members subjected to local buckling under uniform compression. In *Proceeding of 5th International Colloquium on Structural Stability*, SSRC, Brazilian Session, Rio de Janeiro.

Sharp, M.L. (1993). *Behaviour and Design of Aluminium Structures.* McGraw-Hill, Inc.

TALAT (1995). Training in Aluminium Application Technologies. Edited by Friedrich Ostermann.

Chapter 2

RESISTANCE OF MEMBERS

Günther Valtinat

Technical University Hamburg-Harburg

1. INTRODUCTION

Before we go into the details of the calculation and construction according to the ENV 1999-1-1: 1998 (EUROCODE 9) I want to show you in some slides the advantages of the material aluminium and furthermore some constructions in aluminium.
Afterwards we go into the new european standard EUROCODE 9 which has been prepared by an international group of experts from many european countries over the last ten years. The code is now out as ENV 1999-1-1: 1998 for two years and we started with the conversion phase as it is under way for the other EUROCODES. The final dead line for the end of the work is envisaged 2004/2005. Then the phase will come during which the individual national codes and the EUROCODE 9 will be allowed to be applied simultaneously in the european countries. After that time all national codes will be taken from the market. CEN/TC 250, which is responsable for this procedure, has fixed a period around 2009/2010 for all EUROCODES.
The Advantages of the material aluminium may be summarized as follows:

- very light material
- attractive appearance
- multifunctional profiles and products
- multifunctional material
- easy to work in the workshop
- ecxcellent weldability
- high ultimate strength capacity
- excellent durability
- high and very high protection against corrosion
- simple surface treatment with additional painting
- high resistance against brittle fracture also in deep temperature
- compeltely recycable

The EUROCODE 9 gives information and guidance to nearly all above mentioned items.

Before we go into the code I will show you some examples of buildings which have been constructed in aluminium. I do also refer to the section of Prof. Mazzolani's lectures in this book, he collected pictures and slides of hundreds of buildings in aluminum with many specific details.

At first overhead traffic sign bridges for the german motorways:

Figure 1a: Local overhead traffic sign bridge for a four lane street

Figure 1b: Local overhead traffic sign bridge for a one way street with 3 lanes

Figure 1c: Overhead traffic sign cantelever bridge for a motorway

Figure 2a: TRELEMENT construction for kindergardens, kiosks, porter's lodges

Figure 2b: TRELEMENT construction for kindergardens, schools, training centers

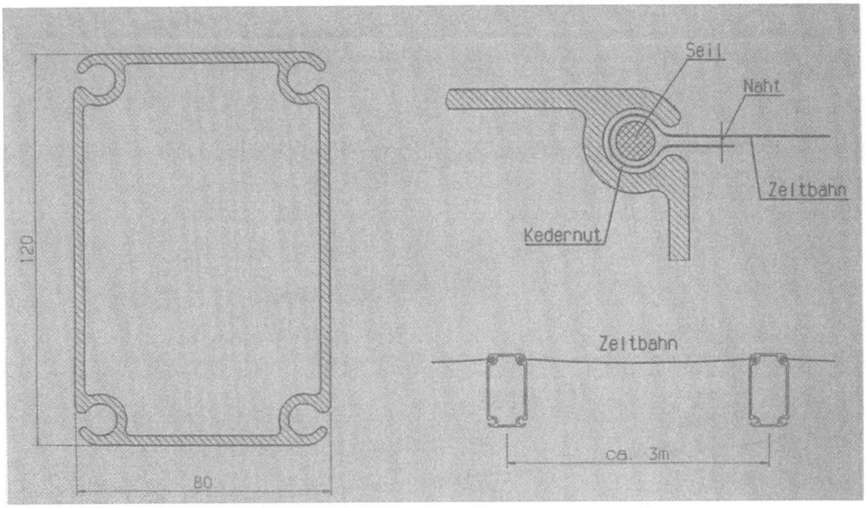

Figure 3a: Keder connections for tents

Figure 3b: Profile with prepared Keder connections for a tent on a wine festival in Udine

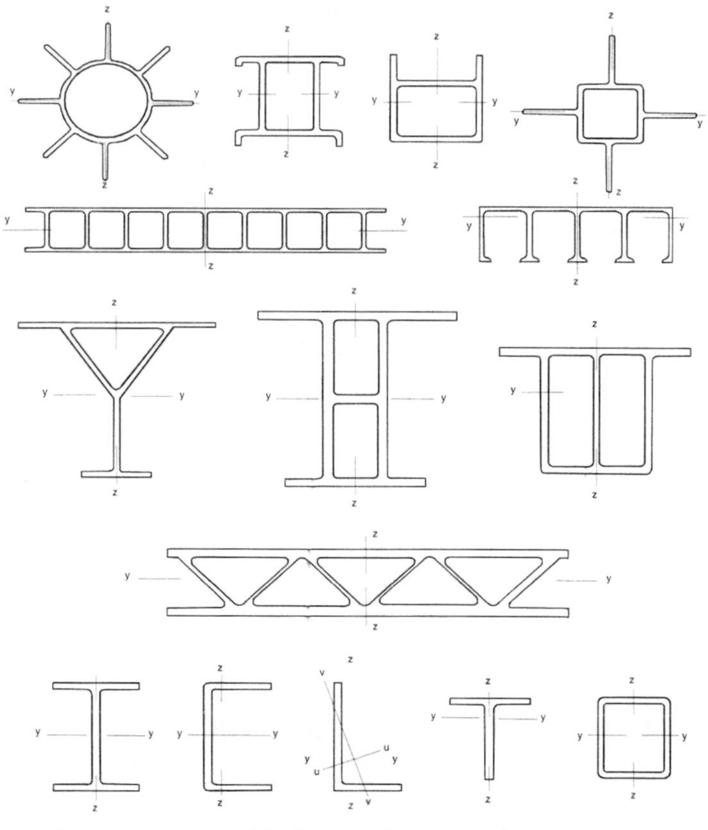

Figure 4: Multifunctional open and hollow sections made from aluminium alloys

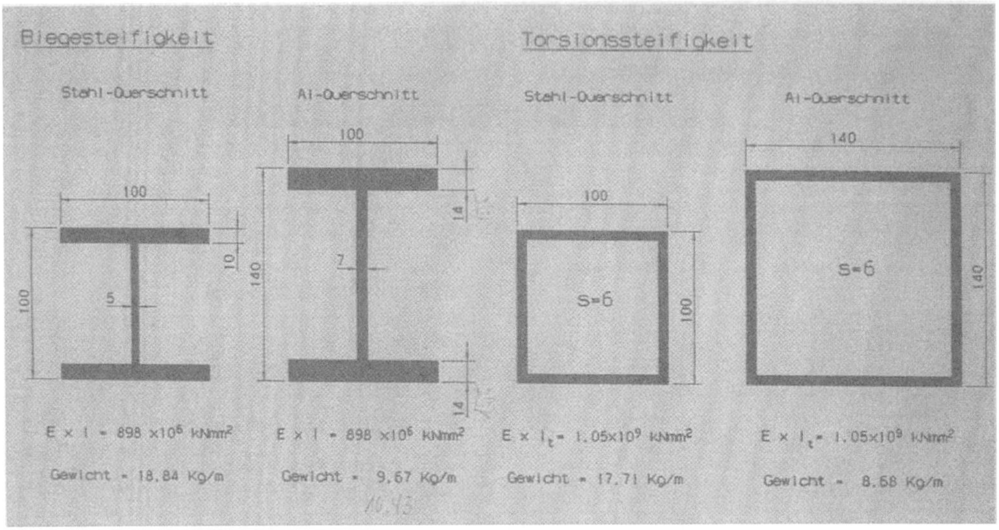

Figure 5: Comparison of the rigidity of steel and aluminium profiles

The figures above show some fields of applications of aluminium profiles. Figure 4 gives an impression about the variety of different multifunctional open and hollow sections which can be produced by extrusion.

The left profile in figure 5 is a steel I-profile 100 x 100 mm which has a certain bending rigidity. The second left profile consists of aluminium and has the same bending rigidity EI as the steel profile. This could be reaches since in this case the depth of the profile was not restricted. So the aluminium profile is 40 % higher than the steel profile but the weigth is only half as much.

An analog calculation of the torsional rigidity could be developed with a hollow section as both the profiles on the right hand show.

The german company Peter Mayer Leichtbau PML in Singen has developed a constructional systems for bridges in aluminium. The advantages are the lightweight of the members, every piece can be carried by one person only, the parts can very easily be mounted and connected by only two people, no crane is needed. All connections are bolted, therefore it is possible to dismantle the construction very easily, no corrosion protection is necessary.

The figures 6a to c show thew aspect view and some constructional details of a pedestrian bridge with a span of about 14 m.

The figure 7 shows the cross section of an aluminium bridge construction type „S“, the figures 8a and b show the aspect view and a cross section of an aluminium bridge construction type „L“.

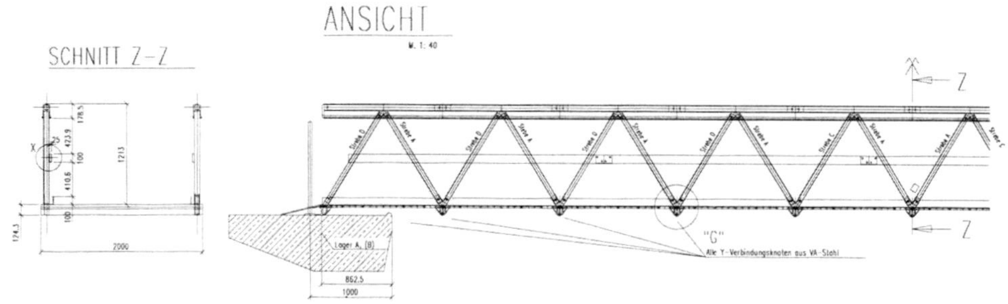

Figure 6a: Aspect view of the PML aluminium bridge system

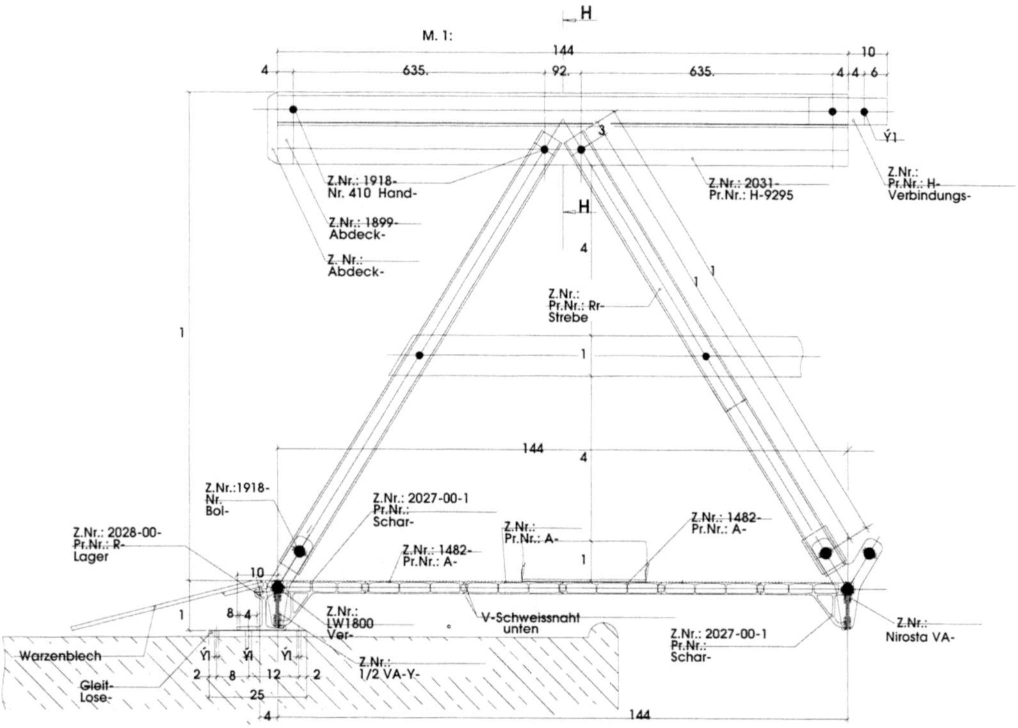

Figure 6b: Detail of the aspect view of the PML aluminium bridge system

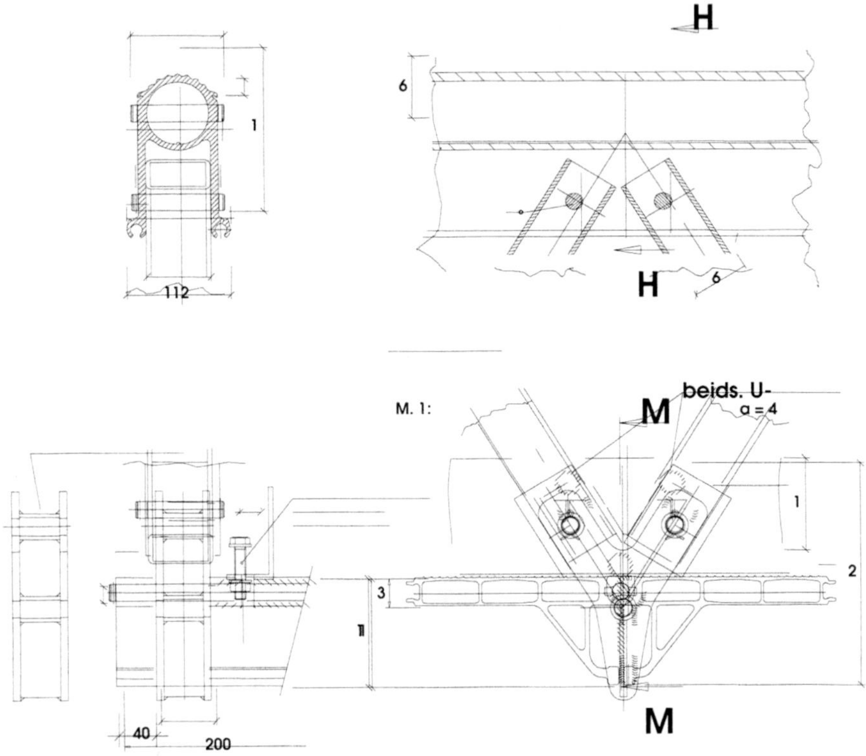

Figure 6c: Constructional details of the PML aluminium bridge system

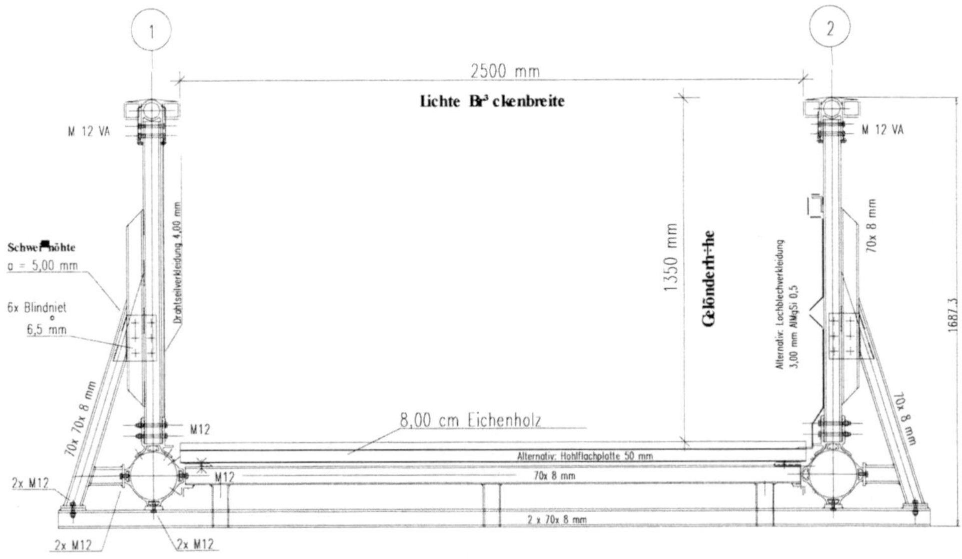

Figure 7: Cross section of the PML aluminium bridge system „S“

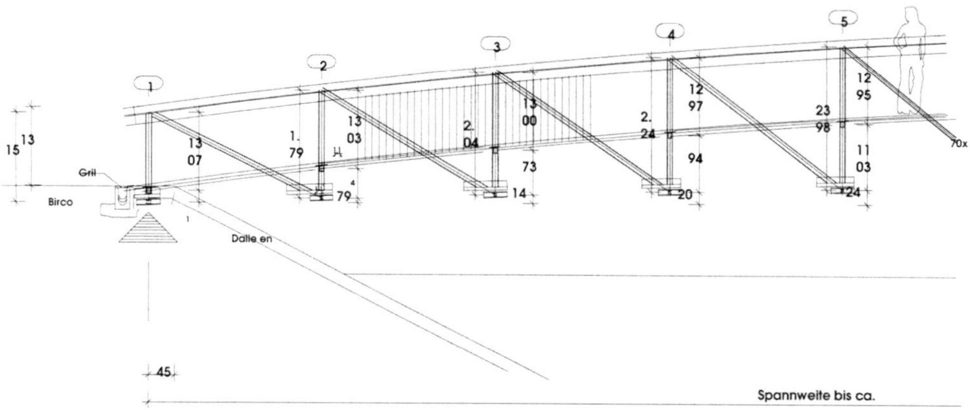

Figure 8a: Aspect view of the PML aluminium bridge system „L“

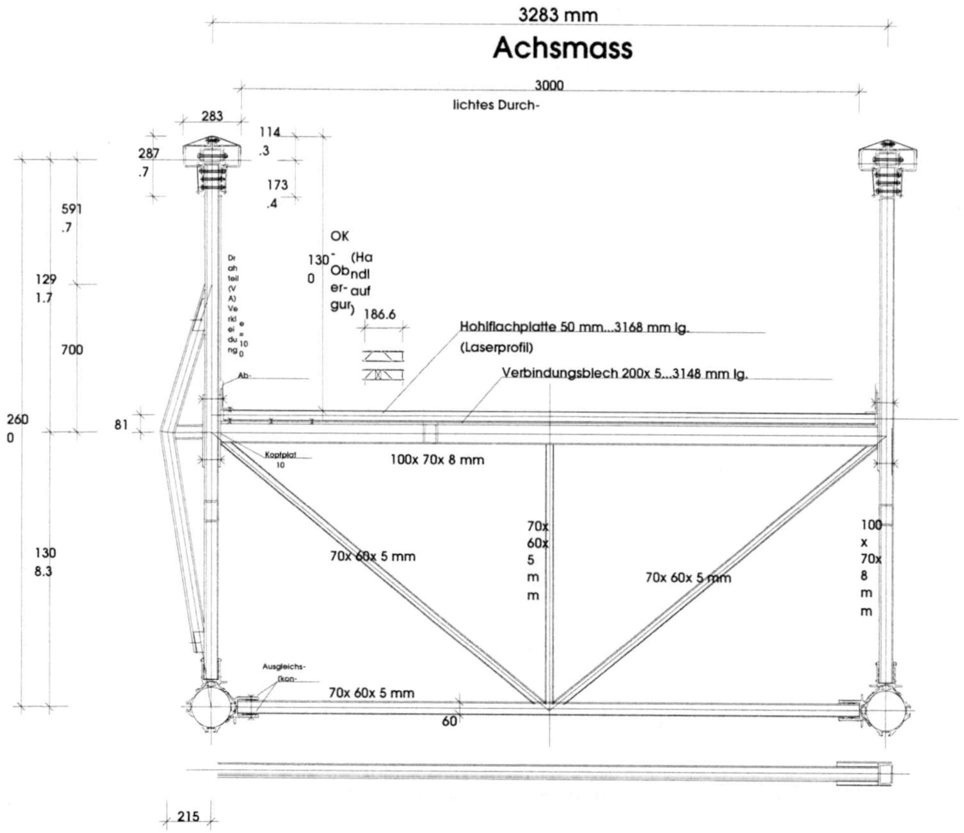

Figure 8b: Cross section of the PML aluminium bridge system „L“

2. MATERIAL, STRESS-STRAIN-DIAGRAMS, MECHANICAL DATA

During the preparation period of the EUROOCODE 9 the CEN/TC250/SC 9 evaluated from a list of international aluminium alloys a group of those which were mostly use in the european countries, whose aluminium market is very much dominated by the american, canadian and french aluminium companies. Mike Bayley (Great Britain), Serge Bompard (France) and Reinhold Gitter (Germany) were the members who selected the aluminium alloys in table 1 with the new (american) and old (european) designation.

Alloy designation		Designation according to DIN 1725	Form of product	Durability [1)]
Numerical	Chemical symbols			
EN AW-3004	AW-Al Mn1Mg1	AlMn1Mg1		k. A.
EN AW-3005	AW-Al Mn1Mg0,5	AlMn1Mg0,5		k. A.
EN AW-3103	AW-Al Mn1	AlMn1	SH,ST,PL,ET	A
EN AW-5005 En AW-5005A	AW-Al Mg1	AlMg1		k. A.
EN AW-5029	AW-Al Mg2Mn0,8	AlMg2Mn0,8		k. A.
EN AW-5052	AW-Al Mg2,5		SH,ST,PL	A
EN AW-5083	AW-Al Mg4,5Mn0,7	AlMg4,5Mn	SH,ST,PL,ET,SEPER/B DT FO	A
EN AW-5454	AW-Al Mg3Mn		SH,ST,PL	A
EN AW-5754	AW-Al Mg3	AlMg3	SH,ST,PL,FO	A
EN AW-6060	AW-Al MgSi	AlMgSi0,5	ET,EP,ER/B,DT	B
EN AW-6061	AW-Al Mg1SiCu	AlMg1SiCu	SH, ST, PL, ET,EP,ER/B,DT	B
EN AW-6063	AW-Al Mg0,7Si	no information	ET,EP,ER/B,DT	B
EN AW-6005	AW-Al SiMg(A)	AlMgSi0,7	EP	B
EN AW-6082	AW-Al Si1MgMn	AlMgSi1	SH,ST,PL,ET,EP,ER/B, DT,FO	B
EN AW-6106	AW-Al MgSiMn	no information		k. A.
EN AW-7020	AW-Al Zn4,5MgCu	AlZn4,5Mg1	SH,ST,PL,ET,SEPER/B DT	C

Key:
- SH - sheet (0,2 to 3,0 mm thick)
- ST - strip
- PL - plate (3,0 tois 25 mm thick)
- ET - extruded tubes
- EP - extruded profiles
- SEP - simple extruded profiles
- ER/B - extruded rods and extruded bars
- DT - drawn tubes
- FO - forgings

1) The durability classes are: A: excellent; B: sufficient /good; C: fair; [2], k. A.: No information.

Table 1: Wrought aluminium alloys for structures and engineering works, forms of products and durability

We know two groups of the material aluminium, the first one is the heat treatable material und the second is the non heat treatable material. Heat treatable alloys belong to the 7xxx- and 6xxx-series, where zinc and silicium are the alloying elements. The most important advantage of these alloys is the heat treatability or the cold treatability to increase the strength by sometimes more than 100 %. The strengthening depends on the temperature (ca. 150 °C to 200 °C) and the time (ca. 2 hours and more) under which the treatment takes place. A cold hardening is also possible but the result is not so high and the time needed is much longer. The disadvantage of the strengthening is that a later heat input such as welding reduces the strength around the weld down to the original untreated condition. This occurs especially when welding is necessary to connect parts to members.To avoid the loss of strength due to welding it is advisable to develop special profiles by extrusion which allow welding in regions with less strain in service (see figure 4). CEN/TC250/SC9 to lower the strength of the heat affected zone according to the table 2. Here I only give the reduction factors ρ_{HAZ} due to welding, later in the lectures by Prof. Soetens you will be informed about the verification of the carrying capacity of welds an other connections.

For all alloys suppplied as extrusions, sheet, plate, drawn tubes and forging in the O and F condition, $\rho_{HAZ} = 1{,}0$.			
Extrusions, sheet, plate, drawn tubes and forging in 6xxx and 7xxx alloys in the T4, T5 and T6 condition:			
Alloys series	Condition	ρ_{WEZ} (MIG-welding)	ρ_{WEZ} (TIG-welding)
6xxx	T4	1,0	-
	T5	0,65	0,60
	T6	0,65	0.50
7xxx	T6	0,80 (A)[1)] 1,0 (B)[1)]	0,60 (A)[1)] 0,80 (B)[1)]
Sheet, plate or forging in 5xxx, 3xxx and 1xxx alloys (in the work hardened (H) condition):			
Alloys series	Condition	ρ_{WEZ} (MIG welding)	ρ_{WEZ} (TIG welding)
5xxx	H22	0,86	0,86
	H24	0,80	0,80
3xxx	H14, 16,18	0,60	0,60
1xxx	H14	0,60	0.60

[1)] Definitions: The values A are valid for tension stresses perpendiculiar to the butt or fillet weld, the values B are valid for all other cases, i. e. a longitudinal stress, a transverse compressive stress or a shear stress.

Table 2: HAZ softening factor ρ_{HAZ} for HAZ material

The 5xxx-serie aluminium alloys usually is not heat treatable but it can also be delivered in a certain hardened condition (H).

Independent of the fact whether the material is heat treated or not the stress-strain-law of the aluminum consists always of a linear part and a non linear part. Dependent on the material and the condition it has a more or less long linear part until the diagram rounds into the elasto-plastic range. The higher the strength of the alloy the similar the stress-strain-diagram is to the steel stress-strain-law, which means that the transition from the elastic part to the plastic one is relatively sharp. The figures 10a and b show stress-strain-laws of different aluminium alloys and of the corresponding heat affected material.

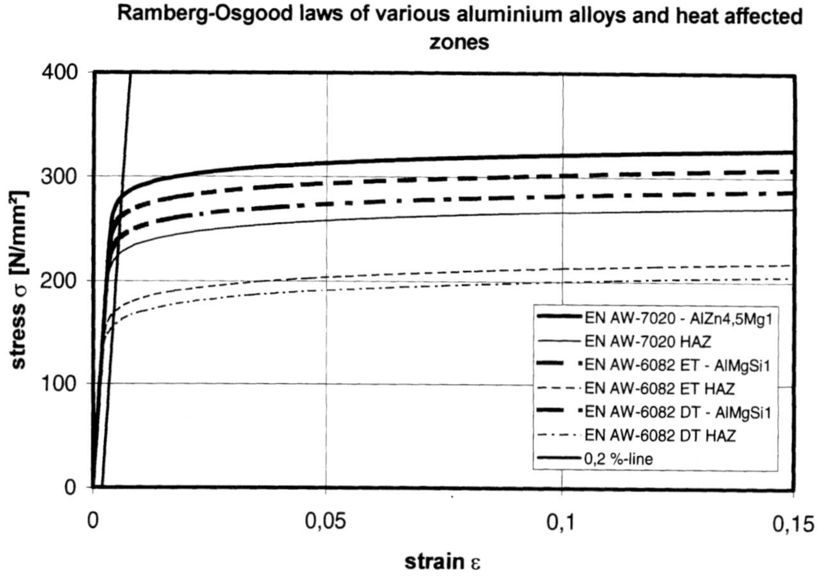

Figure 10a: Stress-strain-laws of high capacity heat treated aluminium materials and of their strength reduced material due to welding and heat affecting

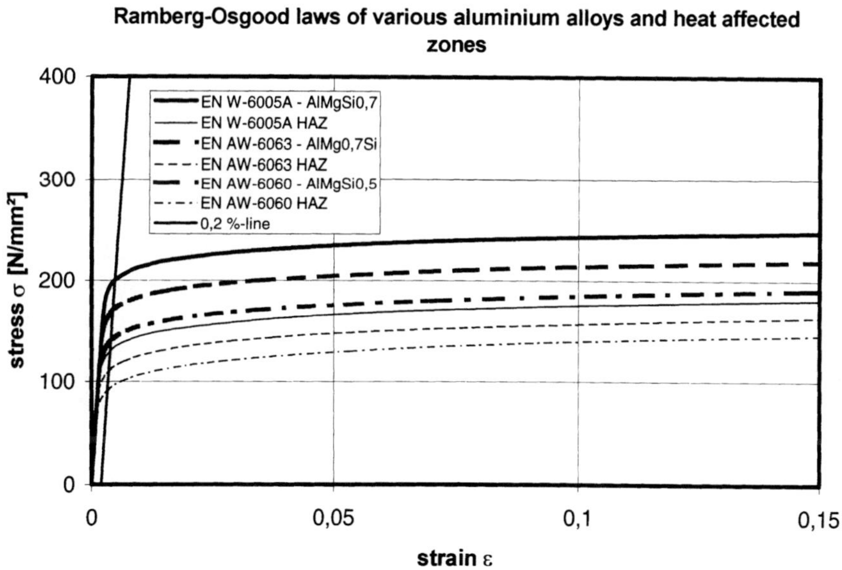

Figure 10b: Stress-strain-laws of medium capacity heat treated aluminium materials and of their strength reduced material due to welding and heat affecting

The stress-strain-laws of material from the weld itself as well as from the HAZ-zone behave very similar to that one of the weak aluminium alloys. Therefore it is possible to apply the same formulae as for the base material. These formulae will be shown later.
It needs well educated engineers and usually finite element calculations to make use of the non linear portion of the stress-strain-law. Therefore for simplification and for a better understanding compared with steel the 0,2 %-limit - similar to that of high duty steels - has been developed. In the all day design this limit can be used as the yield limit similar to the steel S235. The non linearity until the 0,2 %-limit is so small that the little deviation can be neglected.
The $f_{0,2}$-value is the so called "yield" limit strength of aluminium, the f_u-value is the ultimat limit strength. The $f_{0,2}$-value is that stress from where unloading results in a residual strain of 0,2 %. This can be seen in the figures 11a and b, it is where the 0,2 %-line meets the stress-strain-diagrams. All the values are given in chapter 3 of the ENV 1999-1-1:1998 for a great variety of aluminium alloys.
The stress-strain-law of an aluminium alloys can be given in different forms:

- a bilinear diagram (see figure 11a)
- a trilinear diagram (see figure 11b)
- a multilinear diagram
- a potential diagram
- a non linear diagram such as the Ramberg-Osgood law (see figure 11c)

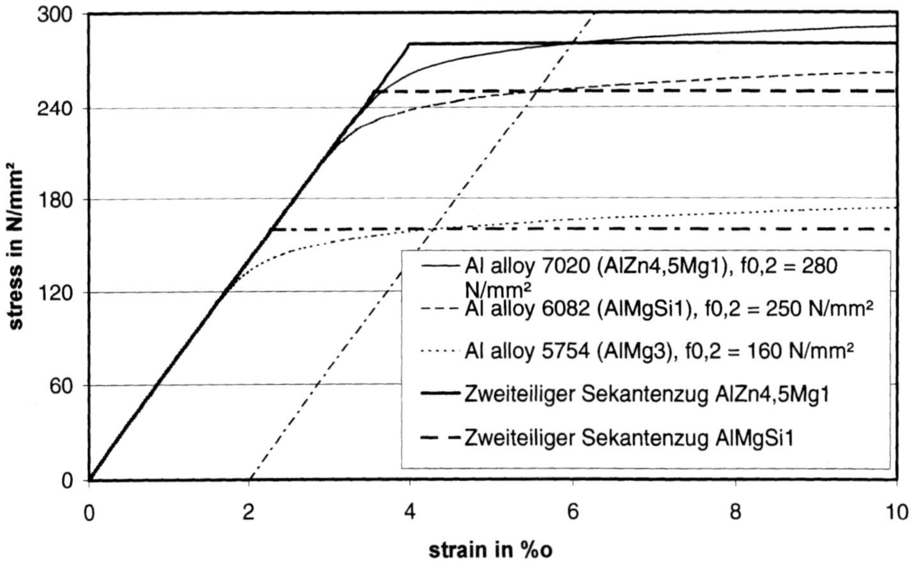

Figure 11a: Bilinear stress-strain-laws for different aluminium alloys

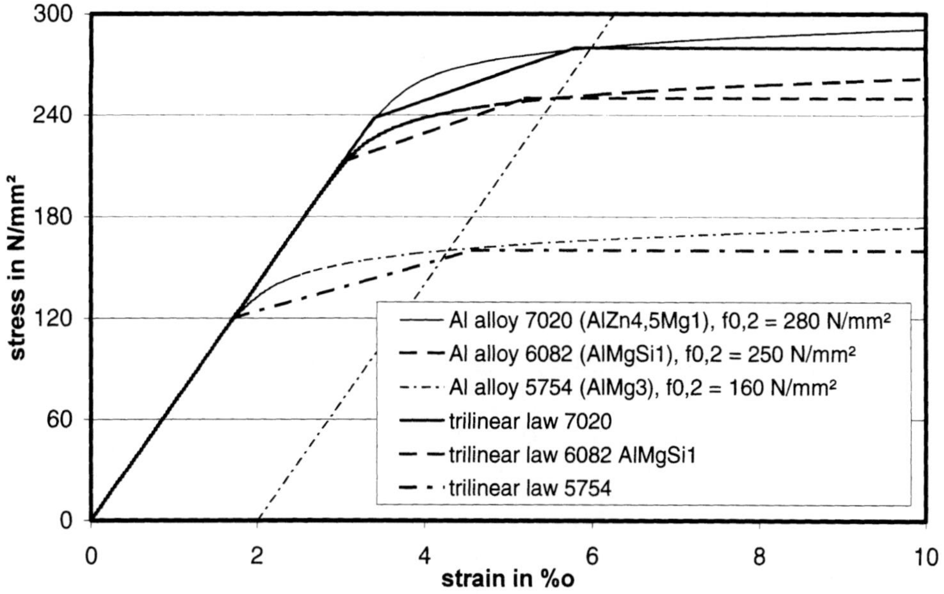

Figure 11b: Trilinear stress-strain-laws for different aluminium alloys

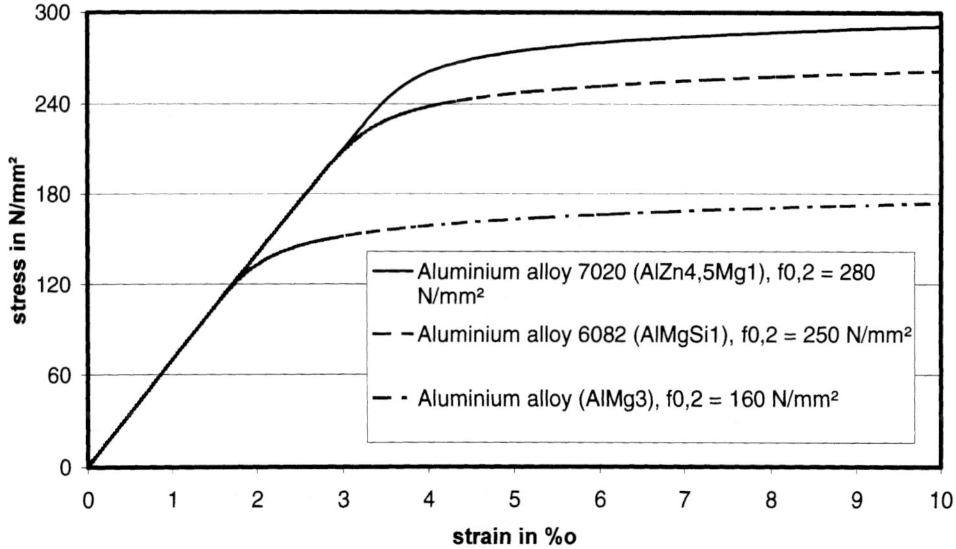

Figure 11c: Non linear stress-strain-laws for different aluminium alloys Ramberg-Osgood-law

After the discussion of the different ways to apply stress-strain-law in the design hereafter we give the basic material data for aluminium alloys:

- Young's modulus E = 70.000 N/mm^2
- shear modulus G = 27.000 N/mm^2
- contraction coefficient μ = 0,3
- linearer thermal expansion coefficient α_{th} = 23 10^{-6} 1/°C
- specific gravity ρ = 2.700 kg/m^3.

The Ramberg-Osgood-law is the internationally accepted stress-strain-law for aluminium alloys (see Chapter 1, Section 2.10.2.). It covers as well the strong heat treated alloys such as the 7xxx- and 6xxx-series as the weaker non heat treated alloys of the 5xxx-series and also the heat affected material and the weld material itself. It consists of a linear part σ/E and a non linear part $0{,}002^*(\sigma/B)^n$ which governs the law in higher stress ranges:

$$\varepsilon = \frac{\sigma}{E} + 0{,}002\left(\frac{\sigma}{\mathrm{B}}\right)^n$$

where

E is the Young's modulus of aluminium 70.000 N/mm²
B is a factor, to be derived by tests, usually it is equal to the $f_{0,2}$-limit
n is an exponent, to be derived by tests.

As Mazzolani has shown the exponent n may be taken also from the following formula:

$$n = \frac{\ln 2}{\ln\left(f_{0,2} / f_{0,1}\right)}$$

Another proposal by Steinhardt was to take n as the value of the $f_{0,2}$-limit in N/mm², divided by 10.

Taking into account the data for B and n we receive the following formula:

$$\varepsilon = \frac{\sigma}{E} + 0{,}002\left(\frac{\sigma}{f_{0,2}}\right)^n$$

Figure 12 may indicate the influence of the n-value taken as 20, 25 and 30 on the σ-ε-curves for the material 6082, it can be demonstrated that this influence mainly occurs in the high stress-strain range, but the stress values differ not more than 4 %.

Additionally to the wrought material the EUROCODE 9 covers also other materials such as cast aluminium and material for rivets and bolts. In bolted connections it is advisable to take either aluminium bolts or stainless steel bolts or hot dip galvanized steel bolts. Never take uncoated or black steel bolts because of contact corrosion problems. Nuts and washers should be of the same material and corrosion protection.

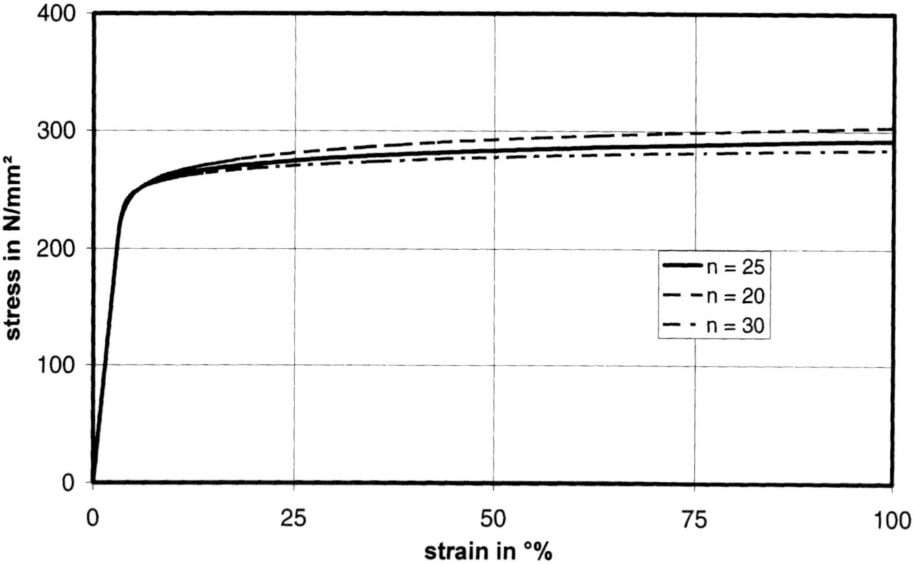

Figure 12: Comparison between stress-strain-laws with different n-values

Now we want to give you some material data of aluminium alloys as they are tabled in the EUROCODE 9. The values in table 3.1 are given for sheet, strip and plate, the values in table 3.2 are given for extruded profiles, extruded tube, extruded rod/bar und drawn tube.
In the EUROCODE 9 there are a lot of more strength data given, such as for materials for connectors and for castings, for forgings and electrically welded tubes etc..
The minimum values of elongation given in tables 3.1 and 3.2 are for information only.

Alloy	Temper	Thickness in mm		$f_{0,2}$ 0,2% Proof stress N/mm²	f_u Ultimate strength N/mm²	A_{50} Minimum elongation %
		over	up to			
EN AW-3103	H14	0,2	25	120	140	2
	H18	0,2	4	145	160	1
EN AW-5052	H12	0,2	4	160	210	4
	H14	0,2	2	180	230	3
EN AW-5454	O/H111	0,2	8	85	215	12
	H24/H34	0,2	25	200	270	5
EN AW-5754	O/H111	0,2	100	80	190	12
	H24/H34	0,2	25	160	240	6
EN AW-5083	O/H111	0,2	50	125	275	11
		50	80	115	270	14
	H24/H34	0,2	25	250	340	4
EN AW-6061	T4	0,4	12,5	110	205	12
	H24/H34	0,4	12	140	290	6
EN AW-6082	T4	0,4	12	110	205	14
	T6	0,4	6	260	310	6
		6	12,5	255	300	9
	T651	12	100	240	295	8
EN AW-7020	T6	0,4	12,5	280	350	7
	T651	12,5	40			6[1)]

[1)] Based on A not A_{50}
NOTE: The minimum elongation values do not apply across the whole range of thickness given, but generally to the thinner materials. Higher values of elongation usually apply to thicker material. For actual minimum values see ENs and prENs listed in 1.3.1.3.

Table 3.1: Minimum guaranteed values of 0,2% proof strength $f_{0,2}$ and ultimate tensile strength f_u for wrought aluminium alloys - Sheet, strip and plate

Alloy	Product-form	Temper	Dimension t wall thickness or thickness mm	$f_{0,2}$ 0,2% Proof stress N/mm²	f_u Ultimate strength N/mm²	A Minimum elongation %
EN AW-5083	ET, EP,ER/B	F,H112	$t \leq$ **20**	110	**260**	12
	DT	H12 H22 H32	$t \leq 10$	200	280	6
		H14 H24 H34	$t \leq 5$	235	300	4
EN AW-6060	EP,ET,ER/B	T5	$t \leq 5$	**120**	**160**	8
	EP		$5 < t \leq 25$	100	140	8
	ET,EP,ER/B	T6	$t \leq$ **15**	**140**	**170**	8
	DT		$t \leq 20$	160	215	12
EN AW-6061	ET,EP,ER/B,DT	T6	$t \leq$ **20**	240	**260**	**8**
EN AW-6063	EP,ET,ER/B	T5	$t \leq 3$	**130**	**175**	8
	EP		$3 < t \leq 25$	110	160	**7**
	ET,EP,ER/B	T6	$t \leq$ **10**	170	215	**8**
	DT		$t \leq 20$	190	220	10
EN W-6005A	EP/O	T6	$t \leq 5$	225	270	8
			$5 < t \leq 10$	215	260	8
			$10 < t \leq 25$	200	250	8
	EP/H	T6	$t \leq 5$	215	255	8
			$5 < t \leq 10$	200	250	8
EN AW-6082	EP,ET,ER/B	T4	$t \leq 25$	**110**	**205**	14
	EP/O, EP/H	T5	$t \leq 5$	230	270	8
	EP/O,EP/H	T6	$t \leq 5$	250	290	8
	ET		$5 < t \leq 25$	260	310	10
	ER/B	T6	$t \leq 20$	250	295	8
			$20 < t \leq 150$	260	310	8
	DT	T6	$t \leq 5$	255	310	8
			$5 < t \leq 20$	240	310	10
EN AW-7020	EP/ER/B,DT,ET	T6	$t \leq$ **15**	**280**	350	10

Key: EP - Extruded profiles; EP/O - Extruded open profiles; EP/H - Extruded hollow profiles; ET - Extruded tube; ER/B - Extruded rod and bar; DT - Drawn tube

Note 1: Where values are quoted in **bold**, greater thicknesses and/or higher mechanical properties may be permitted in some forms, see ENs and prENs in 1.3.1.3.

Note 2: Where minimum elongation values are given in **bold**, higher minimum values may be given for some forms or thicknesses.

Note 3: Where an extruded product employs thicknesses across the thickness range given above, the highest value given may be used provided the manufacturer can support the value by an appropriate quality assurance certificate.

Table 3.2: Minimum guaranteed values of 0,2% proof strength $f_{0,2}$ and ultimate tensile strength f_u for wrought aluminium alloys - Extruded profiles, extruded tube, extruded rod/bar und drawn tube

3. MEMBERS

3.1. General

The verification of all members:

tension members,
girders and
compression members

must be treated differently, depending whether they are non welded or welded. In case of welds there may be welds in the longitudinal direction and welds in the transverse direction of the member. The calculation for the unwelded members ist very similar to steel. In case of the welded members we have taken notice of the reduction in capacity due to the welds.
Aluminium elements shall be proportioned such that the basic design requirements for the **Ultimate Limit State** (section 5 in the EUROCODE 9) are satisfied. While the loads are enhanced by the partial safety factors γ_F and the combination coefficient φ, whose values are given in EUROCODE 1 and depend on the kind of the load (permanent load, live load, exceptional load etc.) the resistance of the cross section, of the member or of the whole structure implies the partial safety factor γ_M for the material. The most important values of the γ_M-factor are:

γ_{M1} =1,10 for all sections, members or structures and for all internal forces and

γ_{M2} =1,25 for all connections.

The **Serviceability Limit States** for structural aluminium (section 4 in the EUROCODE 9) are:

- deformations or deflections that negatively influence the use of the structure
- deformations or deflections that cause damage to the finish and the non structural elements
- deformations or deflections that negatively influence the appearance of the structure
- shape distortions especially of thin walled members
- vibrations.

Details such as limits for vertical deflections of girders are given in EUROCODE 9.

In the following sections we will show, how to handle the individual cases.

3.2. Non welded and welded tension members

A simple tension bar non welded can be treated as in steel. The design tension resistance $N_{t,Rd}$ of the cross section, taken as the smaller of two values corresponding respectively to:

1. general yielding along the member: $N_{t,Rd} = A_g f_0/\gamma_{M1}$
   where
   A_g is either the gross section or a reduced cross section
   f_0 is the limiting stress, usually the yield stress $f_{0,2}$

2. local failure at a critical section such as connenction: $N_{t,Rd} = A_{net} f_a/\gamma_{M2}$
   where
   A_{net} is the net section area, taking into consideration any deduction due to holes, weldings etc.
   f_a is the limiting stress, usually the ultimate stress f_u.

A simple example may show the design procedure according to general yielding along the member:

Tension bar with the tension load T without and with welds.

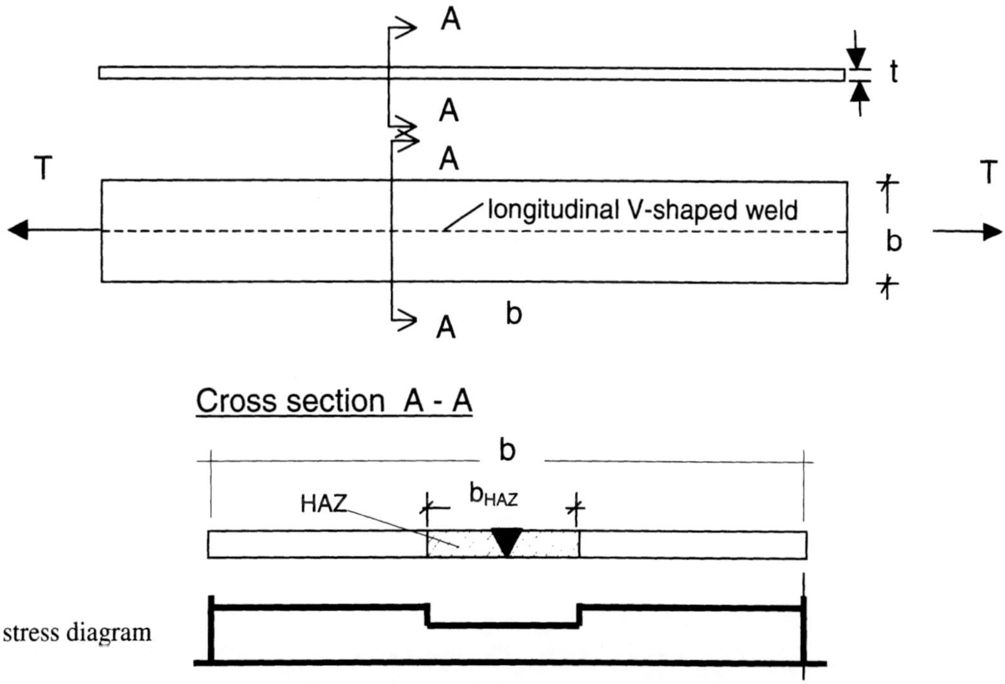

Figure 13: Tension bar without and with weld

The basic idea behind the verification in the ultimate limit state is that every fibre is strained up to the same and constant level ε. Each fibre has its individual stress taken from its corresponding stress-strain-curve at the level of ε. Base material fibres have a higher stress, weld material fibres have a lower stress. The total internal force results from the integral over the force in each fibre of the cross section and must be in equilibrium with the external load T on the bar. This is as long valid as each fibre has enough elongation capacity.

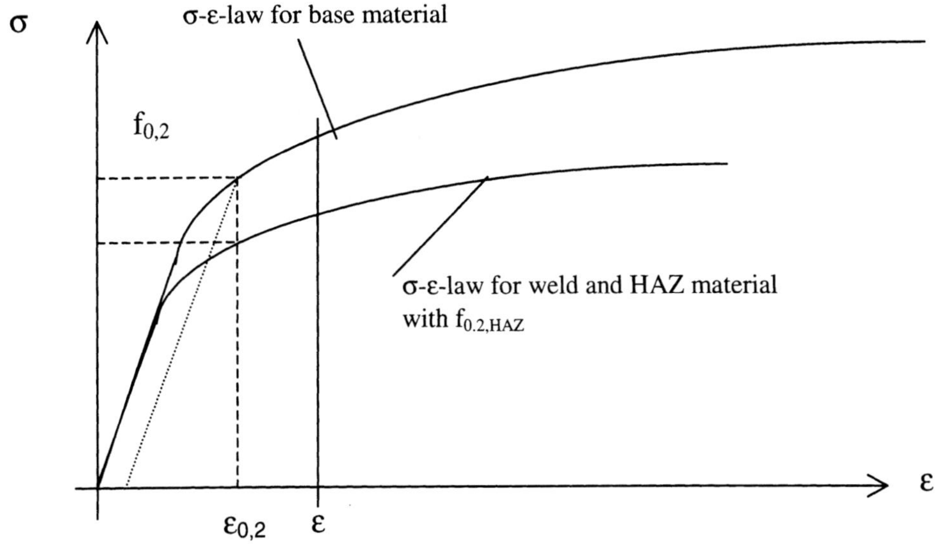

Figure 14: Stress-strain.law for non welded and welded material

Resistance of tension members (in case of no weld take $b_{HAZ} = 0$ and $\rho = 1,0$):

Total cross section: $A = b \cdot t$

HAZ-cross section: $A_{HAZ} = b_{HAZ} \cdot t$

Reduced cross sectional area: $A_{red} = b \cdot t - b_{HAZ} \cdot t \cdot (1 - f_{0.2,HAZ}/f_{0.2}) = b \cdot t - b_{HAZ} \cdot t \cdot (1-\rho)$

with the reduction factor $\rho = f_{0.2,HAZ}/f_{0.2}$

design tension resistance: $N_{t,Rd} = A_{red} \cdot f_{0.2}/\gamma_{M1}$.

3.3. Non welded and welded cross sections for bending

A cross section for bending can be treated as in steel if the basic design idea discussed above for the tension bar is applied. This basic design idea behind the verification in the ultimate limit state for bending is that every fibre is strained according to the elongation diagram with respect to bending which is a basic assumption and has been developed by Bernoulli many many years ago. It says that the plane of the cross section remains a plane after bending and is always perpendicular to the neutral bending axis. Each fibre has its individual elongation, its individual stress can be taken from its corresponding stress-strain-curve at the level of ε. Base material fibres have a higher stress, weld material fibres have a lower stress, if these fibres have the same elongation and if this elongation is above the linear part of the stress-strain-diagram. The total internal bending moment results from the integral over the force in each fibre of the cross section multiplied with the distance between the fibre and the neutral axis. This internal bending moment must be in equilibrium with the external bending moment on the girder. This is as long valid as each fibre has enough elongation capacity. The design moment of resistance $M_{c,Rd}$ of the cross section is:

1. $M_{c,Rd} = f_0 \, \alpha \, W_{el} / \gamma_{M1}$

   where

   α is the shape factor of the cross section

   W_{el} is the elastic modulus of the gross section. Depending on the section the shape factor α is related to the class of the cross section. There are the four classes 1, 2, 3 and 4. Class 1 is able to develop the full plastic moment and additional rotation without local buckling. Class 2 is able to develop the full plastic moment but without rotation, otherwise local buckling has to be anticipated. Class 3 is able to develop the full elastic moment. Class 4 is able to develop only a partial elastic moment, otherwise local buckling occurs (details see EUROCODE 9).

An example may show the design procedure to evaluate the bending resistance of a cross section without and with welds:

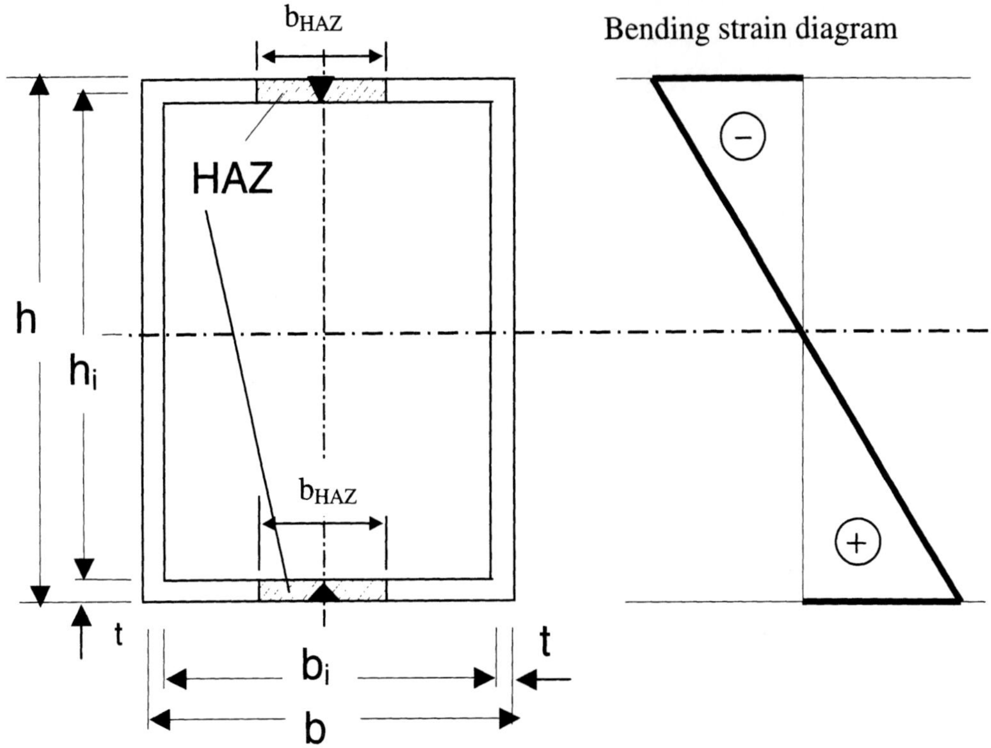

Figure 15: Rectangular hollow section without and with welds in the flanges.

$$I = \frac{1}{12} b \cdot h^3 - \frac{1}{12} b_i \cdot h_i^3$$

$$I_{red} = \frac{1}{12} b \cdot h^3 - \frac{1}{12} b_i \cdot h_i^3 - \frac{2}{12} b_{HAZ} \cdot t^3 + \frac{2}{12} b_{HAZ} \cdot t^3 \cdot \rho - 2b_{HAZ} \cdot t \cdot a^2 + 2b_{HAZ} \cdot t \cdot a^2 \cdot \rho$$

$$= \frac{1}{12} b \cdot h^3 - \frac{1}{12} b_i \cdot h_i^3 - \frac{2}{12} b_{HAZ} \cdot t^3 \cdot (1-\rho) - 2b_{HAZ} \cdot t \cdot a^2 \cdot (1-\rho)$$

with a = 0,5 (h - t)

$$W_{Rd,bottom} = \frac{I_{red}}{z_{bottom}} \quad \text{and} \quad W_{Rd,top} = \frac{I_{red}}{z_{top}}$$

with $z_{bottom} = h/2$ and $z_{top} = h/2$

$$S = \frac{1}{8} b \cdot h^2 - \frac{1}{8} b_i \cdot h_i^2$$

The verifications are then for the bending resistance:

$$M_{E,d} \leq M_{c,Rd}$$

and for the shear resistance:

$$\tau = \frac{Q \cdot S}{I \cdot 2t} \quad \text{nearly equal to} \quad \frac{Q \cdot S_{red}}{I_{red} \cdot 2t}$$

For members under tension and bending see EUROCODE 9.

Very often the welds are not in the top and in the bottom flange but in the webs. This is shown in the figure 16.

The verification of this problem can be develoed with the following formulae for the moment of inertia I and the reduced moment of inertia I_{red}:

$$I = \frac{1}{12} b \cdot h^3 - \frac{1}{12} b_i \cdot h_i^3$$

$$\begin{aligned} I_{red} &= \frac{1}{12} b \cdot h^3 - \frac{1}{12} b_i \cdot h_i^3 - \frac{1}{12} b_{HAZ}^3 \cdot t + \frac{1}{12} b_{HAZ}^3 \cdot t \cdot \rho - 2 b_{HAZ} t a^2 + 2 b_{HAZ} t a^2 \rho \\ &= \frac{1}{12} b \cdot h^3 - \frac{1}{12} b_i \cdot h_i^3 - \frac{1}{12} b_{HAZ}^3 \cdot t \cdot (1 - \rho) - 2 b_{HAZ} t a^2 (1 - \rho) \end{aligned}$$

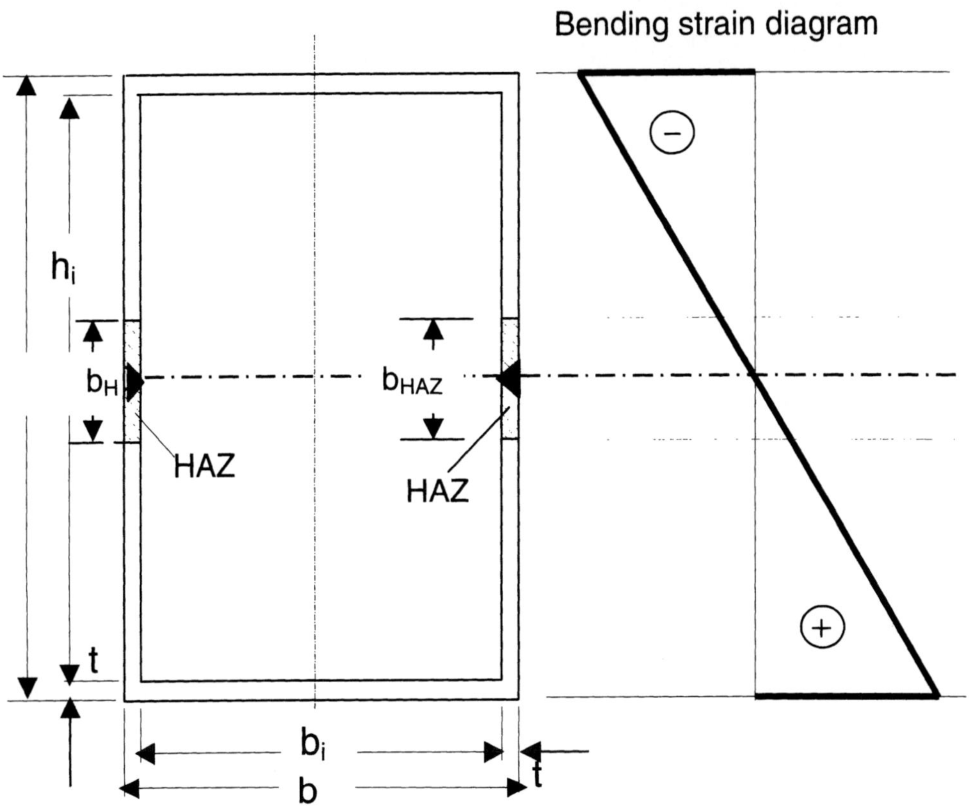

Figure 16: Rectangular hollow section without and with welds in the webs

Since the distance of the welds to the neutral axis is very small the contribution of the HAZ zones ti I can be neglected which can easily be evaluated from the two formulae above. The section modulus is equal for the extreme bottom and top fibre and results from:

$$W_{R,bottom} = W_{R,top} = \frac{I_{red}}{z_{bottom}} = \frac{I_{red}}{z_{top}} \quad \text{(the cross section is symmetrical)}$$

$$S_{red} = 0,125\ b \cdot h^2 - 0,125 \cdot b_i \cdot h_i^2 - b_{HAZ} \cdot t \cdot (1 - \rho)$$

The static moment with respect to the neutral axis can be evaluated with or without the simultaneous account of the heat affected zones in the two geometrical values I and S, the difference can be neglected. But it is necessary to take into consideration either the reduced thickness t_{red} of the web or the reduced shear resistance $\tau_{R,d}$:

The verification of the shear stress in the neutral axis (identical to the HAZ) is as follows:

$$\tau_{E,d} = \frac{Q \cdot S_{red}}{I_{red} \cdot 2t_{red}} \approx \frac{Q \cdot S}{I \cdot 2t_{red}} \leq \tau_{R,d}$$

or

$$\tau_{E,d} = \frac{Q \cdot S_{red}}{I_{red} \cdot 2t} \approx \frac{Q \cdot S}{I \cdot 2t} \leq \tau_{R,d,HAZ}$$

where

$$\tau_{R,d} = \frac{f_{0.2}}{\sqrt{3}\,\gamma_{M1}}$$

$$\tau_{R,d,HAZ} = \frac{\rho f_{0.2}}{\sqrt{3}\,\gamma_{M1}}$$

and

ρ can be taken from table 2.

For members under tension and bending interaction formulae have to be employed, see EUROCODE 9.

3.4. Members under compression and bending (beam columns) explained with a thin walled rectangular hollow section

The beam column below with a compression load $N_k = 300$ kN and a transverse load due to wind $w_k = 12$ kN/m shall be verified for buckling around the strong and around the weak axis of the cross section (see figure 17). The material is 6082 T6 = AlMgSi1 T6 with an 0,2-limit strength of $f_{0,2,k} = 260$ N/mm², an ultimate limit strength of $f_{u,k} = 310$ N/mm² and a reduction factor due to welding of $\rho_{HAZ} = 0{,}65$ (according to ENV 1999-1-1: 1998). The buckling length of the beam column is L = 4.000 mm.

The rectangular hollow section 300x200x6x6 (see figure 18) has end plates at the bottom and at the top of the member which are welded with fillet welds a = 6 mm (see figure 19).

The verification of the beam-column follows now. It is necessary to have the EUROCODE 9 (ENV 1999-1-1: 1998) at hand because in the following verification the chapters from 5.3 to 5.9 of that standard are applied.

The ultimate design resistance with respect to buckling f_s depends on the non dimensional slenderness according to ENV 1999-1-1:1998, 5.8.4.1(1):

$$f_s = \rho \cdot \eta \cdot k_1 \cdot k_2 \cdot f_0$$

where

ρ is the buckling coefficient taken from ENV 1999-1-1: 1998, figure 8 or the following formulae:

η = 1,0 for cross sections of classes 1, 2 or 3

= A_e/A for cross sections of class 4 with $A_e = A - A_c(1-\square_c)$ and with A_c = cross sectional area of that part of the cross section which belongs to class 4

k_1 is a factor to allow for the asymmetry of the cross section, given in ENV 1999-1-1: 1998, table 5.5,

k_2 is a factor to allow for the weakening effects of welding, given in ENV 1999-1-1: 1998, table 5.5,

f_0 = $f_{0,2}$ = is the characteristic strength for bending and for yielding of the total cross section due to tension and compression, given in ENV 1999-1-1: 1998, 5.3.5(1).

The complete interaction formula for the ultimate design resistance of hollow sections under normal force N_{Ed} and under biaxial bending $M_{y,Ed}$ and $M_{z,Ed}$ is given in ENV 1999-1-1: 1998, 5.9.4.2(3) (equation 5,46) as follows:

$$\left(\frac{N_{Ed}}{\chi_{min} \cdot \omega_x \cdot N_{Rd}}\right)^{0,8} + \frac{1}{\omega_0}\left[\left(\frac{M_{y,Ed}}{M_{y,Rd}}\right)^{1,7} + \left(\frac{M_{z,Ed}}{M_{z,Rd}}\right)^{1,7}\right]^{0,6} \le 1,0$$

The individual data will be explained later.

Since the section is a hollow section it is not necessary to verify the strut against lateral torsional buckling.

Not given but necessary data may be taken individually corresponding to the material and to ENV 1999-1-1: 1998.

N_k = 300 kN (permanent load)

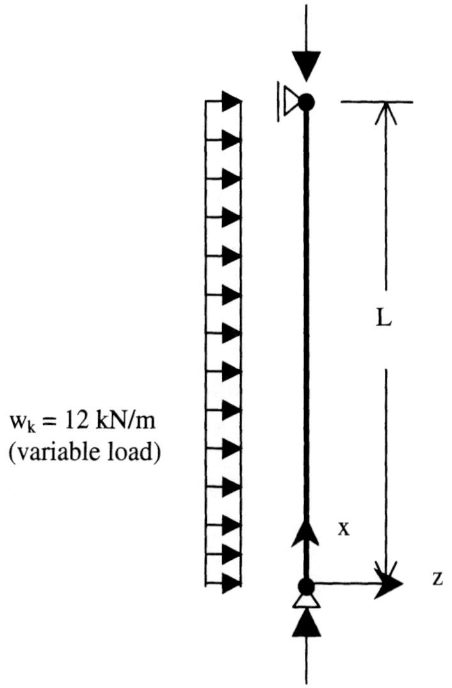

N_k = 300 kN (permanent load)

Figure 17: Beam column with a compression load N_k and a transverse load w_k System and characteristic loads

Effects of the design loads:

$N_{Ed} = \gamma_F \cdot N = 1{,}35 \cdot 300 = 405$ kN (assumption: permanent load)

$M_{y,Ed} = \gamma_F \cdot w \cdot L^2/8 = 1{,}5 \cdot 12 \cdot 4^2/8 = 36$ kNm (assumption: variable load).

For the verification it is necessary to find the data for the design resistance:

κ_y, N_{Rd}, $M_{y,Rd}$
κ_z, und $M_{z,Rd}$.

The allocation of the cross section to a corresponding cross sectional class shall follow ENV 1999-1-1: 1998, 5.4.3, it is necessary to evaluate the slenderness parameter β. The chapter 5.4.3 reads as follows:

5.4.3. Slenderness parameters

(1) The susceptibility of an unreinforced flat element to local buckling is defined by the parameter β, which has the following values:

a) flat outstand or internal elements with no stress gradient: $\beta = b/t$

b) internal element with a stress gradient that results in a neutral axis at the centre: $\beta = 0{,}4\ b/t$ or $\beta = 0{,}4\ d/t$

c) for any other stress gradients: $\beta = g\ b/t$ or $\beta = g\ d/t$

What is the stress distribution over the cross section in the centre of our strut? This stress distribution should at first be evaluated with the non reduced cross sectional data.

The non reduced cross sectional data (see figure 18) and stress distribution (see figure 20) of the profile 300x200x6x6 are

$A = 5.856\ mm^2$
$e_u = 150\ mm$
$e_o = 150\ mm$
$I_y = 7.576 \cdot 10^4\ mm^4$ $I_x = 4.053 \cdot 10^4\ mm^4$
$W_y = 505 \cdot 10^3\ mm^3$ $W_z = 405 \cdot 10^3\ mm^3$
$i_y = 114\ mm$ $i_z = 83\ mm$

EN AW-6082 T6 (AlMgSi1)
$f_{0,2,k} = 260\ N/mm^2$

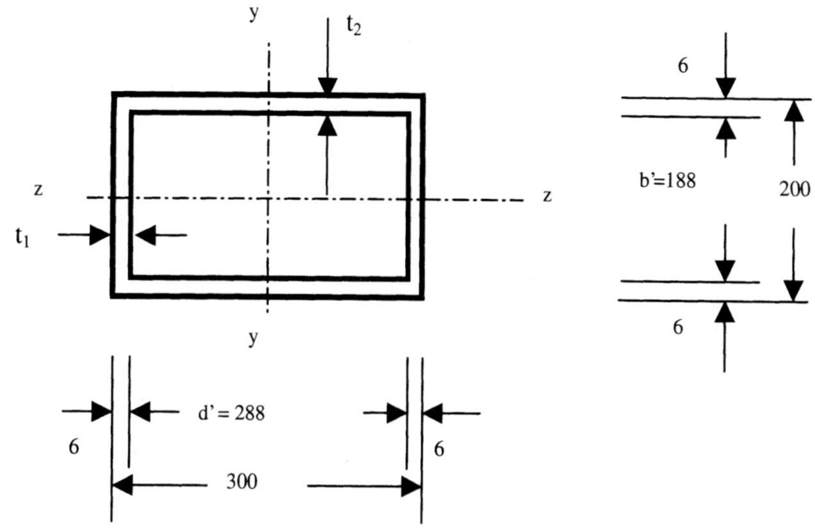

Figure 18: Cross section (dimensions in mm)

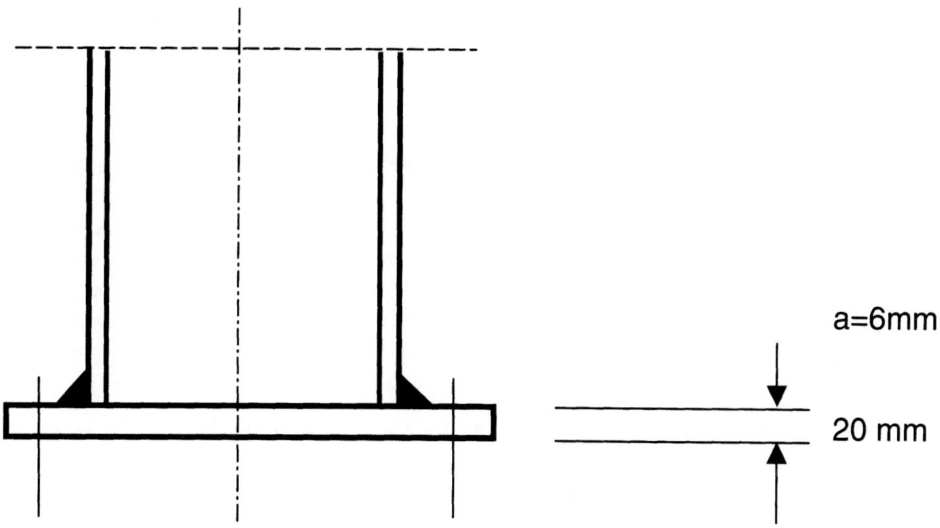

Figure 19: End plate at the bottom of the beam column

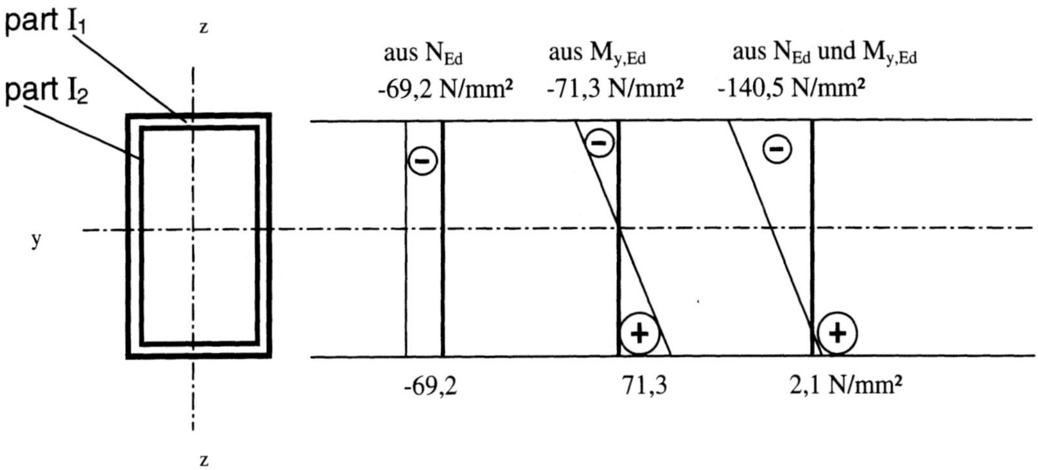

Figure 20: Stress distribution according to the internal forces N_{Ed}, $M_{y,Ed}$ and to the combination of N_{Ed} und $M_{y,Ed}$

Classification of the elements of the cross section according to ENV 5.9.2

(1)P Classification of cross sections for members with combined bending and axial forces is made for the loading components separately according to ENV 5.4. No classification is made for the combined state of stress.

(2) A cross section can belong to different classes for axial force, major axis bending and minor axis bending. The combined state of stress is taken care of in the interaction expressions in ENV 5.9.3 and 5.9.4. These interaction expressions can be used for all classes of cross section. The influence of local buckling and yielding on the resistance for combined loading is taken care of by the capacities in the denominators and the exponents which are functions of the slenderness of the cross section.

Part I_1: σ resulting from N, M and the combination of N and M, according to ENV 5.4.3(1)a) follows:
$\beta_{1,N} = \beta_{1,M} = \beta_{1,N/M} = b'/t = 188/6 = 31{,}3$

Part I_2: σ resulting only from N, according to ENV 5.4.3(1)a) follows:
$\beta_{2,N} = d'/t = 288/6 = 48$

σ resulting only from M, according to ENV 5.4.3(1)b) follows:
$\beta_{2,M} = 0{,}4\ d'/t = 19{,}2$

σ resulting from the combination of N and M, according to ENV 5.4.3(1)c) follows: $\beta_{2,N/M} = g \cdot d'/t$

The value of g results from:
$\psi = \sigma_{min}/\sigma_{max} = +2{,}1/-140{,}4 = -\ 0{,}014$ as follows:
$g = 0{,}7+0{,}3 \cdot \psi = 0{,}7-0{,}3 \cdot 0{,}014 = 0{,}7$ and therefore:
$\beta_{2,N/M} = 0{,}7 \cdot 288/6 = 33{,}6$.

Element classification according to ENV 5.4.4(1) and (2) and ENV table 5.1, bottom line (internal elements):

$$\varepsilon = \sqrt{\frac{250}{260}} = 0{,}981 \rightarrow \beta_3 = 22 \cdot \varepsilon = 21{,}6.$$

The values $\beta_{1,N} = \beta_{1,M} = \beta_{1,N/M} = 31{,}3$ are greater than 21,6, therefore and because we have a heat treated, unwelded material the cross sectional class 4 results.

The value $\beta_{2,N} = 48$ is greater than 21,6 therefore and because we have heat treated and unwelded material the cross section class 4 results.

The value $\beta_{2,M} = 19{,}2$ is smaller than 21,6 and greater than 16 therefore and because we have heat treated, unwelded material the cross section class 3 results.

The value $\beta_{2,N/M} = 33{,}6$ is greater than 21,6 therefore and because we have heat treated and unwelded material the cross section class 4 results.

Now we have to pay attention to ENV 5.4.4(3) to (5). The normal force due to the permanent load creates a constant stress distribution and requires therefore a class 4 cross section in each element. For the first verifications we do neither take notice of the stress distribution due to the moment $M_{y,Ed}$ nor of the stress distribution due to the combined action of N_{Ed} and $M_{y,Ed}$. We therefore allocate the total cross section to class 4.

Buckling resistance around y-y-axis (strong axis) under pure normal force N

Now follow the evaluation of the effective cross section according to ENV 5.4.5(1).

Reduction value ρ_c according to ENV 5.4.5(3)c) for heat treated and non-welded material.

Evaluation of the design resistance of the member

a) $N_{Rd,eff} = A_{eff} \cdot f_{0,2,k}/\gamma_{M1}$, (see analog formula ENV 1999-1-1:1998, 5.8.6, page 75).

Part I_1: $\beta_{1,N} = 31{,}6$, $\beta/\varepsilon > 21{,}6$

$$\rho_{c,1} = \frac{32}{31{,}3/0{,}981} - \frac{220}{(31{,}3/0{,}981)^2} = 1{,}003 - 0{,}216 = 0{,}787$$

$$t_{1,eff} = \rho_{c1} \cdot t = 0{,}787 \cdot 6 = 4{,}72 \text{ mm.}$$

Part I_2: $\beta_{2,N} = 48,\ \beta/\varepsilon = 48{,}95 > 22$

$$\rho_{c,2} = \frac{32}{48/0{,}981} - \frac{220}{(48/0{,}981)^2} = 0{,}654 - 0{,}092 = 0{,}562$$

$$t_{2,eff} = \rho_{c,2} \cdot t = 0{,}562 \cdot 6 = 3{,}37 \text{ mm.}$$

Effective cross sectional values due to $t_{1,eff}$ = 4,72 mm and $t_{2,eff}$ = 3,37 mm and corresponding design resistance of the member:

A_{eff}	= 3.813 mm²	A_{eff}	= 3.813 mm²
e_u	= 150 mm	e_{li}	= 100 mm
e_o	= 150 mm	e_{re}	= 100 mm
$I_{y,eff}$	= $5.385 \cdot 10^4$ mm^4	$I_{z,eff}$	= $2.439 \cdot 10^4$mm^4
$W_{y,eff}$	= $359{,}0 \cdot 10^3$ mm³	$W_{z,eff}$	= $243{,}9 \cdot 10^3$ mm³
$i_{y,eff}$	= 118,8 mm	$i_{z,eff}$	= 80,0 mm

$$N_{Rd,eff} = (A_{eff} \cdot f_{y,k}/\gamma_{M1}) = 3.813 \cdot 260/(1{,}1 \cdot 10^3) = 901{,}3 \text{ kN}$$

Evaluation of all data for the calculation of the buckling resistance of the beam column according to ENV 5.8.4.1(1)

Buckling around the y-y-axis (strong axis):
Calculation of the slenderness λ_y and the buckling factor κ_y:

$$\lambda_y = \frac{4000}{118{,}8} = 33{,}67 \approx 34$$

$$\lambda_1 = \pi \sqrt{\frac{E}{\eta f_{0,2}}}$$

$\eta = A_{eff}/A = 3.813/5.856 = 0{,}651$ for class 4

$$\lambda_1 = \pi \sqrt{\frac{70.000}{0{,}651 \cdot 260}} = 63{,}88 \approx 64$$

$$\bar{\lambda}_y = \lambda_y / \lambda_1 = 0{,}527$$

from ENV table 5.6: $\alpha = 0{,}20$ und $\bar{\lambda}_0 = 0{,}1$

$\phi = 0{,}5\ [1+\alpha\,(\bar{\lambda} - \bar{\lambda}_0) + \bar{\lambda}^2]$, with $\bar{\lambda} = \bar{\lambda}_y$ follows
$\phi = 0{,}5\ [1+0{,}2(0{,}527-0{,}1)+0{,}527^2] = 0{,}682$

and the buckling factor κ_y according to ENV 5.8.4:

$$\kappa_y = \frac{1}{\phi + \sqrt{\phi^2 - \bar{\lambda}^2}} = \frac{1}{0{,}682 + \sqrt{0{,}682^2 - 0{,}527^2}} = 0{,}898.$$

Verification of the load carrying capacity under normal load with respect to buckling around the y-y-axis according to the equation ENV (5.46) in 5.9.4.2(4), hollow sections and tubes with $\omega_x = \omega_0 = 1{,}0$:

$$\left(\frac{405}{0{,}898 \cdot \omega_x \cdot 901{,}3}\right)^{0{,}8} = 0{,}575 < 1{,}0$$

Buckling resistance around the y-y-axis (strong axis) under the combined action of normal force N_{Ed} and bending $M_{y,Ed}$

Now follows the evaluation of the effective cross section according to ENV 5.4.5(1):

Reduction value ρ_c according to ENV 5.4.5(3)c) for heat treated non-welded material

Evaluation of the design resistance of the member for the bending moment $M_{y,Rd}$

a) $M_{y,Rd,eff} = W_{y,el,eff} \cdot f_{y,k}/\gamma_{M1}$, according to ENV 5.6.2.1, equation (5.14) and table 5.3.

Part I_1: $\beta_{1,N} = 31{,}6$, from $\beta/\varepsilon > 22$ follows:

$$\rho_{c,1} = \frac{32}{31{,}3/0{,}981} - \frac{220}{(31{,}3/0{,}981)^2} = 1{,}003 - 0{,}216 = 0{,}787$$

and therefore

$$t_{1,eff} = \rho_{c1} \cdot t = 0{,}787 \cdot 6 = 4{,}72 \text{ mm}.$$

Part I_2: $\beta_{2,N} = 19{,}2$, from $16 < \beta/\varepsilon = 19{,}58 < 22$ follows

$\rho_{c,2} = 1{,}0$ and therefore

$$t_{2,eff} = \rho_{c,2} \cdot t = 1{,}0 \cdot 6 = 6{,}0 \text{ mm} = t_2.$$

Effective cross sectional values due to $t_{1,eff}$ = 4,72 mm, $t_{2,eff}$ = 6,0 mm and corresponding design resistance of the member:

A_{eff}	= 5.359 mm²	A_{eff}	= 5.359 mm²
e_u	= 150 mm	e_{li}	= 100 mm
e_o	= 150 mm	e_{re}	= 100 mm
$I_{y,eff}$	$= 6.499 \cdot 10^4$ mm^4	$I_{z,eff}$	$= 3.895 \cdot 10^4$ mm^4
$W_{y,eff}$	$= 433{,}3 \cdot 10^3$ mm³	$W_{z,eff}$	$= 389{,}5 \cdot 10^3$ mm³
$i_{y,eff}$	= 110,1 mm	$i_{z,eff}$	= 85,25 mm

$$M_{y,Rd} = W_{y,el,eff} \cdot f_{y,k}/\gamma_{M1} = 433{,}3 \cdot 10^3 \cdot 260/(1{,}1 \cdot 10^6) = 102{,}42 \text{ kNm}$$

according to ENV, 5.6.2.1 (1), equation (5.14) and table 5.3.

Verification of the load carrying capacity under combined actions of the normal force N_{Ed} and bending moment $M_{y,Ed}$ with buckling around the y-y-axis according ENV 5.9.4.2(4), equation (5.46) hollow sections and tubes with $\omega_x = \omega_0 = 1{,}0$:

$$\left(\frac{405}{0{,}898 \cdot \omega_x \cdot 901{,}3}\right)^{0,8} + \frac{1}{\omega_0}\left[\left(\frac{36}{102{,}42}\right)^{1,7}\right]^{0,6} = 0{,}575 + 0{,}344 = 0{,}919 < 1{,}0.$$

Buckling resistance around the z-z-axis (weak axis) under normal force N_{Ed}

For the verification of the buckling capacity around the z-z-axis there is no bending moment but only the normal load N_{Ed}. All parts of the cross section are in class 4, as evaluated before.

Evaluation of the buckling parmeters λ_z und κ_z

$$\lambda_z = \frac{4.000}{80} = 50$$

$$\overline{\lambda}_z = \frac{\lambda_z}{\lambda_1} = \frac{50}{64} = 0{,}781$$

from ENV table 5.6: $\alpha = 0{,}2$ and $\overline{\lambda}_0 = 0{,}1$

$$\phi = 0{,}5\,[1 + 0{,}2(0{,}781 - 0{,}1) + 0{,}781^2] = 0{,}873$$

$$\kappa_z = \frac{1}{0{,}873 + \sqrt{0{,}873^2 - 0{,}781^2}} = 0{,}791.$$

Verification of the load carrying capacity due to buckling around the z-z-axis under normal force N_{Ed} only according to the ENV equation (5.46) in 5.9.4.2(4), hollow sections and tubes:

$$\left(\frac{405}{0{,}791 \cdot \omega_x \cdot 901{,}3}\right)^{0{,}8} = 0{,}636 < 1{,}0$$

Result: The sufficient buckling strength of the beam column could be verified.

Verification of the carrying capacity of the MIG-welded foot plate connection of the beam column

The cross section at the foot plate connection is a MIG-welded cross section, therefore the reduction in strength due to welding has to be taken into account by the factor $\rho_{HAZ.}$ Since at the end of the beam column there is no bending moment the stress distribution results only from the normal force and is constant. The extent of the heat affected zone into the cross section is so small ($0{,}5 \cdot b_{HAZ}$ = 20 mm, see ENV 5.5.3(3) and ENV figure 5.6), that in this welded zone it is not necessary to simultaneously take into account any local buckling effect. Therefore the welded cross section is taken as a class 1 cross section.

The verification is made with the non-reduced cross sectional area A = 5.856 mm² according to ENV 5.5.2(1) and (3) and ENV 5.8.6(1):

$N_{Ed} \leq N_{a,Rd} = \rho_{WEZ}\, A\, f_{0,2}/\gamma_{M2}$ whereby $\rho_{HAZ} = 0{,}65$ and $\gamma_{M2} = 1{,}25$

$N_{a,Rd} = (0{,}65*5.856*260/1{,}25)/10^3 = 791{,}7$ kN.

Verification:

$$\frac{N_{Ed}}{N_{a,Rd}} = \frac{405}{791{,}7} = 0{,}51 < 1{,}0$$

The partial safety factor with respect to the resistance $\gamma_{M2} = 1{,}25$ has been taken, because we did the verification for a welded connection.

Chapter 3

RESISTANCE OF PLATED STRUCTURES

Torsten Höglund

Department of Structural Engineering
The Royal Institute of Technology, Stockholm, Sweden

Abstract. The design of plated structures involves check of local buckling as well as distorsional buckling of stiffened plates. Local buckling behavior of cross section elements is discussed in section 1. Depending on the width to thickness ratio, the behavior is different, which is considered for in Eurocode 9 by introducing cross section classes. Especially slender webs are treated. In section 2 thin-walled girders in bending are treated and in section 3 stiffened plates in compression. Section 4 deals with orthtropic plates. In section 5 the shear buckling resistance of plate girders with webs without intermediate stiffeners, webs with transversal and longitudinal stiffeners and trapezoidal corrugated webs are discussed. The design rules in Eurocode 9 and some background information are given.

1. LOCAL BUCKLING BEHAVIOR OF ALUMINIUM ELEMENTS

1.1. Introduction

For slender webs in beams and plate girders loaded in axial compression or bending there is a substantial post-buckling strength provided that, after buckling, web strips supported by the flanges can take over stresses not carried in the middle of the web. The post-critical behaviour of slender plates subjected to compression is very pronounced in that sense that the critical stress according to classical plate buckling theory may be exceeded before the ultimate resistance is reached. The design method in Eurocode 9 is based on the effective thickness concept.

1.2. Slenderness parameter and cross section classes

The resistance of a cross-section element in compression is generally limited by local buckling. The buckling load depends on the slenderness of the cross-section element, which is normally determined by the ratio of the width divided by the thickness ($\beta = b/t$). In many cases the more general parameter for slenderness, λ, is used

$$\lambda = \sqrt{\frac{f_o}{\sigma_{cr}}} \tag{3.1.1}$$

where f_o is the 0,2-limit and σ_{cr} is the elastic buckling stress for a perfect flat plate without initial buckles or residual stresses. λ is proportional to b/t and $\sqrt{f_o / E}$ and depends on the loading and the boundary conditions, e.g. the connection to other cross sectional elements. In Eurocode 9 there is a bar on top of the notation, $\bar{\lambda}$. In this chapter this bar is omitted, as there is no risk of mistake.

For a long plate, width b_w, thickness t_w, simply supported along the edges the critical buckling stress σ_{cr} is

$$\sigma_{cr} = k_\sigma \frac{\pi^2 E}{12(1-\nu^2)} \frac{t_w^2}{b_w^2} \tag{3.1.2}$$

where k_σ is the so-called buckling coefficient. For an axially loaded long plate with simply supported edges, Figure 1, $k_\sigma = 4.0$. The slenderness parameter can now be determined by inserting expression (3.1.2) into (3.1.1).

$$\lambda_w = \sqrt{\frac{f_o}{\sigma_{cr}}} = \sqrt{\frac{f_o}{k_\sigma \frac{\pi^2 E}{12(1-\nu^2)} \frac{t_w^2}{b_w^2}}} = 0.526 \frac{b_w}{t_w} \sqrt{\frac{f_o}{E}} \quad \text{for } k_\sigma = 4 \text{ and } \nu = 0{,}3 \tag{3.1.2}$$

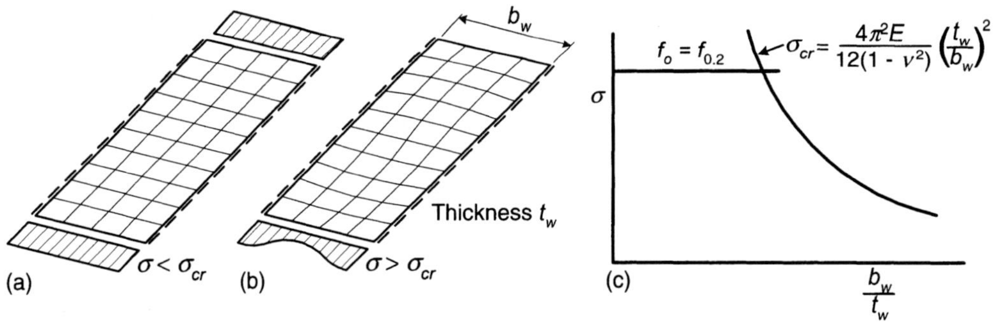

Figure 1: Local buckling of plate in compression

The behaviour of an element in compression depends on the slenderness ratio.

a. If the slenderness ratio of the element is small $(\beta < \beta_1)$ no buckling occurs. The average stress is equal to or even larger than the ultimate strength of the material in tension, cf. Figure 2(a).

b. If the slenderness ratio is somewhat larger $(b_1 < \beta < \beta_2)$ buckling occurs after the compressed element has been plastically deformed to a strain, which is more than about twice the strain corresponding to the 0.2 proof stress f_o $(\varepsilon \approx 1\%)$, Figure 2(b).

c. If the slenderness ratio is further increased $(\beta_2 < \beta < \beta_3)$, buckling occurs once the 0,2 proof stress has been reached and plastic deformation has started. See Figure 2(c).
d. If the slenderness ratio is large $(\beta > \beta_3)$, then buckling occurs before the average stress in the compressed part of the section has reached the 0,2 proof strength. Figure 2(d).

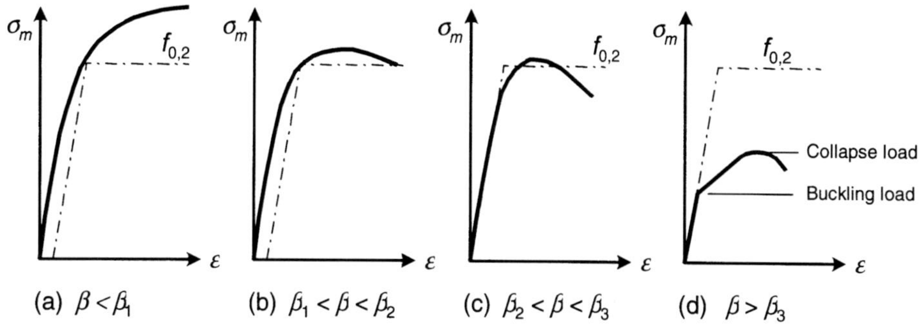

Figure 2: Principle relationships between mean stress σ_m and compressive strain ε for different slenderness β

Failure normally does not occur when some cross sectional element starts to buckle, but after redistribution of stresses and yielding. See Figure 1(b) and 2(d)

The division of cross-sections into four classes for members in bending corresponds to the different behaviour as above. Class 1 and 2 cross sections have compact cross-section elements that behave according to a and b. Class 3 cross sections have semi-slender cross section elements and behave according to c. Class 4 cross sections have one or more slender section elements that behave according to d. See below.

For a member in axial compression, actually only two classes are of interest - non-slender sections with class 1, 2 or 3 cross section elements and slender sections with one or more elements that behave according to d. The division of non-slender sections into three classes is of interest for beams and plate girders in bending only, where redistribution of stresses within the cross section may occur for class 1, 2 and 3 cross sections.

1.3. Behaviour of slender plates

A slender plate with four edges, all simply supported, can carry an ultimate load that is greater than the critical buckling load according to the theory of elasticity for perfectly flat plates. Figures 3(a) and 3(b) show the difference between a plate with two free edges (Figure a) and a plate supported along four edges, (Figure b).

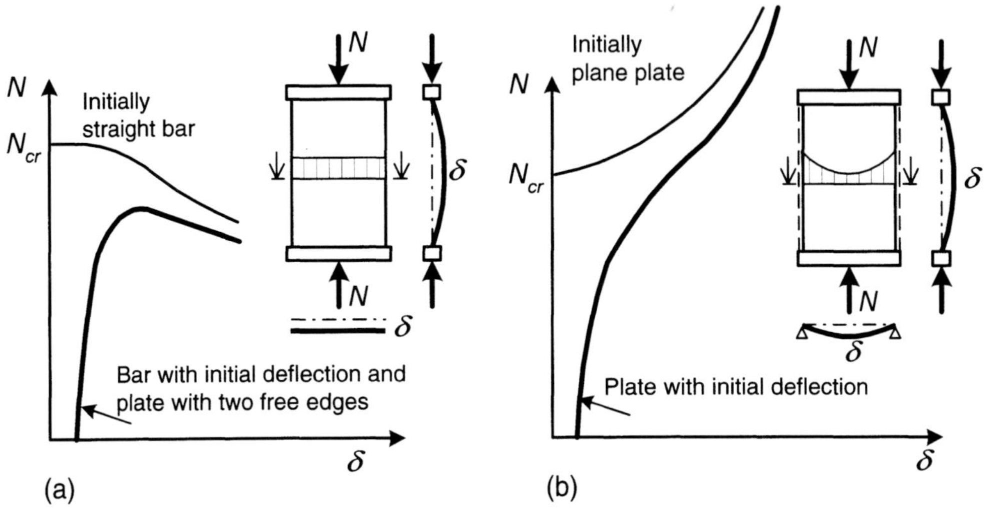

Figure 3: Principle relationship between maximum deflection δ and load N, with and without initial deflections, for compressed bar and plate with two free edges and (b) plate supported along all four edges

A plate with two longitudinally unsupported edges will buckle in the same way as a compressed bar. Hereby, the plate will deform to a surface of single curvature (disregarding small disturbances along the unsupported edges). Every vertical strip will have the same strain and deflection.

After buckling, a plate with four edges, all simply supported, will deform to a surface of double curvature. When compression increases, the strip in the centre will behave in a different way compared to one along the edge. The strip along the edge will remain straight and the compression will result in increased stresses. The central strip, on the contrary, will deflect without any particular increase in stress.

This description of the behaviour of a rectangular plate with simply supported edges, applies only under the condition that the compressed edges are still straight after the plate has buckled. In a long plate that buckles into a number of half-waves, Figure 1(b), the node lines act as straight loaded edges.

The behaviour is also similar for a rectangular plate with three supported edges and one free edge (an outstand element), where the last edge is parallel to the direction of the load. The same condition applies here as in the previous case, if the loaded edges stay straight after buckling and if they are parallel and in the same plane. The compression flange of an I-beam is an example of this case. Each half flange acts as a plate supported along three edges and, after buckling into a number of half-waves, the node lines correspond to the straight, loaded edges.

When the edge of the flange buckles, redistribution of stresses occur with an increase of stresses close to the web and a decrease of stresses at the edges of the flange, see Figure 4(a).

For a single sided flange, as in a channel section, Figure 4(b), the behaviour is usually different. When the free edge of the flange buckle, the ends of the beam rotate and a redistribution of stresses from the free edge to the connected edge does not occur if the load still acts in the centre of the cross section (pinned ends). Therefore, the buckling stress usually is an upper limit for the resistance of an unsymmetrical cross section with outstands.

For sections composed entirely of radiating outstands, such as angles, tees and cruciforms, local and torsional buckling are closely related. Usually, only torsional buckling (lateral-torsional for tees) needs to be checked.

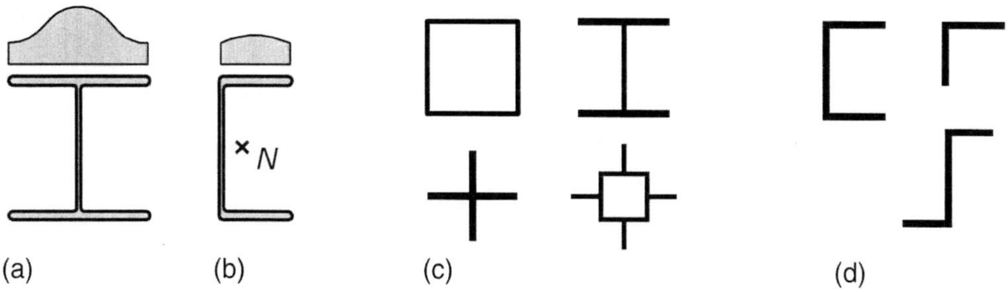

Figure 4: (a) Stress distribution after buckling of a flange of an I-section and a flange of a channel section with pinned ends, (b) Cross sections with stable post-buckling behaviour and (c) Cross sections with no post-buckling strength

The load exceeding the buckling load is usually called post-buckling range (post buckling strength, post critical range). A stable post buckling range exists for an internal cross section element for most load cases such as axial load, bending moment and transverse load. For cross section elements with one free edge, a stable post buckling range exists in flanges of I-profiles and composite cross sections as in Figure 4(c), but normally not in flanges of channels, angles and Z-sections as in Figure 4(d).

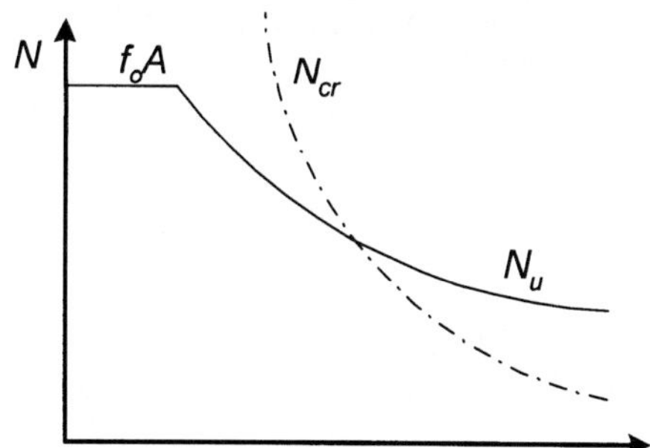

Figure 5: Principle relationship between slenderness parameter λ, resistance N_u and buckling load N_{cr}

The resistance in the post buckling range is of importance for slender plates. The post buckling range increases with increasing slenderness ratio, see Figure 5.

1.4. Effective cross sections

The relation between compression u and axial load N for a slender plate is illustrated in Figure 6. This is characterised by the use of the term effective area. The effective area, as in many regulations for cold-formed structures, can be determined by multiplying the effective width with the gross thickness of the plate, or, as in Eurocode 9, by multiplying an effective thickness with the gross width.

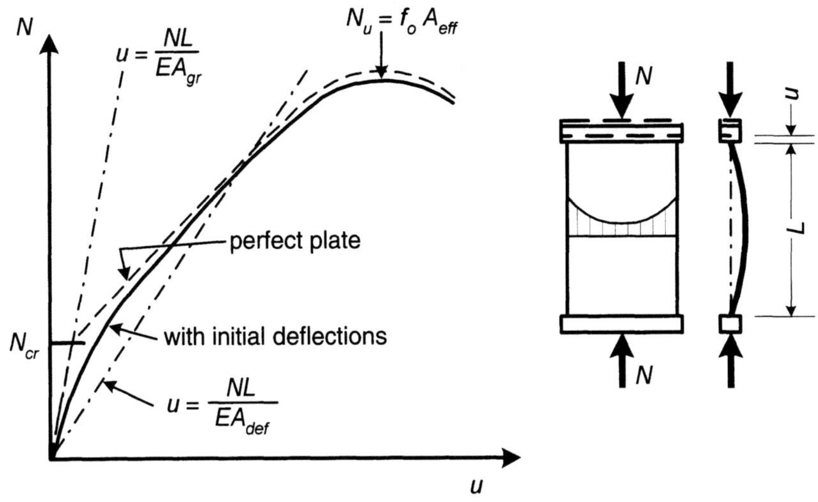

Figure 6: Principle relationship between compression u and axial load N

A_{gr} area of gross cross section, used for calculation of deformation prior to buckling.
A_{def} area of effective cross section used for calculation of deformation. A_{def} depends on the load level
A_{eff} area of effective cross section used for calculation of resistance.

In some recently adopted national standards the effective thickness is used since it leads to simple calculations. In Eurocode 9, the effective thickness is used for calculation of resistance. In many cases an effective width concept is used for calculation of stiffness. See section 5.

Effective cross section for calculation of axial compression force resistance is defined as

$$N_u = A_{eff} f_o \tag{3.1.4}$$

where N_u is the ultimate load for the cross section element.

The axial deformations can be calculated from the relationship

$$u = \frac{N L}{E A_{def}} \tag{3.1.5}$$

where A_{def} depends on the load level.

The definition of A_{def} is different from that for A_{eff}, and, in principle, $A_{def} \neq A_{eff}$ even close to failure, see next section.

1.5. Calculation technique for class 4 cross sections

The resistance of many structures is independent of the overall bending deflection, as in the case of a short, centric compressed column. The ultimate load is governed only by *local buckling*, i.e. of A_{eff}. For a beam in bending, the resistance is determined by the section modulus W_{eff} for the effective cross section.

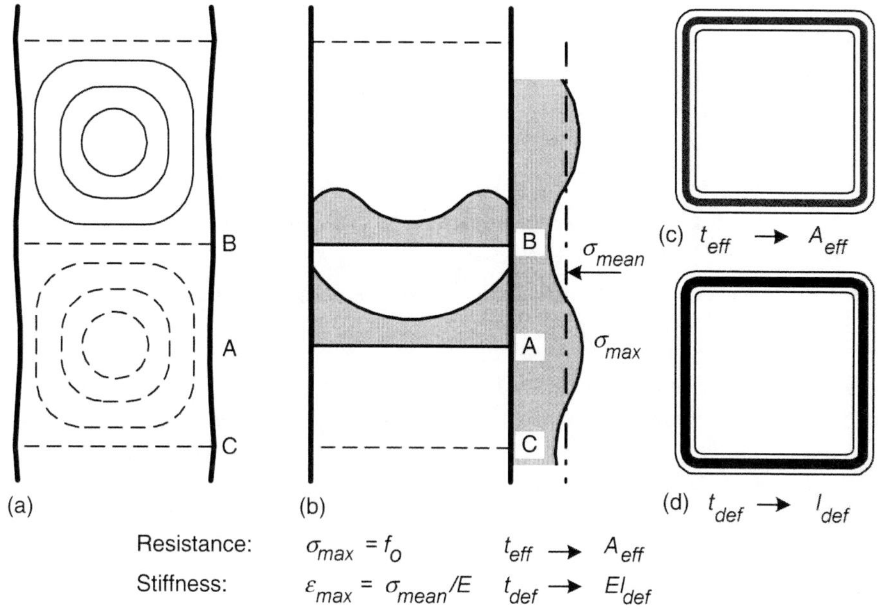

Figure 7: Stress distribution in a buckled plate and effective cross sections

The resistance for other structures may also depend on the *bending stiffness*. One example is the resistance of a long column, which depends on the bending stiffness as well as the strength of the most compressed cross section element. The bending stiffness is determined by the deformation of the compressed cross section element expressed by I_{def}.

These two effective cross sections are not the same. The effective cross section for stiffness is normally larger than the effective cross section for strength, which is explained in conjunction to Figure 7. The left figure shows one flange of a quadratic hollow section loaded in compression. The flange is buckled in several half waves and, as the section is quadratic, the length of the buckles is the same as the width of the flanges. In the figure is also shown the stress distribution in a section through the middle of the buckle, section A-A, and also through the node line between two buckles, section B-B. These stress distributions differ quite a lot in a slender flange. In the section through the buckle the stresses are largest at the borders. In the section between two buckles the stresses are levelled out to a certain extent. This means that the stresses (and the corresponding strains) vary along the column from a maximum at the centres of the buckles to a minimum at the nodes.

Of importance for the axial *strength* is the *maximum stress*. Of importance for the *axial stiffness* is the *mean strain* along the whole column. The effective area for stiffness is then a function of the mean strain along the column, which is proportional to the mean stress, whereas the effective area for strength is given by the maximum stress.

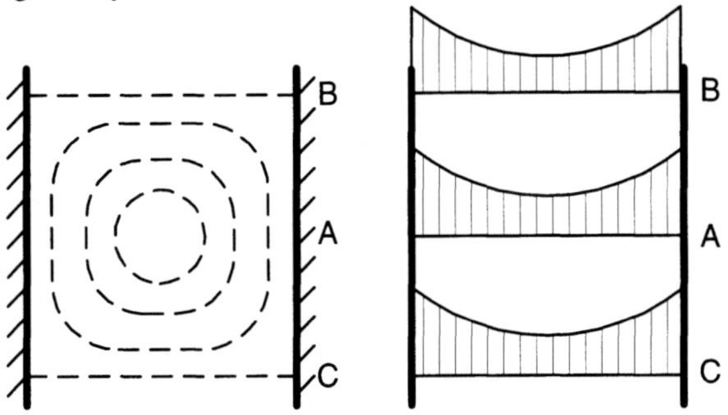

Figure 8: Stress distribution in a plate with straight edges

One condition for the development of the stress distribution in Figure 7 is that the longitudinal edges are free to deflect inwards when local buckling occurs. This is the case in a hollow section. In an inner panel of a wide plate with stiffeners, adjacent panels prevent the edges to deflect inwards. Then the stress distribution through the nodes is very similar to the stress distribution through the buckles, Figure 8. The maximum stresses are also smaller in a plate with straight edges than in a hollow section.

The calculation procedures according to Eurocode 9 for different loading conditions are presented in Section 2, 3, 4 and 5.

1.6. Calculation of deflections of beams with class 4 cross section

To calculate the deflection of a beam with class 4 cross section is very complicated, due to the fact that the effective stiffness varies with the load level and then also along the beam. In Eurocode 9, a simplified procedure is used, which mean that only the second moment of area I_{eff} for the effective cross section in the ultimate limit state need to be calculated besides I_{gr} for the gross cross section.

The calculation of elastic deflection should generally be based on the properties of the gross cross section of the member. However, for slender sections it may be necessary to take reduced section properties to allow for local buckling.

Advantage may be taken from reduced stress levels for class 4 sections to calculate the effective thickness, using the following fictitious second moment of area I_{fic}, constant along the beam

$$I_{fic} = I_{gr} - \frac{\sigma_{gr}}{f_o}(I_{gr} - I_{eff}) \tag{3.1.6}$$

where:

I_{gr} = second moment of area of the gross cross section
I_{eff} = second moment of area of the effective cross section in the ultimate limit state, with allowance for local buckling
σ_{gr} = maximum compressive bending stress in the serviceability limit state, based on the gross cross section (positive in the formula).
f_o = characteristic strength for bending and overall yielding

In Figure 9 the deflection δ is given as a function of the load level σ_{gr} / f_o. The deflections δ based on I_{gr}, I_{eff} and I_{fic} are compared. The curve based on I_{fic} is similar to curves from tests. Note that deflections are usually calculated at the serviceability limit state only.

For a perfect beam, with cross section elements without initial buckles, for loading less then corresponding to the critical buckling stress σ_{cr}, the deflection should be that calculated for the gross cross section I_{gr} (dotted line below σ_{cr} / f_y in Figure 9). However, there are always imperfections (initial buckles, non-linear material and residual stresses) that result in larger deflections. Therefore, the deflections based on the fictitious second moment of area I_{fic} are in good agreement with tests.

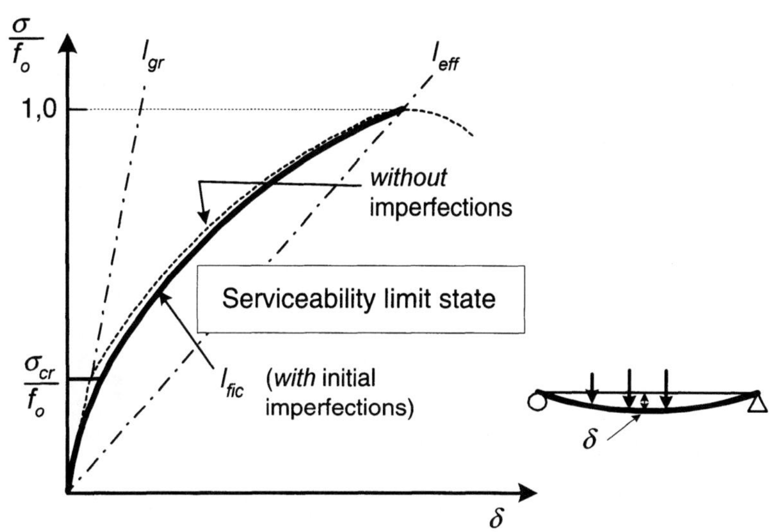

Figure 9: Load-deflection curves

1.7. Fatigue due to breathing

When a beam with a slender web is loaded in compression or bending, because of initial deflection or buckling, the web will have out-of-plane deformations. When the beam is unloaded, the web will return to the original position. If the load is cyclic, this phenomenon will repeat, which is called "breathing". Due to the deflection of the web, plate-bending stresses are created. There is a risk of initiating fatigue cracks at the flange-to-web junction and along the stiffeners. Possible types of cracks due to breathing are shown in Figure 10.

The stresses in the web are not only dependent on the load, but also on the slenderness of the web. This makes the fatigue problem of web breathing complicated. Plate bending stresses normal to panel boundaries are shown to be the primary cause of fatigue cracks.

For girders in bending, fatigue assessment for type 1 crack is automatically accomplished by the fatigue assessment for type 2 cracks. Type 1 cracks occur at a toe on the web side of the fillet weld connecting the web to the compression flange and type 2 cracks are observed at the toe on the web side of the fillet weld connecting the vertical stiffener to the web. See Figure 10(a). Breathing does not very much influence type 3 cracks, initiated at the fillet weld connecting the web to the tension flange, as the web deflects very little on the tension side of the web.

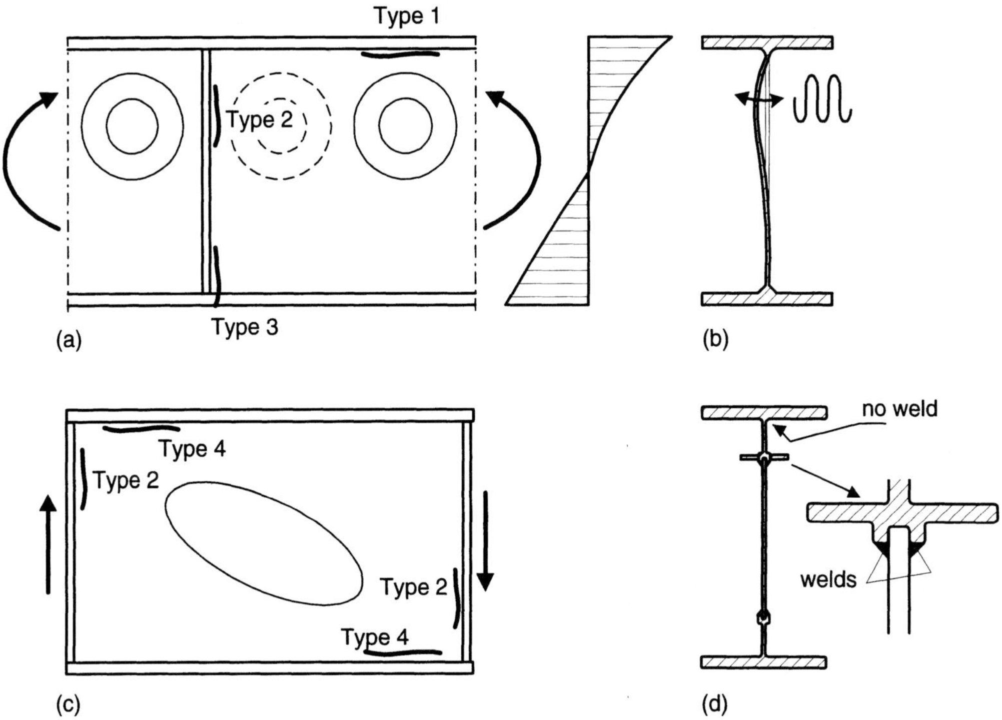

Figure 10: Breathing (a) Beam in bending (b) Web breathing (c) Web in shear (d) Web with stiffener

For girders in shear the bending stresses due to breathing acts together with in-plane shear stresses and tensile membrane stresses due to tension field action in the post buckling stage of a slender web. To avoid type 4 cracks, see Figure 10(c), due to breathing of a web in shear, it has usually been recommended that the maximum applied in-plane shear stress (at the ultimate limit state for fatigue) should be below the shear buckling stress for the simply supported condition along the four edges. However, another method, shown to give not so conservative result, is to base the fatigue check on stresses calculated for the effective section. For shear loading, the shear stress is then calculated as

$$\tau_{fat} = \frac{V_{fat,Ed}}{h_w t_w \rho_v} \qquad (3.1.7)$$

where ρ_v is the reduction factor for shear buckling.

The above findings and recommendations are based on research on steel plate girders. No tests have been found on aluminium plate girders. Due to the smaller elastic modulus of aluminium the web deflection might be larger than in a corresponding steel girder. On the other hand, the resulting bending stresses normal to the panel boundaries are smaller than in steel as the elastic modulus is smaller. As the fatigue resistance of aluminium is less than for steel, fatigue due to breathing might be more severe then in steel. However, the fatigue strength of thin-walled aluminium plate girders can be increased substantially by the use of extruded T-shaped flanges, where there are no welds at the junction between the flange and the web. See Figure 10(d). Furthermore, stiffeners can be added to the web of the T-shaped flange profile to increase the elastic buckling stress of the web.

2. THIN-WALLED ALUMINIUM GIRDERS IN BENDING

2.1. Introduction

For slender webs in beams and plate girders loaded in bending there is a substantial post-buckling strength provided that, after buckling, web strips supported by the flanges can take over stresses not carried in the middle area of the compressed part of the web. In many design codes post buckling behaviour is accounted for with the effective width concept. In Eurocode 9 the effective thickness concept is used, which gives almost he same result but is simpler to use. Even in beams with non-slender cross sections the final failure is due to local buckling, however in a different way. To account for the different behaviour for different slenderness elements, cross section classes are introduced in most modern design codes as Eurocode 9.

2.2. Behaviour of beams in bending

Yielding and local buckling

The ultimate limit state of a beam can occur in different ways depending upon the geometry of the beam (the span L, the b/t ratio of the individual parts etc.), the loading and support conditions and the type of connection. Failure is most often accompanied by local buckling of compressed cross section parts. Exceptions are compact cross sections, such as solid rectangular and circular sections and beams made by material with small ductility where tensile failure may occur.

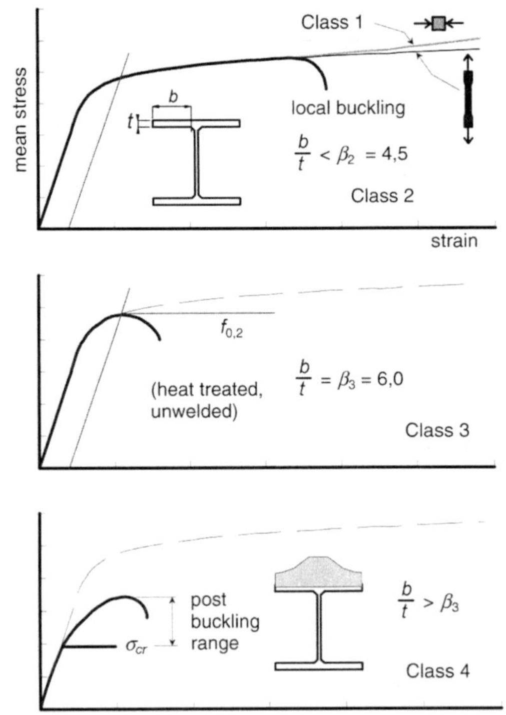

Figure 11: Stress-strain curves for compression flange in different cross section class

Compact cross section (class 1)

In the case of a beam with compact cross-section, in which local buckling or flexural-torsional buckling are not likely to occur, the beam experiences the inelastic range after reaching the limiting elastic moment $M_{el} = W_{el} f_o$ until the ultimate moment M_u is reached. This moment cannot be defined (as it is for steel structures) as the full plastic moment $M_{pl} = W_{pl} f_o$. In fact, due to the hardening behaviour of the σ–ε law of aluminium, a limiting curvature has to be defined corresponding to the limit of strain in the outermost fibre of the cross section. The increase in strength, M_u - M_{el}, obtained in this phase can be quantified through a relation

$$M_u = \alpha \ M_{el} \tag{3.2.1}$$

which defines in a general form a shape factor $\alpha = \alpha_1$ which is not solely dependent upon the cross-sectional geometry, as is usual, but also depends upon the

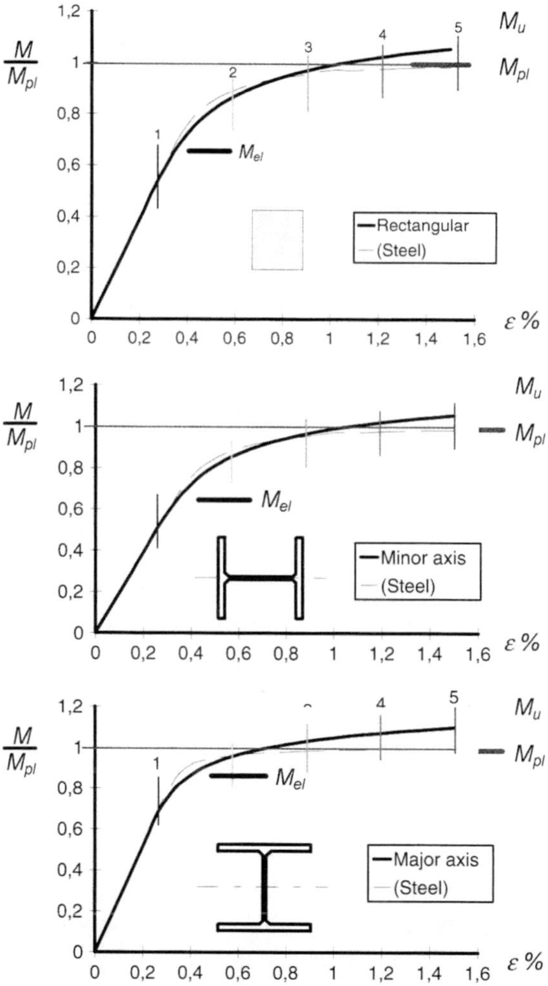

Figure 12: Moment-strain curves for compact cross sections

parameters in the σ–ε law and on the definition of the limiting curvature. Figure 12 show that $M_u > M_{pl}$ if ε is larger than about 1%. Then $\alpha_1 > M_{pl} / M_{el}$.

In this case the rotational capacity of the cross-section, which characterise the flexural ductility of the beam, allows redistribution of the internal forces, and it is therefore possible to carry out a limit analysis of the whole structure.

Non-slender cross-section (class 2)

In the case of open profiles, local buckling phenomena are most likely to occur in the compressed region of the cross-section and cause a decrease in the M–ε curve of the beam. This unstable behaviour is dependent upon the $\beta = b/t$ ratio. If the decreasing portion of the curve occurs after the ultimate moment M_u is reached, the beam keeps the same maximum load-carrying capacity, but the rotational capacity of the cross-section is reduced why redistribution of the internal forces is limited. Note that using M_{pl} as the ultimate resistance corresponds to a strain at the extreme fibres of about 1% for rectangular cross sections as well as for mayor and minor axis bending of H cross-sections. See Figure 12. Also note that the difference between the curves for aluminium with its hardening behaviour of the σ–ε law and steel is not that large.

Slender cross-section (class 3)

If the decreasing portion of the curve occurs before the ultimate moment M_u is reached (class 3 cross-section, see Figure 11), the load-carrying capacity of the beam is affected by local buckling phenomena to a higher degree if the b/t ratio is large. The bending moment resistance M_u can reach and to some extent exceed the elastic moment M_{el} but not to more than the plastic moment M_{pl}. Also ductility decreases to the extent that redistribution of internal forces cannot be considered.

Thin-walled cross section (class 4)

One or more cross section elements are so slender that the moment corresponding to classic plate buckling theory $M_{cr} = W_{el}\sigma_{cr}$ is less than the yield moment $M_{el} = W_{el} f_o$. However, after buckling, redistribution of stresses within the section can occur and the loading can be increased until yielding occurs in some area of the cross section. See Figure 11. The post buckling range is more pronounced for a slender section for which the difference between M_{el} and M_{cr} is large. See Section 1 for further information on behaviour of slender elements in compression.

2.3. Beams without web stiffeners

Classification of cross sections

Based on the above, four classes of cross sections are defined, as follows:

Class 1 cross sections are those which can form a plastic hinge with the rotation capacity required for plastic analysis.

Class 2 cross sections are those which can develop their plastic moment resistance, but have limited rotation capacity.

Class 3 cross sections are those in which the calculated stress in the extreme compression fibre of the member can reach its proof strength, but local buckling is liable to prevent development of the full plastic moment resistance.

Class 4 cross sections are those in which it is necessary to make explicit allowances for the effects of local buckling when determining their moment resistance or compression resistance.

Stress distribution, shape factor and slenderness limits for the four cross section classes are summarised in Table 1 for unwelded sections. For welded sections the same formulae are applicable, except that W_{pl}, W_{el} and W_{eff} are reduced to allow for HAZ softening. See 3.7.

Class	Actual and nominal stress distribution	Shape factor	Slenderness limits
1	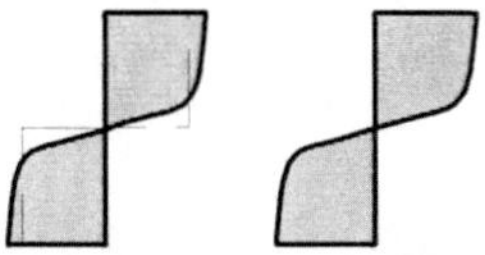	$\alpha \geq \frac{W_{pl}}{W_{el}}$ (See Eurocode 9, Annex G)	$\beta \leq \beta_1$
2	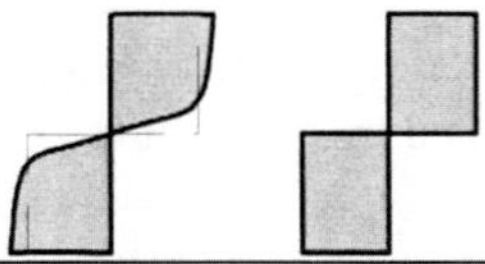	$\alpha = \frac{W_{pl}}{W_{el}}$	$\beta_1 < \beta \leq \beta_2$
3	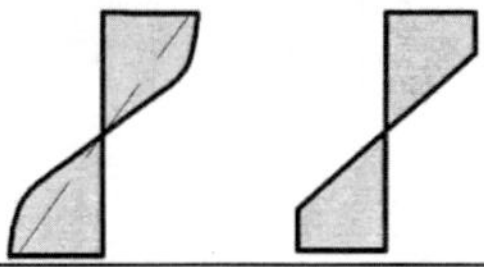	$\alpha = 1 + \frac{\beta_3 - \beta}{\beta_3 - \beta_2}\left(\frac{W_{pl}}{W_{el}} - 1\right)$	$\beta_2 < \beta \leq \beta_3$
4	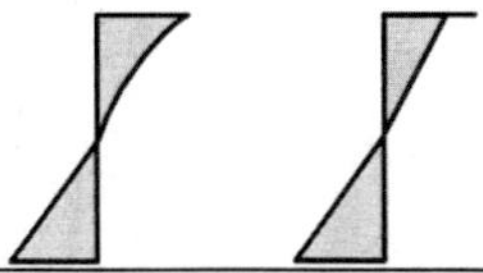	$\alpha = \frac{W_{eff}}{W_{el}}$	$\beta > \beta_3$

Table 1: Stress distribution and shape factor for cross section class 1 to 4

The classification of a section depends on the proportions of each of its compression elements. The compression elements include every element of a cross section that is either totally or partially in compression.

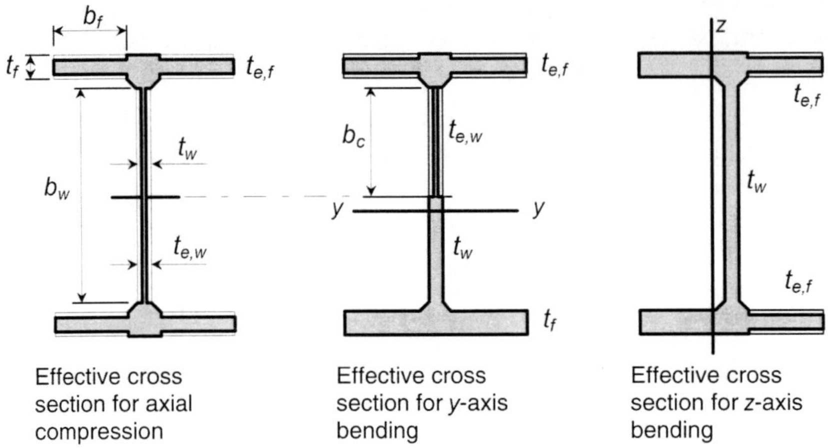

Figure 13: Effective cross sections

The various compression elements in a cross section (such as a web or a flange) can, in general, be in different classes. The cross section should be classified by quoting the least favourable class of its compression elements.

The cross section class depends on the loading. Therefore, a cross section can belong to different classes for *y*- and *z*-axis bending as well as axial force. See examples in Figure 13. No cross section class or effective cross section is defined for the combined action of normal force and bending moment, but for the separate actions. Combinations of forces are dealt with by interaction formulae.

Cross section elements

The following basic types of thin-walled elements are identified in the classification process:

a) flat outstand element
b) flat internal element
c) curved internal element.

These elements can be unreinforced, or reinforced by longitudinal stiffening ribs or edge lips or bulbs (see Figure 14).

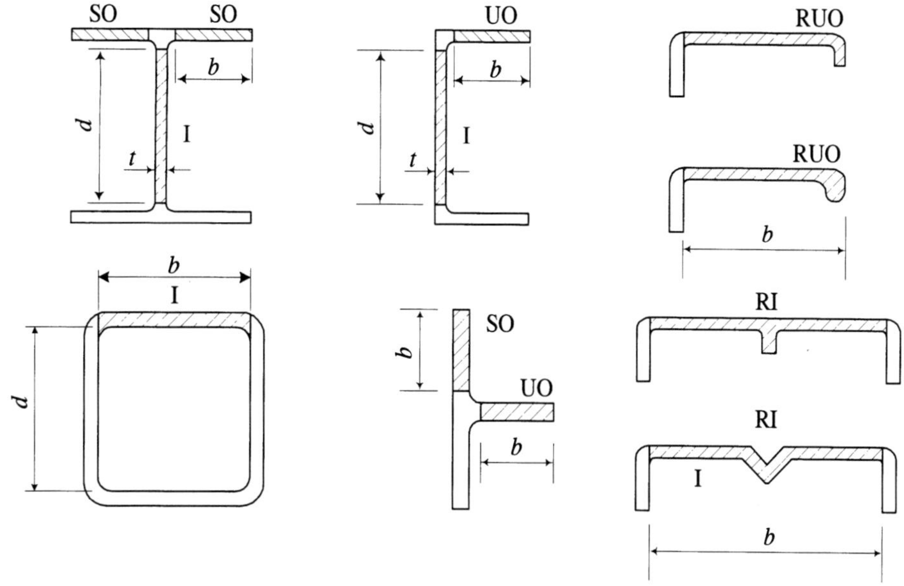

(a) ---------------- Unreinforced ----------------   (b) Reinforced

Key SO: symmetric outstand
UO: unsymmetric outstand
I: internal

RI: reinforced, internal
RUO: reinforced unsymmertical outstand

Figure 14: Types of element

Slenderness parameter

The susceptibility of an un-reinforced flat element to local buckling is defined by the slenderness parameter β, which has the following values:

a) flat outstand or internal elements with no stress gradient $\beta = b/t$ (3.2.2a)
b) internal element with a stress gradient that results in a neutral axis at the centre $\beta = 0{,}40\ b/t$ (3.2.2b)
c) for any other stress gradients $\beta = g\ b/t$ (3.2.2c)

where b is the width of an element, t is the element thickness and g is the stress gradient coefficient. For vertical elements in Figure 14, d is used instead of b in the expressions for β.

For internal elements and outstand elements with peak compression at root, g is given by the expressions:

$$g = 0{,}70 + 0{,}30\psi \qquad (1 > \psi > -1), \qquad (3.2.3)$$

$$g = 0{,}80/(1 - \psi) \qquad (\psi < -1), \text{ see Figure 15} \qquad (3.2.4)$$

where ψ is the ratio of the stresses at the edges of the plate under consideration, related to the maximum compressive stress.

For outstand elements with peak compression at toe use $g = 1$.

When considering the susceptibility of a reinforced flat element and cylindrical elements to local buckling, see Eurocode 9.

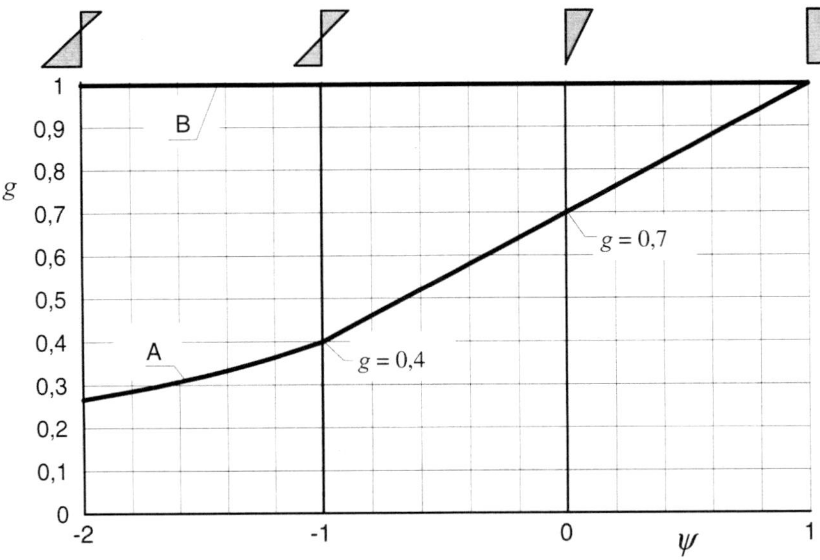

Figure 15: Flat internal elements under stress gradient, values of g. For internal elements or outstands (peak compression at root) use curve A. For outstands (peak compression at toe) use line B

Element classification limits

The classification of elements in cross sections is linked to the values of the maximum slenderness parameter β. For elements in beams:

$$
\begin{array}{rl}
\beta \leq \beta_2 & : \text{class 1} \\
\beta_1 < \beta \leq \beta_2 & : \text{class 2} \\
\beta_2 < \beta \leq \beta_3 & : \text{class 3} \\
\beta > \beta_3 & : \text{class 4}
\end{array}
$$

Values of β_1, β_2 and β_3 are given in Table 2.

	Heat treated, unwelded			Heat treated, welded or non heat treated, unwelded			Non heat treated, welded		
Elements	$\frac{\beta_1}{\varepsilon}$	$\frac{\beta_2}{\varepsilon}$	$\frac{\beta_3}{\varepsilon}$	$\frac{\beta_1}{\varepsilon}$	$\frac{\beta_2}{\varepsilon}$	$\frac{\beta_3}{\varepsilon}$	$\frac{\beta_1}{\varepsilon}$	$\frac{\beta_2}{\varepsilon}$	$\frac{\beta_3}{\varepsilon}$
Outstand	3	4,5	6	2,5	4	5	2	3	4
Internal	11	16	22	9	13	18	7	11	15

Table 2: Slenderness parameters β_1/ε, β_2/ε and β_3/ε

$\varepsilon = \sqrt{250/f_o}$ where fo is in N/mm^2

In the Table, an element is considered welded if it contains welding at an edge or at any point within its width. However, cross-sections of a member that do not contain welding may be considered as unwelded even if the member is welded elsewhere along its length. Note that in a welded element the classification is independent of the extent of the HAZ.

It is permissible to use a modified expression,

$$\varepsilon = \sqrt{250\, z_1/(f_o z_2)}$$

when classifying elements in members under bending, if the elements are less highly stressed than the most severely stressed fibres in the section. In this expression, z_1 is the distance from the elastic neutral axis of the effective section to the most severely stressed fibres, and z_2 is the distance from the elastic neutral axis of the effective section to the element under consideration. See Figure 16. In principal z_1 and z_2 should be evaluated on the effective section by means of an iterative procedure. Normally two steps are sufficient, see Section 6.

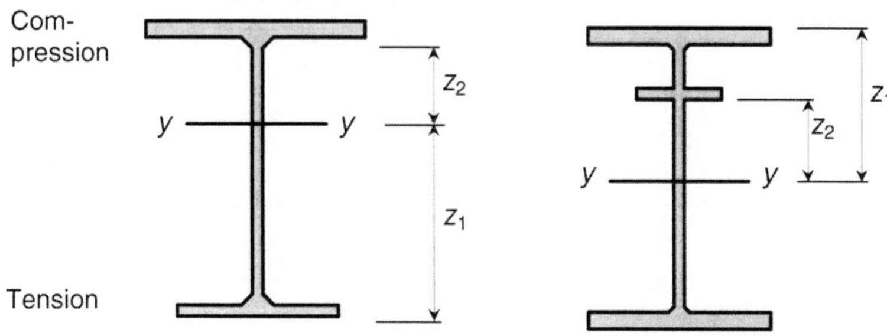

Figure 16: Distances z_1 and z_2

Effective thickness

Local buckling in class 4 members is generally allowed for by replacing the true section by an effective section. The effective section is obtained by employing a local buckling coefficient ρ_c to factor down the thickness. ρ_c is applied to any uniform thickness class 4 element that is wholly or partly in compression. Elements that are not uniform in thickness require special study by the designer.

The coefficient ρ_c is found separately for different elements of the section, in terms of the ratio β/ε, where β is found as in 3.3 and ε is defined under Table 2.

Values of ρ_c are as follows:

$$\rho_c = 1{,}0 \qquad \text{when } \frac{\beta}{\varepsilon} \leq \frac{\beta_3}{\varepsilon}, \tag{3.2.5}$$

$$\rho_c = \frac{A}{\beta/\varepsilon} - \frac{B}{(\beta/\varepsilon)^2} \qquad \text{when } \frac{\beta}{\varepsilon} > \frac{\beta_3}{\varepsilon} \tag{3.2.6}$$

where β_3/ε, A and B are given in Table 3.

Elements	Heat treated, unwelded			Heat treated, welded or non heat treated, unwelded			Non heat treated, welded		
	$\frac{\beta_3}{\varepsilon}$	A	B	$\frac{\beta_3}{\varepsilon}$	A	B	$\frac{\beta_3}{\varepsilon}$	A	B
Outstand	6	10	24	5	9	20	4	8	16
Internal	22	32	220	18	29	198	15	25	150

Table 3: Slenderness parameter β_3/ε and coefficients A and B in the expression for ρ_c

Effective cross section

Symmetric I-girder with class 1, 2 or 3 flanges and class 4 web.
The effective parts of a class 4 cross section are combined into an effective cross section. An example of effective cross section for a symmetric I-girder with class 1, 2 or 3 flanges and class 4 web is given in Figure 17. Note that no iteration process is necessary. The effective thickness of the web is based on the width b_w and $\psi = -1{,}0$. The web thickness is reduced to the effective thickness $t_{w,ef}$ within $b_c = b_w/2$ on the compression side only.

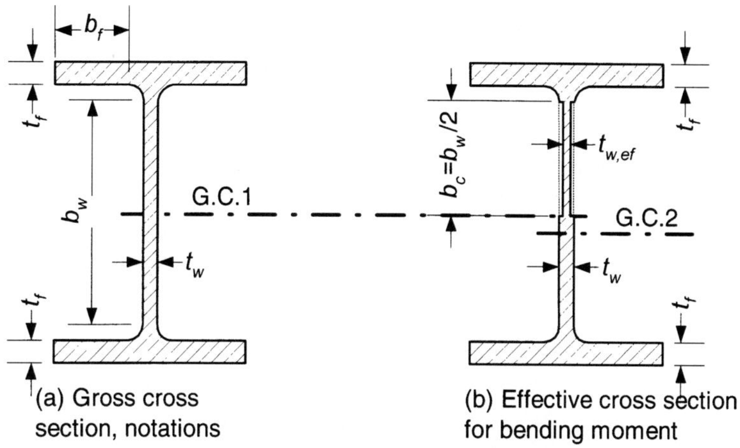

Figure 17: Effective cross section for an I-girder with class 4 web and equal, class 1,2 or 3 flanges

Symmetric I-girder with class 4 flanges and web or I-girder with different flanges.
Calculation steps for a symmetric I-girder with class 4 flanges and web, or an un-symmetric I-girder with class 4 web are given in Figure 18. The effective thickness of the web is based on the width b_w and $\psi = 1 - b_w/b_c$ where b_c is the width of the compression part of the web calculated for the cross section with reduced compression flange but unreduced web (Step 2). The web thickness is reduced to the effective thickness $t_{w,ef}$ within b_c (Step 3).

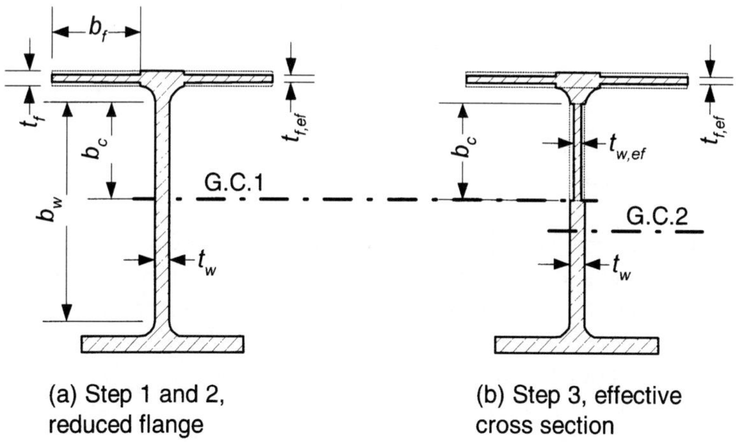

Step 1: Effective thickness of compression flange
Step 2: Centre of gravidity G.C.1 of cross section with reduced compression flange
Step 3: Effective thickness of web part in compression. Effective cross section

Figure 18: Effective cross section for I-girder with different flanges or class 4 flanges

Welded section

In welded sections the effective thickness is obtained using a reduced thickness $t_{eff} = \rho_{haz}\, t$ for the HAZ material or the reduced thickness $t_{eff} = \rho_c\, t$ for class 4 elements, whichever is the smaller. An example of effective section is shown in Figure 19. $\rho_{c,1}$ is based on b_1/t_f for the outstand parts of the compression flange and $\rho_{c,2}$ is based on b_2/t_f for the internal part of the compression flange. b_{haz} and ρ_{haz} are given in the code. For the example in Figure 19(a) $\rho_{haz} < \rho_c$. Then $\rho_{haz}\, t_f$ is the effective thickness within the width $2b_{haz} + t_w$ and $\rho_{c,1}\, t_f$ on the outstands and $\rho_{c,2}\, t_f$ between the webs. In Figure 19(b) $\rho_{haz} > \rho_c$. Then the effective thickness is $\rho_{haz}\, t$ within the width $2s_{weld} + t_w$ where s_{weld} is the leg length of the fillet weld. In order to simplify, the whole flange can be reduced as shown in Figure 19(c).

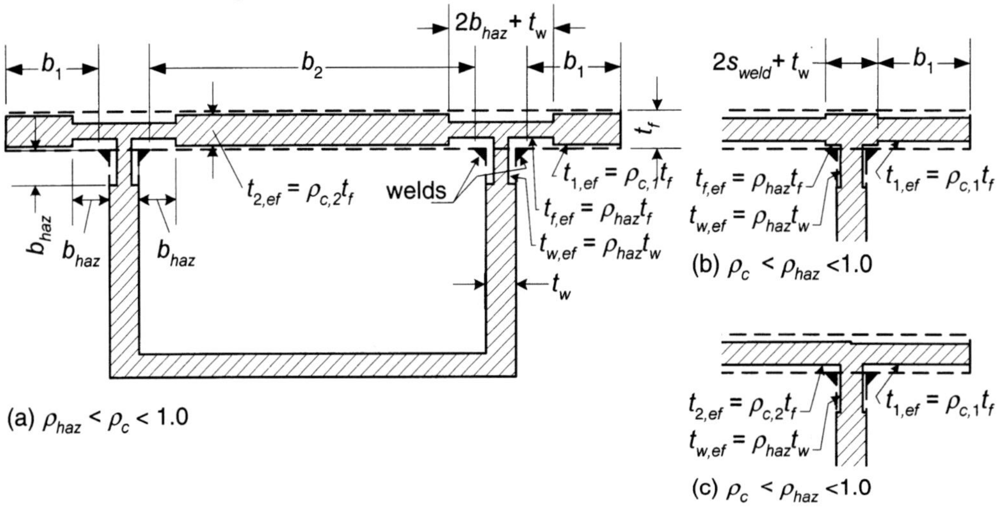

Figure 19: Effective section of welded class 4 hollow cross section made by a channel (= web and bottom flange) welded to a plate (= top flange)

Section with holes

Small holes in the web may be ignored provided that they do not occupy more than 20% of the sectional area of the web. In a section with larger holes the design value of the bending moment resistance M_{Rd} is the lesser of $M_{a,Rd}$ and $M_{c,Rd}$, where $M_{a,Rd}$ is based on the design value of the ultimate strength f_u/γ_{M2} in the net section and $M_{c,Rd}$ is based on the design value of the yield strength f_o/γ_{M1} with no allowance for holes. For both $M_{a,Rd}$ and $M_{c,Rd}$, allowance should be made for HAZ-softening, if any.

$$M_{a,Rd} = W_{net} \frac{f_u}{\gamma_{M2}} \quad \text{in the net section} \tag{3.2.7}$$

$$M_{o,Rd} = \alpha\, W_{el} \frac{f_o}{\gamma_{M1}} \quad \text{at each cross-section} \tag{3.2.8}$$

where W_{net} is the elastic modulus of the net section allowing for holes and HAZ softening, if welded. The stress distribution for $M_{a,Rd}$ and $M_{c,Rd}$ is illustrated in Figure 20 for a class 3 cross section beam with holes in the web.

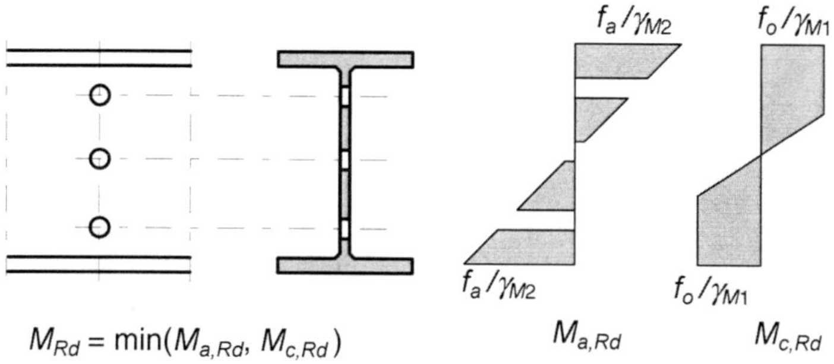

Figure 20: Nominal stress distribution of a section with holes

2.4. Plate girder with web stiffeners

Transverse stiffeners do not influence the bending resistance unless they are very closely spaced (less than $b_w/2$) or there are longitudinal stiffeners as well.

Usually the girder is checked for the maximum moment. However, if the length a between stiffeners is small and the moment gradient is large the moment arising under design load at a distance $0.4a$ from the more heavily stressed end should be checked instead of the maximum moment in the bay.

Longitudinal stiffened girders.

Longitudinal web stiffeners of an aluminium girder can be made in different ways. Some examples are shown in Figure 21. They can be (a) part of the sections welded together, (b) plates or angle sections, (c) trapezoidal section which together with the web plate form a closed stiffener or (d) orthotropic plates made by extrusions welded together - (d) hollow sections or (e) open sections.

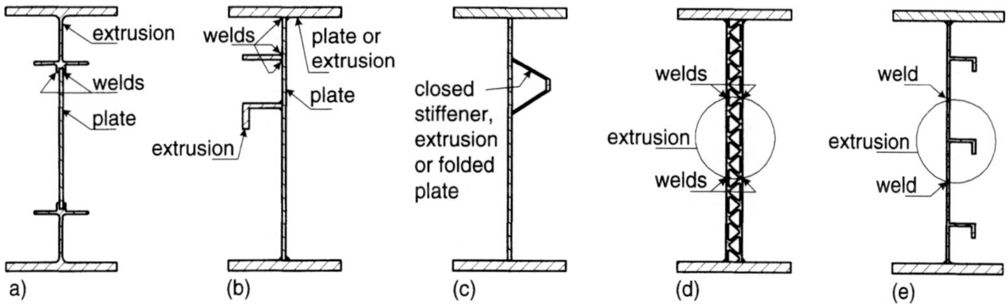

Figure 21: Possible longitudinal stiffeners in plate girders

The orthotropic plates (d) and (e) can be made by MIG welding or by friction stir welding, FSW. All alternatives except (d) can be calculated with the following method.

Plate buckling due to longitudinal compressive stresses may be taken into account by the use of an effective cross-section applicable to class 4 cross-sections, where the effective cross-section properties should be based on the effective areas of the compression elements and their locations within the effective cross-section.

In a first step the effective areas of flat compression sub panels between stiffeners should be obtained using effective thickness according to 3.5. See Figure 22.

Overall plate buckling, including buckling of the stiffeners, is considered as column buckling of a fictitious column consisting of the stiffeners and half the adjacent part of the web. If the stresses change from compression to tension within the sub panel, one third of the compressed part is taken as part of the fictitious column. See Figure 22(c).

The effective thickness of the different parts of the fictitious column section are further reduced in a second step with a reduction factor χ_c obtained from the appropriate column curve relevant for column buckling of the fictitious column as a simple strut out of the plane of the web.

The non-dimensional slenderness parameter λ_c in calculating χ_c is

$$\lambda_c = \sqrt{\frac{A_{st,ef} f_o}{N_{cr}}} \tag{3.2.9}$$

where $A_{st,ef}$ is the effective area of the fictitious column from the first step.

N_{cr} is the elastic buckling load given by the following expression:

$$N_{cr} = 1.05\, E \frac{\sqrt{I_{st}\, t_w^3\, b_w}}{b_1\, b_2} \qquad \text{if } a > a_c \tag{3.2.10}$$

$$N_{cr} = \frac{\pi^2 E I_{st}}{a^2} + \frac{E\, t_w^3\, b_w\, a^2}{4\pi^2 (1-\nu^2)\, b_1^2\, b_2^2} \qquad \text{if } a \le a_c \tag{3.2.11}$$

$$a_c = 4.33 \sqrt[4]{\frac{I_{st}\, b_2^2\, b_2^2}{t_w^3\, b_w}} \tag{3.2.12}$$

where:

I_{st} is second moment of area of the gross cross section of the fictitious column about an axis through its centroid and parallel to the plane of the web.

b_1 and b_2 are distances from longitudinal edges to the stiffener $(b_1 + b_2 = b_w)$.

When calculating I the column consists of the actual stiffener together with an effective width $15t_{st}$ of the web plate on both sides of the stiffener. See Figure 22(d1) and (d2).

In case of two longitudinal stiffeners, both in compression, the two stiffeners are considered as lumped together, with an effective area and a second moment of area equal to the sum of those of the individual stiffeners. The location of the lumped stiffener is the position of the resultant of the axial forces in the stiffeners. If one of the stiffeners is in tension the procedure will be conservative.

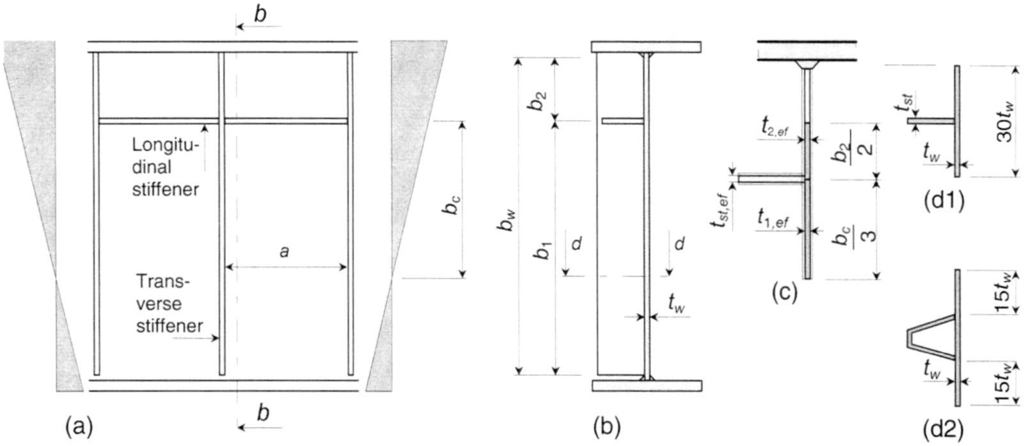

Figure 22: (a) Stiffened web, (b) cross section, (c) effective area of fictitious column, (d) fictitious column cross section for calculation of I_{st} , (d1) open stiffener, (d2) closed stiffener

3. STIFFENED ALUMINIUM PLATES IN COMPRESSION

3.1. Introduction

A direct method is given in Eurocode 9 for elements reinforced with single-sided ribs or lip of thickness equal to the element thickness. This method is first of all applicable to extrusions. In the following sections, a more general procedure is presented, where the design of stiffened elements is based on the assumption that the stiffener itself acts as a beam on elastic foundation. This procedure is applicable to extrusions, plates with welded stiffeners and cold-formed folded stiffeners.

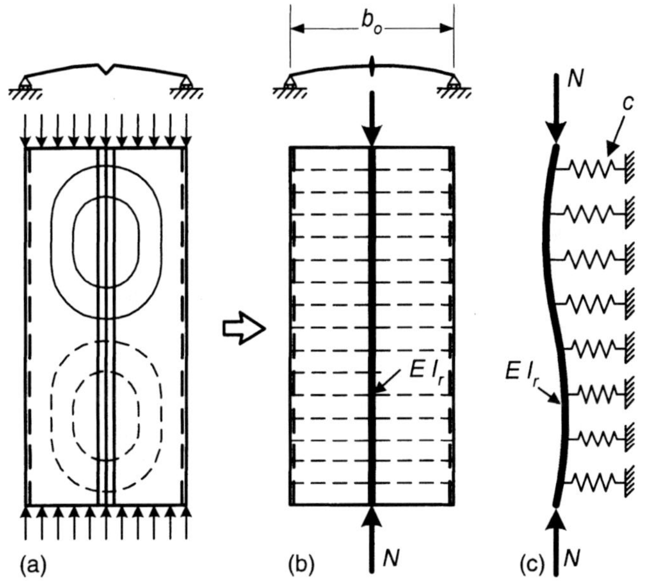

Figure 23: Model of longitudinally stiffened element in compression

3.2. Elements with edge or intermediate stiffeners in compression

Elastic critical load

Design of stiffened elements is based on the assumption that the stiffener itself acts as a bar on an elastic foundation *c*, where the modulus of the foundation is represented by springs corresponding to transverse plate strips supported on the longitudinal support, see Figure 23. Buckling of a bar on an elastic foundation can take place in one or several waves. The buckling load is, Timoshenko and Gere (1961):

$$N_{cr,r} = \left(\frac{n\pi}{l}\right)^2 EI_r + \left(\frac{l}{n\pi}\right)^2 c \qquad (3.3.1)$$

where n is the number of half sine waves, l is the length of the element, EI_r is the bending stiffness of the equivalent stiffener bar and c is the spring stiffness per unite length. For different length that number of n is chosen which gives the smallest buckling load $N_{cr,r}$, see Figure 24. Minimum for $N_{cr,r}$ occur for 1, 2, 3... times the buckling length l_b corresponding to the minimum load where

$$l_b = \pi \sqrt[4]{\frac{EI_r}{c}} \qquad (3.3.2)$$

The minimum load is

$$N_{cr,r} = 2\sqrt{c\, EI_r} \qquad (3.3.3)$$

When the plate is long compared to the buckling length l_b, then $N_{cr,r}$ does not differ very much from the minimum value. Therefore, expression (3.3.1) with $n = 1$ can be used for $l \le l_b$ and expression (3.3.3) for $l > l_b$.

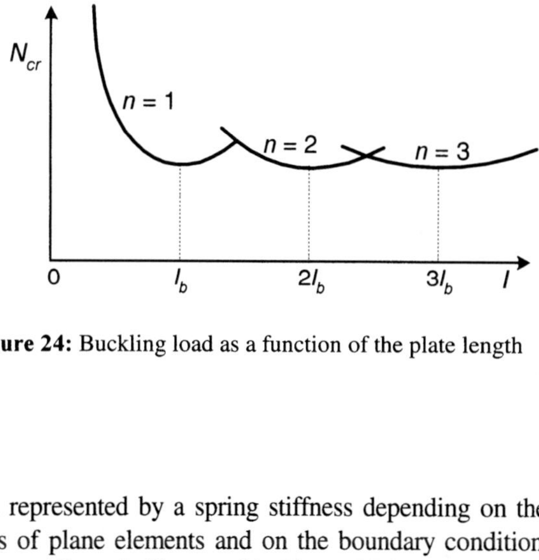

Figure 24: Buckling load as a function of the plate length

Spring stiffness

The elastic foundation is represented by a spring stiffness depending on the transverse bending stiffness of adjacent parts of plane elements and on the boundary conditions of these elements. Determination of the spring stiffness is illustrated in Figure 25 for intermediate and edge stiffeners respectively.

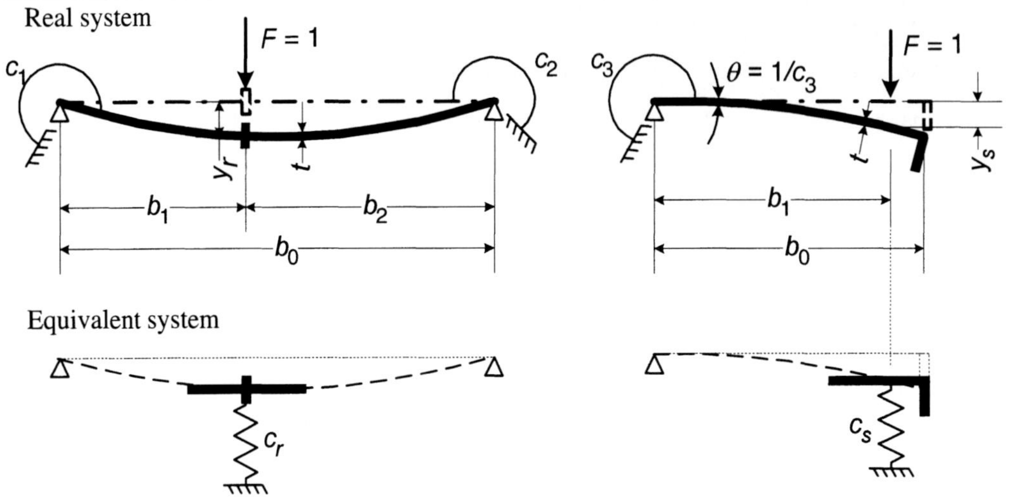

Spring stiffness

$$c_r = \frac{1}{y_r} \quad \text{where}$$

$$y_r = \frac{4(1-\nu^2)\,b_1^2\,b_2^2}{E\,t^3\,(b_1+b_2)}$$

if $c_1 = c_2 = 0$

$$c_s = \frac{1}{y_s} \quad \text{where}$$

$$y_s = \frac{4(1-\nu^2)b_1^3}{Et^3} + \phi\, b_1$$

ϕ is the rotation of adjacent parts of the cross section caused by a spring moment $1 \cdot b_1$

Figure 25: Determination of spring stiffness

Resistance of stiffener

The resistance of the stiffener is given by

$$N_{r,Rd} = \chi\, A_r\, f_o / \gamma_{M1} \tag{3.3.4}$$

where

χ = reduction factor for flexural buckling, to be determined according to expression (3.3.6) and (3.3.7) for a slenderness λ_c according to expression (3.3.5) and an imperfection factor $\alpha = 0{,}2$ and $\lambda_0 = 0{,}6$

A_r = area of effective cross section composed of the stiffener and half the adjacent plane elements (with effective thickness, see Section 3 and 4)

$$\lambda_c = \sqrt{\frac{f_o A_r}{N_{r,cr}}} \tag{3.3.5}$$

$N_{r,cr}$ is the buckling load of the stiffener given by expression (3.3.3) for long elements with large spacing between transverse stiffeners and expression (3.3.1) with $n = 1$ for short element or elements with small spacing between transverse stiffeners.

$$\chi = \frac{1}{\phi + \sqrt{\phi^2 - \lambda_c^2}} \quad \text{but} \leq 1.0 \tag{3.3.6}$$

$$\phi = 0.5(1 + \alpha(\lambda_c - \lambda_0) + \lambda_c^2) \tag{3.3.7}$$

In expression (3.3.1) and (3.3.3)

c = spring stiffness, see Section 3 and 4
I_r = second moment of area of an effective cross section composed of the stiffener itself and part of adjacent plane elements with the effective width $15t$. See Section 3 and 4.
l = member length or spacing between rigid transverse stiffeners.

3.3. Intermediate stiffeners

The design rules given below are valid for elements, which are supported along both edges. The stiffener cross section incorporates the stiffener itself plus the effective portions of adjacent parts of the section. Stiffeners may be grooves or ribs in extrusions or profiles welded to a plate. The effective section is described in Figure 26. The validity of the design expression is limited to two equally shaped stiffeners at maximum.

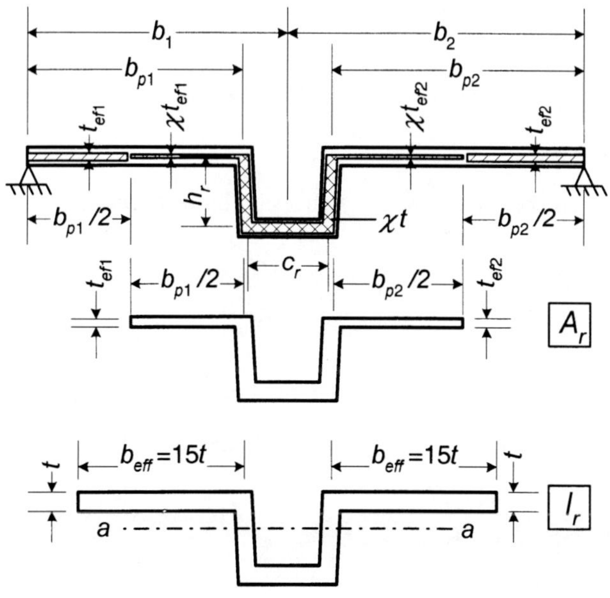

Figure 26: Intermediate stiffeners; final effective cross section and effective cross section for A_r and I_r

The effective area $A_{r,eff}$ of intermediate stiffeners according to Figure 26 may be estimated as follows:

Step 1: Determination of the effective parts of the stiffened element, assuming that the stiffeners provide effective supports to the adjacent plate elements.

Step 2: Consideration of the elastic support of the plate element by determining the buckling stress of the stiffener area, considered being elastically supported as follows:

- Determination of the effective thickness t_{ef1} and t_{ef2} and effective width b_{eff} of the element (Step 1) of the stiffener and the adjacent plane elements.
- The effective cross-sectional area A_r is based on the effective thickness t_{ef1} and t_{ef2}. The moment of inertia is based on the effective width b_{eff} and is referred to the neutral axis a-a of the stiffener section.
- The critical buckling load of the intermediate stiffener with an area A_r is given by expression (3.3.1) or (3.3.3) for a spring stiffness according to Figure 25.
- The reduced effective area $A_{r,eff}$ then will be:

  $$A_{r,eff} = \chi \cdot A_r \tag{3.3.8}$$

  with χ according to expression (3.3.6) and (3.3.7).

One stiffener

For a section with a groove as in Figure 26 the grove is also deformed transversely, why the spring stiffness according to Figure 25 will be $(\nu = 0.3)$

$$c = \frac{Et^3(b_1 + b_2)}{4(1-\nu^2)b_1^2 b_2^2} \approx \frac{0{,}27\,Et^3 s_1}{b_{p1}^2 (s_1 - b_{p1})^2} \tag{3.3.9}$$

for a section with constant thickness along the stiffened element, where

$$s_1 = b_{p1} + b_{p2} + 2h_r + c_r \tag{3.3.10}$$

The critical buckling load, for a long element, will according to expression (3.3.3), be

$$N_{r,cr} = 1{,}05E \frac{\sqrt{I_r t^3 s_1}}{b_{p1}(s_1 - b_{p1})} \tag{3.3.11}$$

Two symmetrical stiffeners

For an element with two symmetrical stiffeners located a distance b_1 from the longitudinal supports, Figure 27(a).

$$y_r = \frac{b_1(s - 2b_1)}{2 \cdot 2D} b_1 + \frac{b_1^2}{2 \cdot 2D}\left(b_1 - \frac{b_1}{3}\right) = \frac{b_1^2(3s - 4b_1)}{12D} \tag{3.3.12}$$

where $D = \dfrac{t^3}{12(1-\nu^2)}$

$$c = \frac{Et^3}{(1-\nu^2)b_1^2(3s-4b_1)} \approx \frac{1.1Et^3}{b_1^2(3s-4b_1)} \tag{3.3.13}$$

For a section with grooves with same thickness as the stiffened element, Figure 27(b):

$$s = 2b_1 + 4s_r + 2c_r + b_2 \tag{3.3.14}$$

For an extrusion or a plate with welded stiffeners, Figure 27(c): $s = b$.

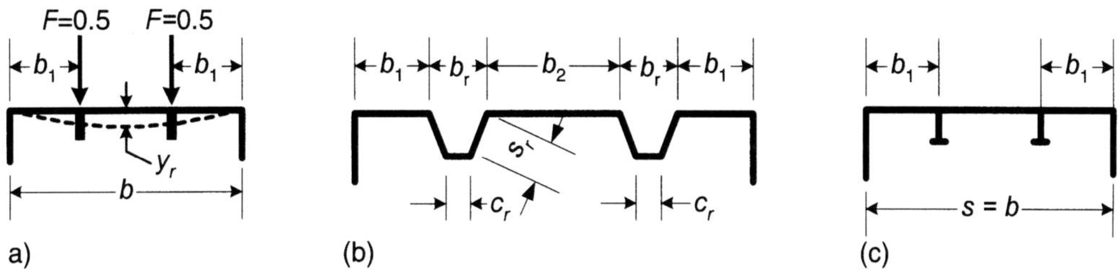

Figure 27: Elements with two symmetrical stiffeners

The critical buckling load, for the two stiffeners in a long element, will be, according to expression (3.3.3)

$$N_{r,cr} = 2.1E\sqrt{\frac{I_r t^3}{b_1^2(3s-4b_1)}} \tag{3.3.15}$$

3.4. Edge stiffeners

Edge stiffeners or edge folds shall only be considered as supports to the plane element, if the angle between the stiffener or fold and the plate deviates from the right angle by not more than 45° and if the width of the edge stiffener is larger than $0{,}2b_p$, with b_p according to Figure 28. Otherwise, they should be ignored.

The design procedure follows the procedure for an intermediate stiffener, except that when calculating the spring stiffness it is now need to take account of deformation of adjacent element of the cross section.

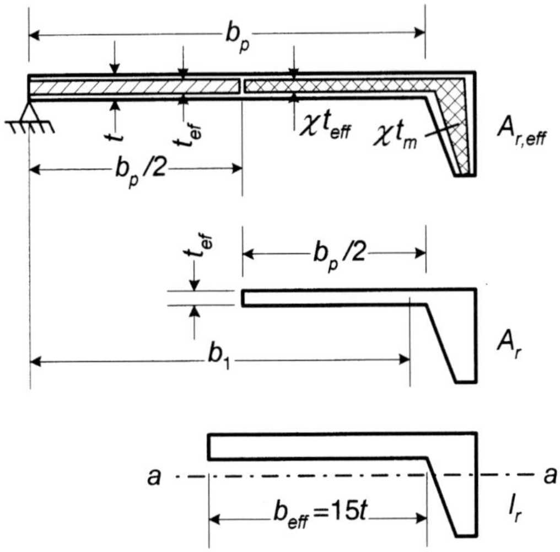

Figure 28: Edge stiffener, notations and effective cross sections for A_r and I_r

The effective area A_{ref} of edge stiffeners according to Figure 28 may be estimated as follows:

Step 1: Determination of the effective edge section with an effective thickness of the plate element assuming the plate to be rigidly supported along both longitudinal edges.

Step 2: Consideration of the elastic support of the plate element by determining the buckling load of the edge section, considered being elastically supported as follows:

- Determination of effective thickness t_{eff} and effective width $b_{eff} = 15t$ of the stiffener and the adjacent plane cross sectional element, cf. Figure 28. (Step 1)
- The effective cross-sectional area A_r is calculated by using the effective thickness t_{eff} , cf. Figure 28
- The second moment of area I_r is based on a cross section with effective width $b_{eff} = 15t$ for internal element and $b_{eff} = 12t$ for outstands, referred to the neutral axis a-a of the effective edge section.
- The critical buckling load $N_{cr,r}$ of the edge stiffener is given by expression (3.3.1) or (3.3.3).
- The reduced effective area A_{ref} then will be $A_{ref} = \chi A_r$ with χ according to expression (3.3.6), expressed by a second reduction of the thickness to $\chi \cdot t_{ef}$ or $\chi \cdot t_{av}$ where t_{av} = average thickness of edge stiffener.

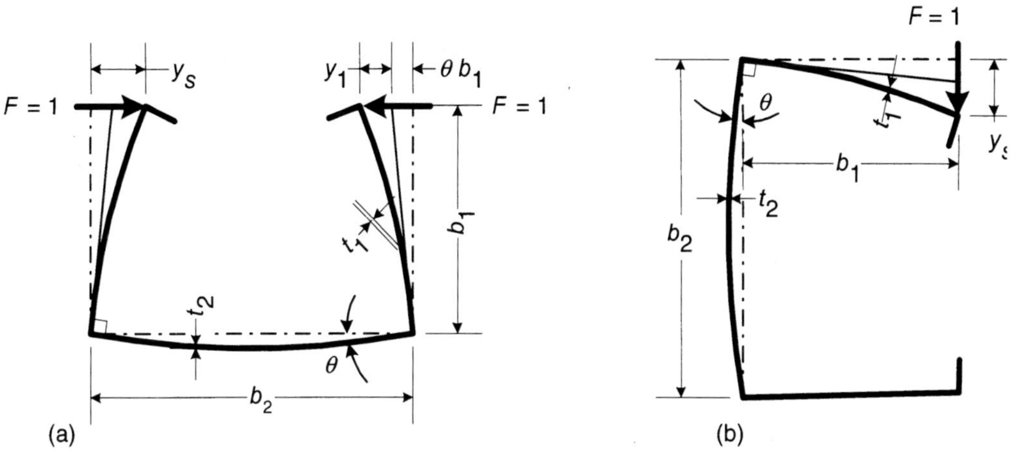

Figure 29: Channel section

For the channel section in Figure 29(a) with compression in both stiffeners the deflection y_s is given by

$$y_s = y_1 + \theta b_1 = \frac{F b_1^3 \, 12(1-\nu^2)}{3 E t_1^3} + \frac{F b_1^2 b_2 \, 12(1-\nu^2)}{2 E t_2^3} \tag{3.3.16}$$

Expression (3.3.16), with $c = F / y_s$, then gives for a section without transverse stiffeners:

$$N_{cr,r} = 1{,}05 E \sqrt{\frac{I_s t_1^3}{b_1^3 \left(1 + \frac{1{,}5 b_2 t_1^3}{b_1 t_2^3}\right)}} \quad \text{both stiffeners in compression} \tag{3.3.17}$$

For a C section in bending, Figure 29(b), there is no force F on the tension flange why the digit 2 in the dominator in the second term in expression (3.3.16) shall be replaced by 3 why

$$N_{cr,r} = 1{,}05 E \sqrt{\frac{I_s t_1^3}{b_1^3 \left(1 + \frac{b_2 t_1^3}{b_1 t_2^3}\right)}} \quad \text{one stiffener in compression} \tag{3.3.18}$$

4. BUCKLING RESISTANCE OF ALUMINIUM ORTHOTROPIC PLATES

4.1. Introduction

Orthotropic plates are commonly used in aluminium structures since they can easily be made by extrusions welded together or by folding a plate. In Figure 30, section (b) can be produced by extrusions, width *w*, or by welding (f) opened, (g) closed or (h) combined stiffeners to a plate, width *b*.

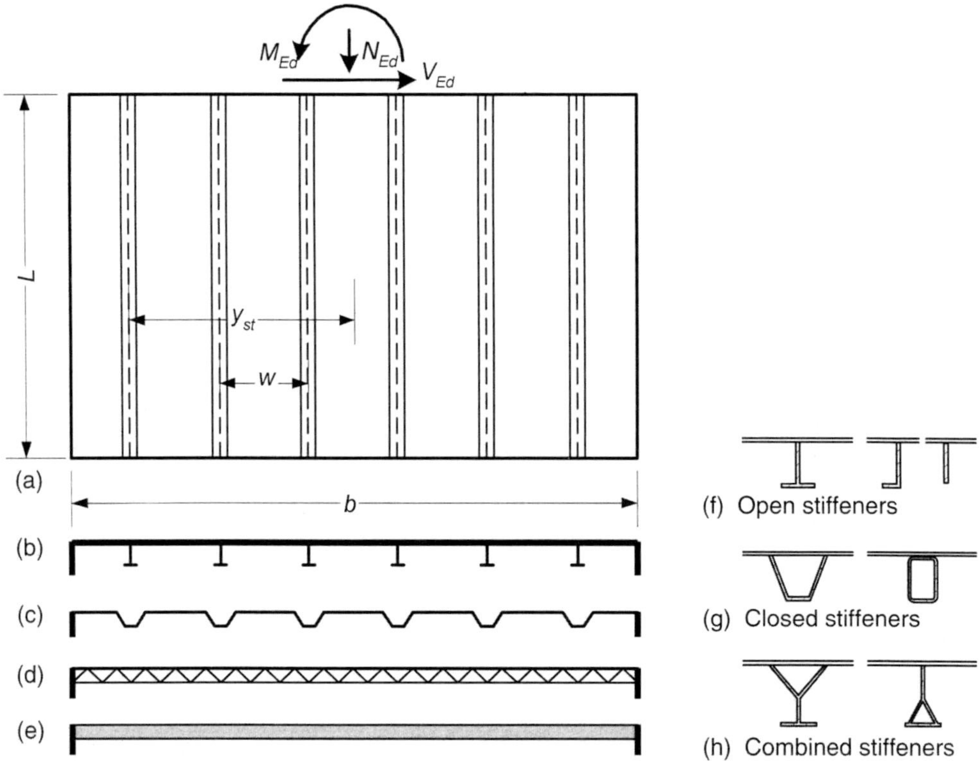

Figure 30: Stiffened plates and types of stiffeners

The folded plate in Figure 30(c) is produced by cold forming. The double-skin plate, Figure 30 (d) is made by multicell hollow extrusions welded together. Figure 30(e) represents a sandwich plate with aluminium plate faces and an insulating core. Sections (b), (c) and (d) are the types of orthotropic plates dealt with here.

Cross section	B_x	B_y	H
1 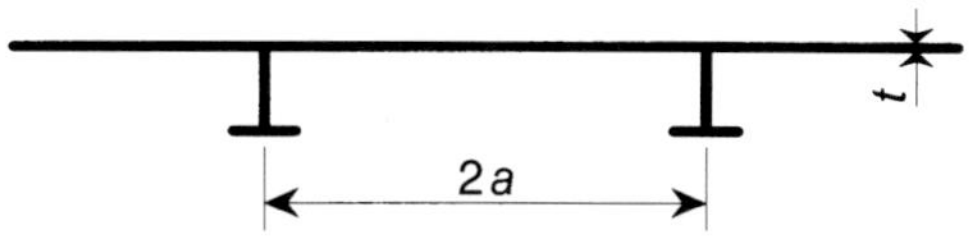	$\frac{EI_L}{2a}$	$\frac{Et^3}{12(1-\nu^2)}$	$\frac{Gt^3}{6}$
2 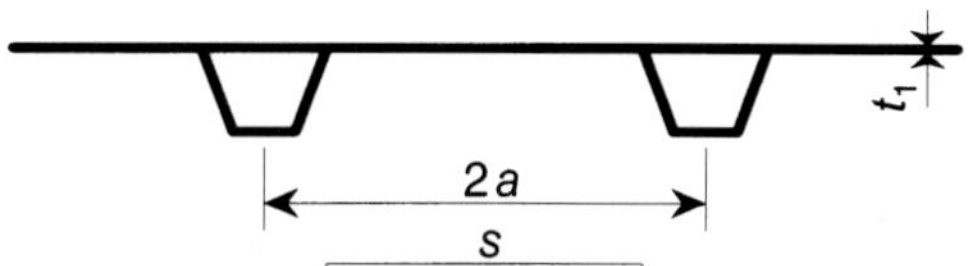	$\frac{EI_L}{2a}$	Eq.(3.4.7)	Eq.(3.4.8)
3 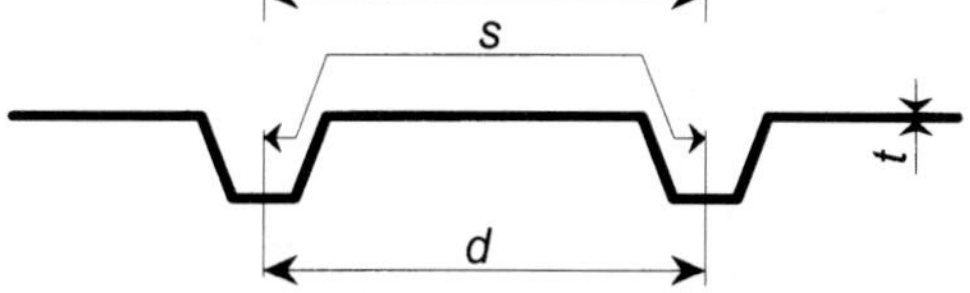	$\frac{EI_L}{d}$	$\frac{d}{s}\ \frac{Et^3}{12(1-\nu^2)}$	$\frac{d}{s}\ \frac{Gt^3}{6}$
4 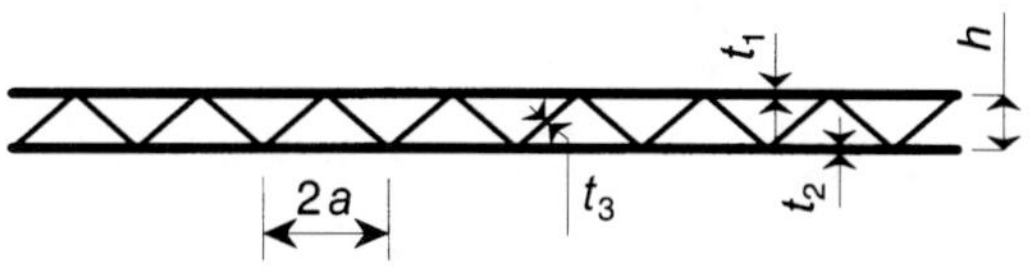	$\frac{EI_L}{2a}$	$\frac{Et_1t_2h^2}{t_1+t_2}$	$\frac{GI_T}{2a}$
5 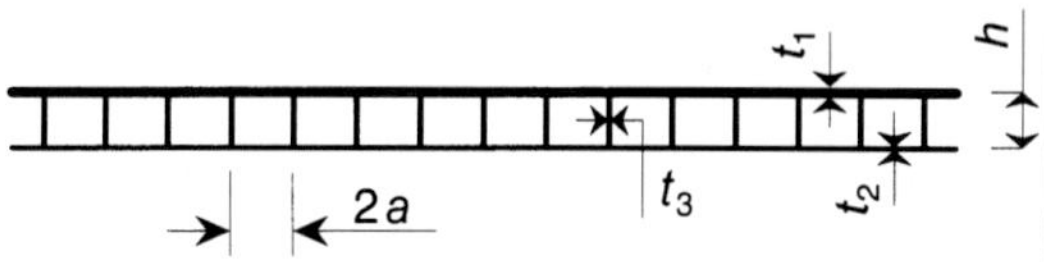	$\frac{EI_L}{2a}$	See Euro-code 9	See Euro-code 9
6 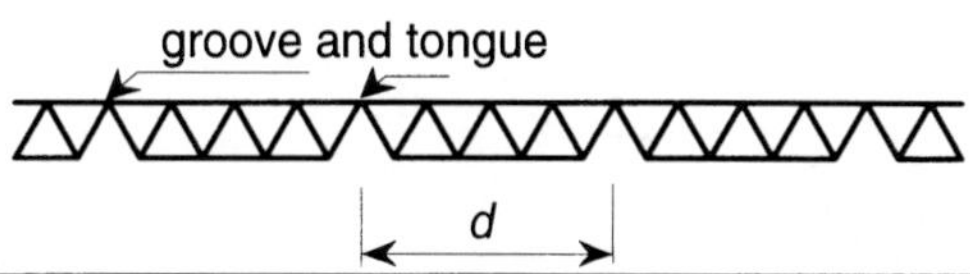	$\frac{EI_L}{d}$	0	$\frac{GI_T}{d}$

I_L is the second moment of area of one stiffener and adjacent plating (within $2a$ or d) in the longitudinal direction

I_T is the torsional stiffness of the same cross section.

Table 4: Flexural and torsional rigidity of orthotropic plates

4.2. Orthotropic plates

Orthotropic plates can have one single layer, cross section 1, 2 and 3 in Table 4, or double layer, cross section 4, 5 and 6. Plates with one single layer have large difference in bending stiffness in longitudinal and transverse direction, $B_y << B_x$. The bending stiffness B_y of section 1 and 3 is less or equal to the bending stiffness of a plane plate, thickness t, $B = Et^3/(12(1-\nu^2))$. Section 2 has larger B_y than B. Deformations according to Figure 31(a) result in a bending stiffness according to expression (3.4.9a). Cross section deformations, Figure 31(b), also reduce the effective torsional stiffness, resulting in expression (3.4.8).

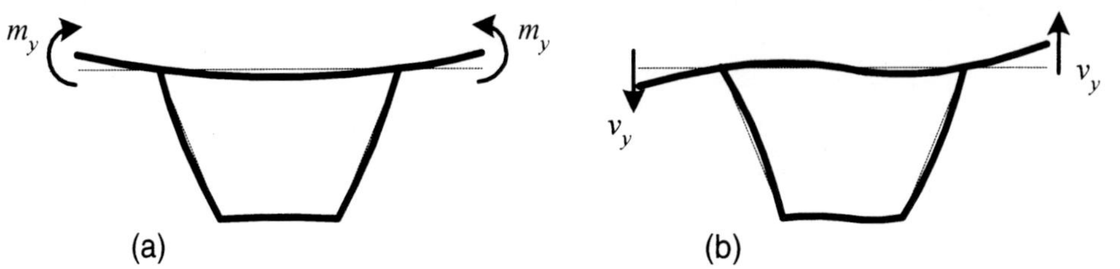

Figure 31: Cross section deformation of trapezoidal stiffeners of (a) transverse bending and (b) shear force

Double skin orthotropic plates, cross section 4 and 5 in Table 4, have similar bending stiffness in all directions. For cross section 5, the effective torsional stiffness is very much influenced by cross section deformations according to Figure 32(c).

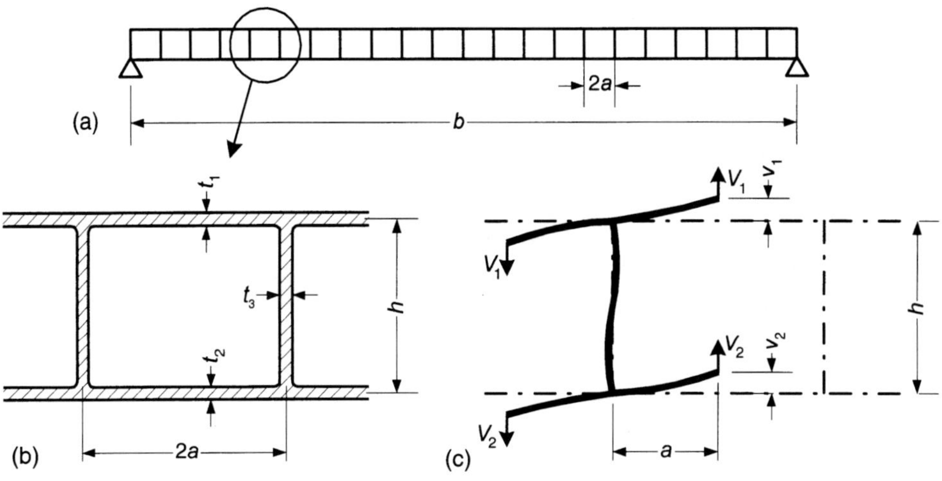

Figure 32: Cross section deformation of a cell elements in torsion

The extrusion technique makes it possible to produce double skin plates (decks) by welding multi hollow section profiler together. However, sharp inside corners in the holes should be avoided. Furthermore, often the number of web in a profile will be kept to a minimum to reduce the number of holes. Therefore large radius or small flat parts a_4 between the webs is preferred as in Figure 33(b). If this flat part is too large, deformations as for multi-cell plate in Figure 33(c) must be checked. No design formula is given in Eurocode 9.

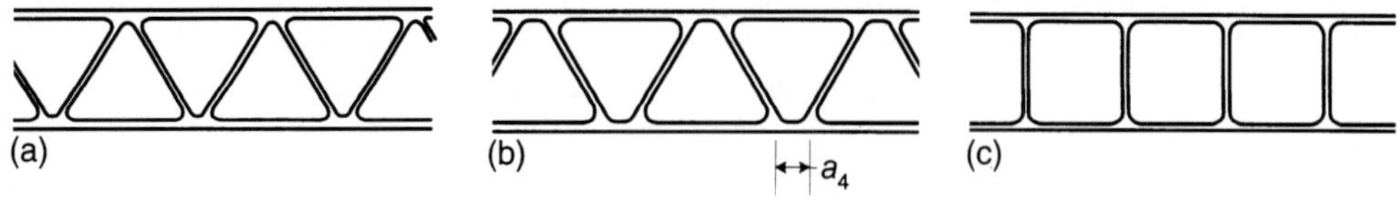

Figure 33: Double skin cross sections

Cross section 6 in Table 4 consists of hollow section profiles put together with tongues and grooves. The plate has no transverse bending stiffness. However, the torsional stiffness is large, which has large influence on the resistance.

4.3. Resistance of orthotropic plate in compression

Buckling of the middle part of a short ortotropic plate is similar as flexural buckling of a single strut, so there is hardly any post buckling strength. In a long panel, under certain conditions, there may be a substantial post-buckling strength. In Eurocode 9 a method is recommended, where no post-buckling strength is allowed for.

The resistance of an orthotropic plate is given by

$$N_{r,Rd} = \chi \, A_{eff} \, f_o \, / \, \gamma_{M1} \qquad (3.4.1)$$

where

χ = reduction factor for flexural buckling, to be determined according to expression (3.4.3) and (3.4.4) for a slenderness λ_c according to expression (3.4.2) and an imperfection factor α = 0,2 and λ_0 = 0,6

A_{eff} = area of effective cross section of the whole panel (with effective thickness, if class 4 sub elements)

$$\lambda_c = \sqrt{\frac{f_o A_r}{N_{cr}}} \qquad (3.4.2)$$

$$\chi = \frac{1}{\phi + \sqrt{\phi^2 - \lambda_c^2}} \quad \text{but} \le 1.0 \qquad (3.4.3)$$

$$\phi = 0.5(1 + \alpha(\lambda_c - \lambda_0) + \lambda_c^2) \qquad (3.4.4)$$

For an orthotropic plate under *uniform compression* the elastic orthotropic buckling load N_{cr} for a simply supported plate is given by Bulson (1970)

$$N_{cr} = \frac{\pi^2}{b}\left[\frac{B_x}{(L/b)^2} + 2H + B_y\,(L/b)^2\right] \qquad \text{when } \frac{L}{b} < \sqrt[4]{\frac{B_x}{B_y}} \qquad (3.4.5)$$

$$N_{cr} = \frac{2\pi^2}{b}\left[\sqrt{B_x B_y} + H\right] \qquad \text{when } \frac{L}{b} \ge \sqrt[4]{\frac{B_x}{B_y}} \qquad (3.4.6)$$

Expressions for B_x, B_y and H for different cross sections are given in Table 4, where Eq. (3.4.7) and Eq. (3.4.8) are given below. For simple flat stiffeners I_L should be based on the effective thickness of the flat outstand element, see Figure 34(c), otherwise on the gross cross section.

$$B_y = \frac{2B\,a}{2a_4 + \dfrac{2a_1 a_3 t_1^3\,(4a_2 t_3^3 + a_3 t_2^3)}{a_3 t_1^3 (4a_2 t_3^3 + a_3 t_2^3) + a_1 t_3^3\,(12a_2 t_3^3 + 4a_3 t_2^3)}} \tag{3.4.7}$$

where $B = \dfrac{E\,t_1^3}{12(1-\nu^2)}$

$$H = 2B + \frac{\dfrac{G\,I_T}{2\,a}}{1 + \dfrac{3{,}3 G I_T a_4^2}{L^2 a\,B}\left[1 + \dfrac{1}{97C_1/L_4 + C_2}\right]} \tag{3.4.8}$$

where

$$C_1 = 4(1-\nu^2)(a_2 + a_3)\,a_1^2 a_4^2\,h^2 t_2 / (3a\,t_1^3) \tag{3.4.9}$$

$$C_2 = \frac{4(a_1 + a_2)^2\,a_1 a_4\,(1 + a_1/a_2 + a_2/a_1 + a^2/(a_1 a_3))}{a_2^3 (3a_3 + 4a_2)}\left(\frac{t_2}{t_1}\right)^3 \tag{3.4.10}$$

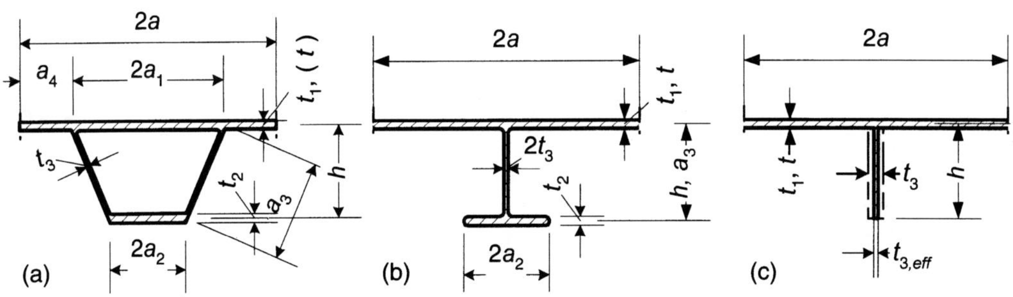

Figure 34: Cross section notations of stiffeners

4.4. Resistance of orthotropic plate in shear

The shear force resistance with respect to ***yielding and local shear buckling*** of flat parts is

$$V_{w,Rd} = 0.7\,\rho_v \sum t_w\,h_w\,f_{ow} / \gamma_{M1} \tag{3.4.11}$$

where ρ_v is found from

$$\rho_v = \eta \qquad \text{if } \lambda_w \le 0.48/\eta \tag{3.4.12}$$

$$\rho_v = 0.48 / \lambda_w \qquad \text{if } \lambda_w > 0.48/\eta \tag{3.4.13}$$

The factor η is dependent on the ultimate-to-yield stress ratio f_{uw} / f_{ow}.

$$\eta = 0.4 + 0.2 f_{uw} / f_{ow} \tag{3.4.14}$$

The slenderness parameter is

$$\lambda_w = 0.35 \frac{b_m}{t_w} \sqrt{\frac{f_{ow}}{E}} \tag{3.4.15}$$

The most slender sub panel with the largest width to thickness ratio b_m / t_w is used to calculate λ_w where b_m is flat width of the sub panel. In a single web plate $\sum t = t$. In a double skin plate $\sum t_w$ is given by $\sum t_w = t_1 + t_2 + t_3$ for cross section 4 in Table 4 and $\sum t_w = t_1 + t_2$ for cross section 5.

The shear force resistance with respect to ***global shear buckling*** is determined according to

$$V_{o,Rd} = \chi_o h_w \sum t_w \frac{f_{ow}}{\gamma_{M1}} \tag{3.4.16}$$

where:

$$\chi_o = \frac{0.60}{0.8 + \lambda_{ow}^2} \quad \text{but not more than } 0.6 \tag{3.4.17}$$

$$\lambda_{ow} = \sqrt{\frac{h_w t_w f_{ow}}{V_{o,cr}}} \tag{3.4.18}$$

The elastic buckling load is given by Bulson (1970)

$$V_{o,cr} = \frac{k_\tau \pi^2}{b} \sqrt[4]{B_x \, B_y^3} \tag{3.4.19}$$

$$k_\tau = 3.25 - 0.567\phi + 1.92\phi^2 + (1.95 + 0.1\phi + 2.75\phi^2)\eta_h \tag{3.4.20}$$

$$\phi = \frac{L}{b} \sqrt[4]{\frac{B_y}{B_x}} \qquad \text{(Valid for } 0 < \varphi < 1{,}2\text{)} \tag{3.4.21}$$

$$\eta_h = \frac{H}{\sqrt{B_x \, B_y}} \qquad \text{(Valid for } 0 < \eta_h < 1{,}5\text{)} \tag{3.4.22}$$

B_y, B_x and H are given in Table 4.

5. SHEAR BUCKLING RESISTANCE OF ALUMINIUM PLATE GIRDERS

5.1. Introduction

For webs in shear there is a substantial post-buckling strength provided that, after buckling, tension membrane stresses, anchored in surrounding flanges and transverse stiffeners, can develop. The post-critical behaviour of slender plates subjected to shear is very pronounced in that sense that the critical stress according to classical plate buckling theory may be exceeded several times before the ultimate resistance is reached. The design method in Eurocode 9 is based on the so-called rotated stress field method in Höglund (1972). Originally it was developed for girders with web stiffeners at the supports only, a structure for which other tension field methods are often very conservative. In this section is presented:

- the rotated stress field method
- design formulae for girders with stiffeners at supports only
- influence of transversal and longitudinal web stiffeners
- influence of rigid flanges
- influence of bending moment
- influence of welds
- comparison with tests
- design of stiffeners including trapezoidal web
- concentrated loads and flange induced buckling.

5.2. Web with stiffeners at supports only

In a pure shear state of stress the absolute value of the principal membrane stresses σ_1 and σ_2 are the same as long as no buckling has occurred ($\tau < \tau_{cr}$, see Figure 35(a)). After reaching the buckling load ($V_{cr} = \tau_{cr} h_w t_w$) the web plate will buckle and redistribution of stresses start. Increased load results in increased tensile stress σ_1 but only slightly, or not at all, increased compressive stress σ_2 .

For a very thin web, after buckling, σ_2 is much less than σ_1 and can be neglected. For the extreme case, where the surrounding flanges are completely prevented to move towards each other by an external structure, Figure 35(b), a so-called ***ideal tension field*** can develop without any compression stresses. For a slender web then

$$\tau = \sigma_1 \sin\varphi \cos\varphi = 0.5\,\sigma_1 \sin 2\varphi \qquad (3.5.1)$$

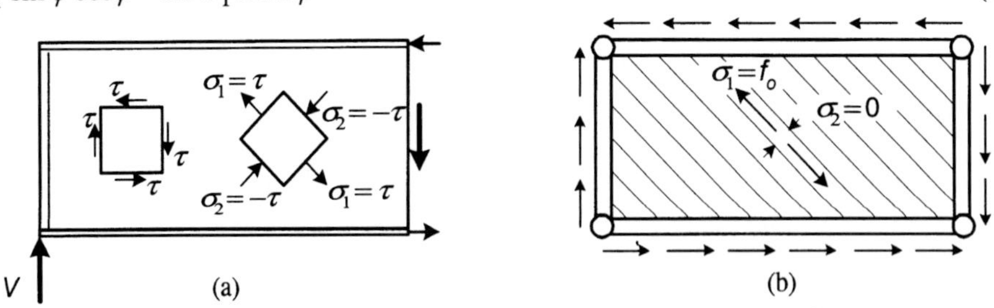

(a) (b)

Figure 35: (a) Pure shear state of stress. (b) Ideal tension field

The direction of the tensile stresses is chosen to give τ = maximum. Setting σ_1 equal to the yield strength f_{ow} of the web, then the ultimate shear stress τ_u is

$$\frac{\tau_u}{f_v} = \frac{0.5 f_{ow}}{f_v} = \frac{\sqrt{3}}{2} \approx 0.87 \text{ for } \varphi = 45^o \qquad (3.5.2)$$

where $f_v = \dfrac{f_{ow}}{\sqrt{3}}$ (3.5.3)

It should be pointed out that this ideal tension field theory is valid only if the flanges are prevented to move towards each other by an external structure, so it is applicable only for some aeroplane components, e.g. a panel in a plate with rigid cross beams and stringers.

In a long beam, with transverse stiffeners at the ends only, nothing but the web prevents the flanges to move towards each other, why the resulting membrane stresses in the transverse direction is zero. Equilibrium for a triangle according to Figure 36(g) gives

$$\sigma_1 = \frac{\tau}{\tan \varphi} \qquad (3.5.4)$$

$$\sigma_2 = -\tau \tan \varphi \qquad (3.5.5)$$

where φ constitutes the direction of the principal stress.

This state of stress has a stress component σ_h in the longitudinal direction

$$\sigma_h = \tau \left(\frac{1}{\tan \varphi} - \tan \varphi \right) = \sigma_1 + \sigma_2 \qquad (3.5.6)$$

Note that $\sigma_h = 0$ if $\varphi = 45^o$. The total longitudinal force in the web is approximately

$$N_h \approx 0.7 \sigma_h h_w t_w \qquad (3.5.7)$$

where the factor 0.7 is added to take account of the fact that, close to the flanges, there is more or less a pure shear state of stress, Figure 36(e) and (d).

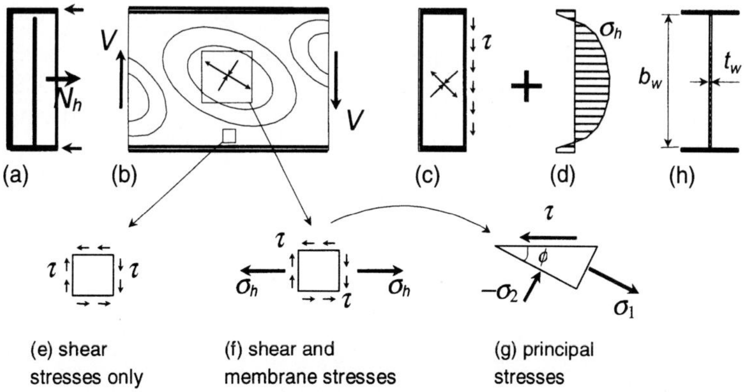

Figure 36: State of stress in ate web of a beam with transverse stiffeners at the ends only (rotated stress field)

This force has to be anchored at the ends of the beam by a transverse short beam called ***rigid end post***, in order to fully develop the rotated stress field. See Figure 36(a). The flanges support this end post, which results in compressive forces in the flanges at the ends of the beam. Design of the end post is presented in Section 9.

The ultimate shear strength of the beam can be derived using the von Mises yield criterion

$$\sigma_1^2 - \sigma_1 \sigma_2 + \sigma_2^2 = f_{ow}^2 \tag{3.5.8}$$

and assuming that, after buckling, the compressive stress remain equal to the shear buckling stress τ_{cr}, but acting in a smaller angle than 45°.

$$\sigma_2 = -\tau_{cr} \tag{3.5.9}$$

Furthermore the slenderness parameter λ_w is introduced

$$\lambda_w = \sqrt{\frac{f_v}{\tau_{cr}}} \quad \text{where} \quad \tau_{cr} = k_\tau \frac{\pi^2 E}{12(1-\nu^2)} \left(\frac{t_w}{b_w} \right)^2 \tag{3.5.10}$$

From expressions (3.5.3), (3.5.4), (3.5.5), (3.5.8), (3.5.9) and (3.5.10) the ultimate strength $\tau_u = \tau$ can be derived as a function of the yield strength f_v in shear and the slenderness parameter λ .

$$\frac{\tau_u}{f_v} = \frac{\sqrt[4]{3}}{\lambda_w} \sqrt{\sqrt{1 - \frac{1}{4\lambda_w^4}} - \frac{1}{2\sqrt{3}\,\lambda_w^2}} \quad \text{for} \quad \lambda_w \geq 1.00 \tag{3.5.11}$$

The large square root in (3.5.11) is close to 1.00 if $\lambda_w > 2.5$. Then

$$\frac{\tau_u}{f_v} \approx \frac{1.32}{\lambda_w} \quad \text{for} \quad \lambda_w > 2.5 \tag{3.5.12}$$

The inclination of the tension stress σ_1 defined by the angle φ is decreased when the ratio τ_u / τ_{cr} is increased. That's why the theory is called the ***rotated stress field*** theory.

In the diagram in Figure 37, τ_u / f_v is given as a function of the slenderness parameter λ_w. The shear stress corresponding to the buckling load, $1/\lambda_w^2$, is also given in the diagram, as well as a line, corresponding to the ideal tension field and some test results on plate girders with stiffeners at the supports only.

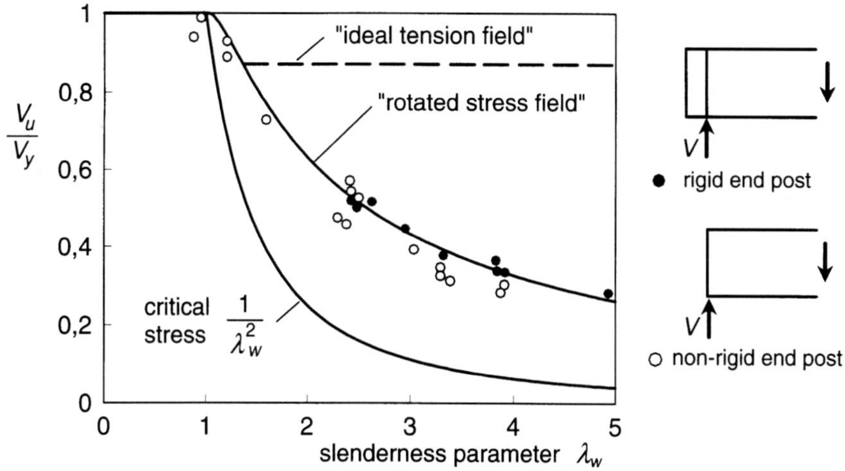

Figure 37: Shear force resistance according to tension field theories and tests

The solid circles are tests on beams with rigid end posts. These tests agree very well with the curve for the rotated stress field theory.

The unfilled circles correspond to beams with only one stiffener at the girder end, ***non-rigid end post***. Such end posts have only limited ability to serve as anchors for the longitudinal membrane stresses, and hence the ultimate load is less than for beams with rigid end posts. However, there is still a substantial post buckling strength.

In Eurocode 9 design formulas for the reduction factor

$$\rho_v = \frac{\tau_u}{f_{ow}} \tag{3.5.13}$$

are given. The value of ρ_v is somewhat reduced compared to the rotated stress field theory to allow for scatter in test results due to initial imperfections and plastic buckling. The shear force capacity is

$$V_w = \rho_v f_{ow} h_w t_w \tag{3.5.14}$$

where h_w is the distance between the inside of the flanges (whereas b_w is the distance between the fillets, see Figure 40(a)).

The shear buckling reduction factor ρ_v is given in Table 5 and in the diagram in Figure 38.

λ_w	Rigid end post	Non-rigid end post
$\lambda_w < 0.48/\eta$	η	η
$0.48/\eta \le \lambda_w < 0.949$	$0.48/\lambda_w$	$0.48/\lambda_w$
$\lambda_w \ge 0.949$	$1.32/(1.66+\lambda_w)$	$0.48/\lambda_w$

Table 5: Reduction factor ρ_v for shear buckling

For small slenderness ratios, $\lambda_w < 0.48/\eta$, strain hardening in shear can take place, giving larger strength than corresponding to initial yielding, $\eta = 1/\sqrt{3} \approx 0.58$. Then, for aluminium, $\eta = 0.4 + 0.2 f_{uw} / f_{ow}$ but not less than 0.7 where f_{uw} is the ultimate strength of the web material.

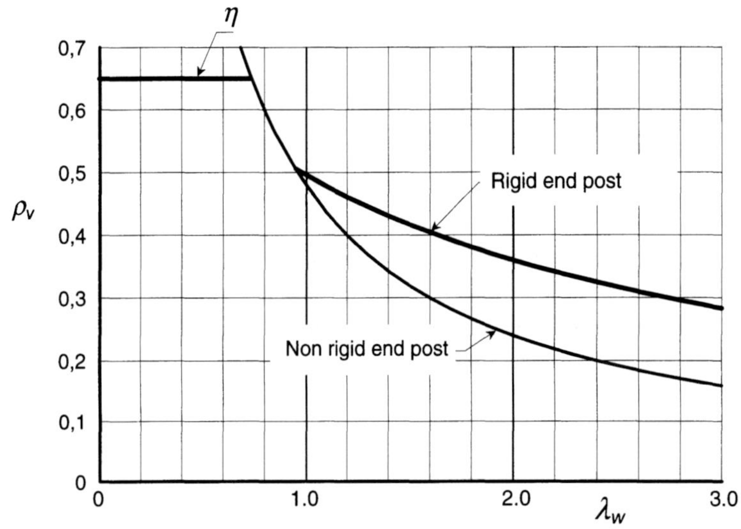

Figure 38: Reduction factor for shear buckling

5.3. Web with transverse stiffeners

Transverse stiffeners, welded to the web, have two main effects on the behaviour and strength of a girder in shear: first, they prevent the web from out of plane deflections, thus increasing the elastic buckling stress, and second, they prevent the flanges from moving towards each other.

If the bending rigidity of the flanges about an axis in the centre of the flange is large, then the prevention of the flanges to move towards each other can be large. See Section 5.

The shear buckling coefficient k_τ is, if simply supports are assumed all around the web panel of length a and width b_w,

$$k_\tau = 5.34 + 4(b_w / a)^2 \quad \text{when} \quad a / b_w \ge 1.00 \tag{3.5.15a}$$

$$k_\tau = 4.00 + 5.34(b_w / a)^2 \quad \text{when} \quad a / b_w < 1.00 \tag{3.5.15b}$$

If the distance between the transverse stiffeners is large, then $k_\tau \cong 5.34$. The slenderness parameter then is

$$\lambda_w = \sqrt{\frac{f_v}{\tau_{cr}}} = \sqrt{\frac{f_o / \sqrt{3}}{5.34 \frac{\pi^2 E}{12(1-\nu^2)} \frac{t_w^2}{b_w^2}}} = 0.35 \frac{b_w}{t_w} \sqrt{\frac{f_o}{E}} \quad \text{for } \nu = 0.3 \tag{3.5.16a}$$

If the distance between the transverse stiffeners is small, or if there are longitudinal stiffeners as well, then

$$\lambda_w = \frac{0.81}{\sqrt{k_\tau}} \frac{b_w}{t_w} \sqrt{\frac{f_o}{E}} \tag{3.5.16b}$$

5.4. Web with longitudinal stiffeners

Longitudinal stiffeners also prevent the web from out-of-plane deflections. If they are rigid enough they divide the web panel between transverse stiffeners into sub-panels which buckle individually. If not, web buckling includes lateral deflection of the longitudinal stiffeners. In design, the method for un-stiffened webs is used but the buckling coefficient is increased.

In Crate and Lo (1948) the buckling coefficient k_τ is given as a function of the stiffness parameter $\gamma_{st} = EI / b_w D$ for ***one centric longitudinal stiffener*** in a long web panel. The theoretical curve can be approximated by

$$k_\tau = 5.34 + \sqrt[3]{\gamma_{st}} \tag{3.5.17a}$$

See Figure 39.

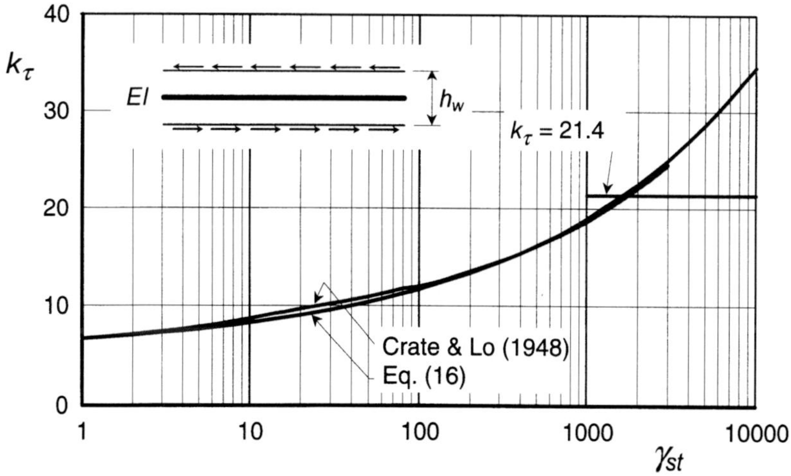

Figure 39: Buckling coefficient k_τ for a plate with one longitudinal stiffener

However, this value of k_τ cannot directly be used in design because the post-buckling strength is much less in a stiffened web than in a web without stiffeners. Therefore, the stiffness I of the stiffener is reduced with a factor of 1/3. Inserting the reduced value $I = I_{st}/3$ and $D = E t_w^3/(12(1-\nu^2))$ into (3.5.17a) and using $\nu = 0.3$ gives

$$k_\tau = 5.34 + 2.1\sqrt[3]{\frac{I_{st}}{b_w t_w^2}} \tag{3.5.17b}$$

In a web with ***rigid transverse stiffeners and a longitudinal stiffener*** with relative stiffness γ_{st}, the coefficient k_τ can be approximated with

$$k_\tau = 5.34 + \frac{4}{(a/b_w)^2} + \frac{3.45\,\gamma_{st}^{3/4}}{(a/b_w)^2} \tag{3.5.18}$$

Formula (3.5.18) are in good agreement with values given in the literature (Klöppel and Sheer, 1960), see Höglund (1995).

Again, after reducing the stiffness of the longitudinal stiffener with a factor of 1/3 and after rearranging we have

$$k_\tau = 5.34 + 4\left(\frac{b_w}{a}\right)^2 + 9.1\left(\frac{b_w}{a}\right)^2\left(\frac{I_{st}}{t_w^3 b_w}\right)^{3/4} \tag{3.5.19}$$

The largest of k_τ according to eqn.(3.5.17b) and (3.5.19) should be used, but τ_{cr} should not be larger than that of the sub-panel giving the smallest value of τ_{cr} e.g. panel with depth b_{w1} in Figure 40(a).

If there are ***several longitudinal stiffeners***, then it is found that eqn.(3.5.17b) and (3.5.19) can be used with I_{st} as the sum of the second moment of area of the individual stiffeners.

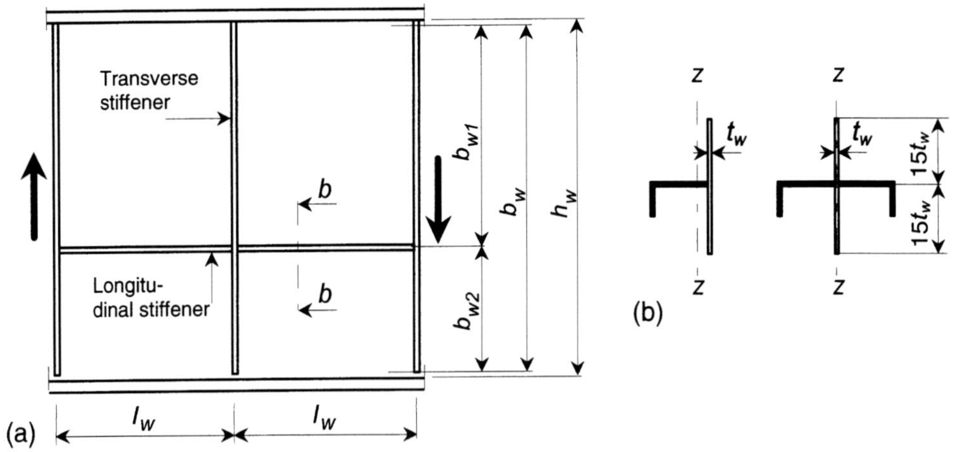

Figure 40: (a) Web with transverse and longitudinal stiffeners. (b) Effective cross section of stiffeners

When calculating the stiffness of the longitudinal stiffener, the second moment of area is based on the stiffener itself and strips of the web, width $15t_w$, on both side of the stiffener. See Figure 40(b).

It is assumed that the increase in τ_{cr} associated with a low value of the aspect ratio, a/b_w, causes an increase in the load-carrying capacity, V_w, which corresponds to the increase in τ_{cr}. Therefore, k_τ given by eqn. (3.5.17b) or (3.5.19) is substituted in eqn. (3.5.10) and eqn. (3.5.16).

5.5. Influence of rigid flanges

As mentioned, transverse stiffeners prevent the web from deflecting and prevent the flanges from approaching to each other except in the vicinity of the stiffeners. In the stage of failure, four hinges, denoted E, H, G, and K, form in the top and bottom flanges, see Figure 41(b). A tension stress field, EHGK, develops in the web. This tension stress field differs from the tension field described for instance by Rockey and Škaloud (1968), in that it is assumed to be developed between the flanges only. Another part of the tension field according to Rockey and Škaloud (1968) correspond to the rotated stress field shown in Figure 41(a). The shear force, V_f, which is transmitted by the tension stress field is obtained from the equation of equilibrium of the flange portion c. This equation gives

$$V_f = \frac{4W_{pl,f} f_{of}}{c} \tag{3.5.20}$$

where $W_{pl,f} = b_f t_f^2 / 4$ is the plastic bending modulus of one flange.

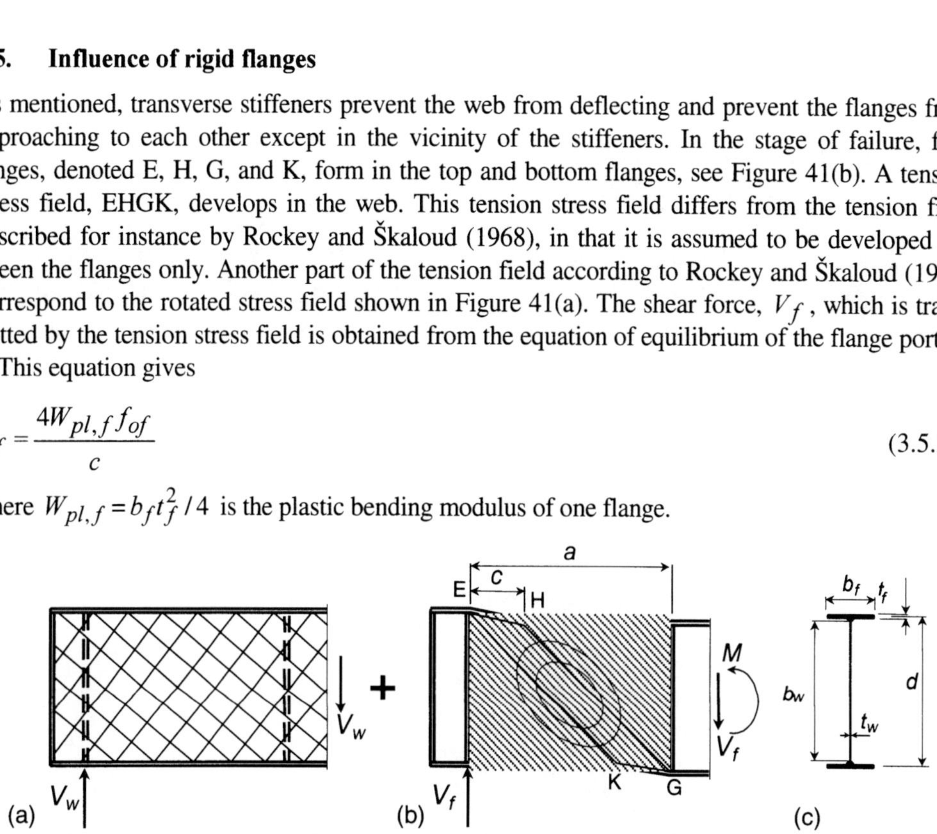

Figure 41: Model of the web in the post-buckling range. (a) Shear force carried by the web by rotated stress field. (b) Shear force carried by truss action, which is dependent on the bending strength of the flanges. (c) Notations for cross section

The tensile stresses in the tension stress field produce a stiffening effect on the web (this effect is favourable to the load-carrying capacity) at the same time as the effective stress increases in the web (this effect is mostly unfavourable). However, it is assumed that the shear resistance of the web, V_w, is not changed by the formation of the tension field between flanges. Then the shear resistance of the girder, V_u, is the sum of the shear resistance of the web, V , and the shear resistance contributed by the flanges, V_f,

$$V_u = V_w + V_f \tag{3.5.21}$$

As has been shown by the test series carried out by Rockey and Škaloud (1968) and Škaloud (1971) the distance, c, between the plastic hinges in the flanges varies from 0.16 to 0.55 times the spacing, a, of the stiffeners, depending on the rigidity of the flanges. The distance c is estimated by means of eqn (3.5.22) for aluminium plate girders

$$c = \left(0.08 + \frac{4.4\, b_f\, t_f^2\, f_{of}}{t_w b_w^2\, f_{ow}} \right) a \tag{3.5.22}$$

5.6. Influence of bending moment

If the girder is subjected to a shear force and at the same time to a small bending moment ($M < M_f$, where $M_f = A_f d\, f_{of}$ is the moment capacity of the flanges, d is the distance between centre of flanges), then it is assumed that the stresses in the web which are caused by the bending moment do not influence that portion, V_w, of the shear force which is resisted by the web. On the other hand, these stresses influence that portion, V_f, of the shear force which is resisted by the tension field between the flanges.

$$V_{red} = V_w + V_f \left(1 - \left(\frac{M}{M_f}\right)^2 \right) \quad \text{when } M \le M_f \tag{3.5.23}$$

This expression corresponds to the curve between A and B in Figure 42.

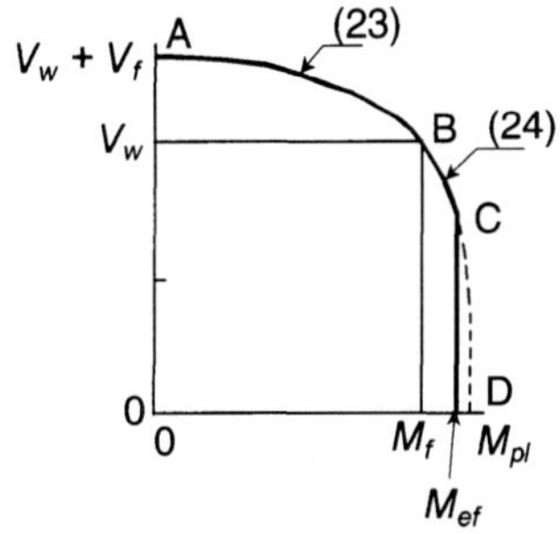

Figure 42: Interaction diagram for a girder subjected to a shear force and a bending moment

If $M > M_f$, then the flanges cannot contribute to the shear capacity of the girder, and the capacity of the web to carry shear forces is reduced. The interaction formula given by Basler (1961) is applied (see the curve BCD in the interaction diagram in Figure 42).

$$M = M_f + (M_{pl} - M_f)\left[1 - \left(\frac{V}{V_w}\right)^2 \right] \tag{3.5.24}$$

where M_{pl} is the plastic bending moment capacity of the girder. Furthermore the bending moment shall not exceed

$$M = W_{ef} f_o \tag{3.5.25}$$

corresponding to the vertical line from C in Figure 42.

5.7. Influence of welds

As shown in Figure 36(e) the membrane stresses in the web close to the flanges are more or less shear stresses only. Then reduced strength in the heat affected zones (haz) due to welding of the web to the flanges, does not seem to influence the shear strength of the web very much. Figure 43(a) shows that the diagonal tensile stress field, in a web with transverse stiffeners, is spread out at the corners of the web with heat affected zones around the borders, which results in smaller stresses in haz, Edlund et al. (2001). The two rows of elements close to the borders are lighter than the elements nearby in Figure 43(a) *with haz*, whereas the blackness is the same in these rows of elements as in the adjacent elements in Figure 43(b) *without haz*. Therefore, in the above design expressions, the yield strength of the parent material shall be used and not the strength in the heat affected zone. Furthermore, it shall be noted that all tests, supporting the design rules, are made on welded girders.

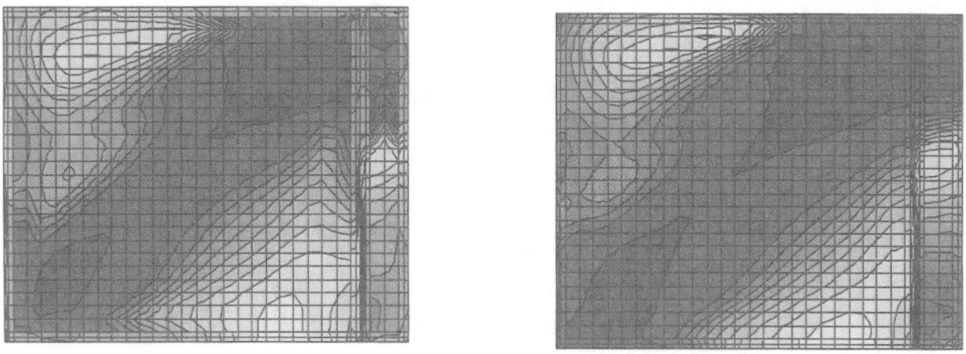

(a) (b)

Figure 43: Von Mises membrane stresses in the web of a beam loaded in shear at a load level close to the ultimate load. (a) Welded girder with heat affected zones in two rows of elements close to the borders. (b) Girder without heat affected zones. The Figures show the right end of the girder in Figure 44, Edlund et al. (2001)

Figure 43 also show that the web between the stiffeners at the end of the girder (forming the end post, see Figure 44) is stressed almost as the adjacent web. The end post is also bent inward as seen in Figure 44.

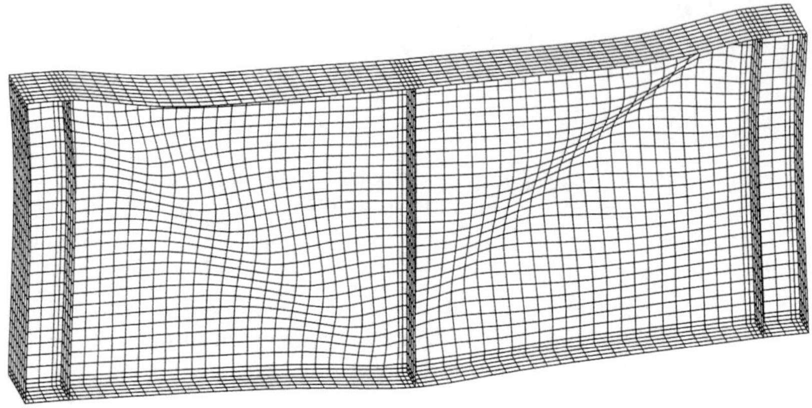

Figure 44: Deformation of a girder loadad with a point load in the middle and supported below the inner of the two stiffeners at the ends, Edlund et al. (2001)

5.8. Comparisons with tests

Many other design methods have been proposed for a plate girder in shear. See e.g. survey in Galambos (1988). Most of these theories start with the elastic buckling load V_{cr} and add a load corresponding to different types of diagonal tension fields. Many of these theories give good results for beams with small web panel aspect ratios a / b_w, but conservative results when the distance between the transverse stiffeners is large, because the contribution from the tension field is small. See Höglund (1995).

As is shown in the comparisons with the tests the rotated stress field method (including V_f) gives good agreement with tests. The method is also much simpler to use.

Comparisons are made for tests on aluminium girders reported in the following references:
The tests cover

- girders with stiffeners at supports only: Benson (1992)
- girders with transverse intermediate stiffeners: Evans and Hamoodi (1987), Evans and Burt (1990), Rockey and Evans (1970), Hamoodi (1983), Seah (1984), Burt (1987), Brown (1990), Benson (1992) and Evans and Lee (1995)
- girders with longitudinal and transverse stiffeners: Hamoodi (1983), Seah (1984)
- girders with trapezoidal corrugated webs, Benson (1992)

The total number of tests is 93. For some of these types of girders the number of tests is few. For instance there are only three tests on aluminium girders with corrugated webs (Benson 1992) and eight tests on longitudinally stiffened webs of aluminium (Hamoodi, 1983 and Seah, 1984)

Especially the few aluminium girders with trapezoidal corrugated webs make the design method for aluminium girders uncertain for this case. The method given in the ENV version of Eurocode 9 is therefore conservative.

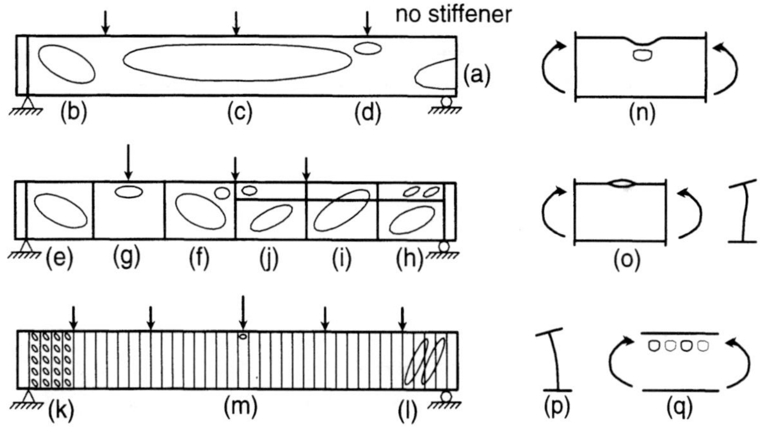

Figure 45: (a) - (n) Web buckling modes and (o) - (q) flange buckling modes

The resistance of plate girder webs depends on many parameters, for instance depth-to-thickness ratio h_w / t_w and web stiffener arrangements. In the test reports failure modes illustrated in Figure 45 have been observed.

The appraise covers test beams with stiffened and non stiffened webs failed due to shear buckling (failure modes b, e, f, h, i, j, k and l) and over-all web buckling caused by transverse forces and curvature induced transverse stresses in the web (failure mode c).

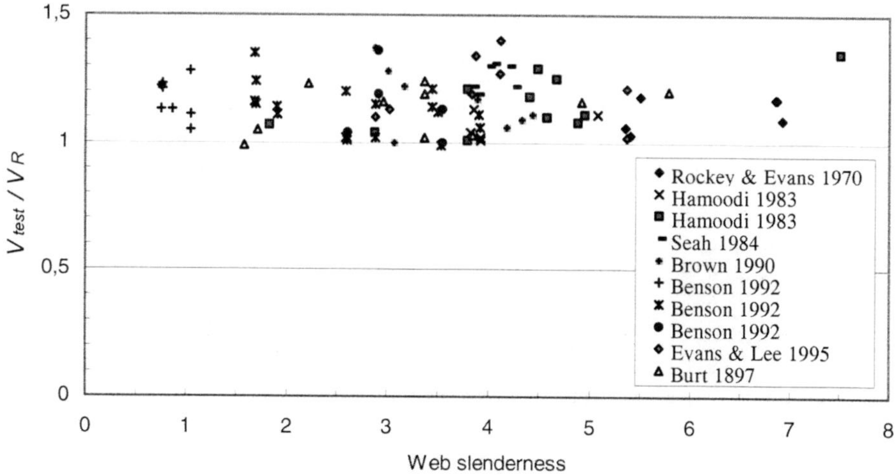

Figure 46: Comparison of test result and calculation according to Eurocode 9, V_{test} / V_R as function of web slenderness ratio λ_w

Test results on girders with flat webs are summarised in the diagram in Figure 46. A special buckling mode called overall web buckling has been found in aluminium girders without intermediate stiffeners [9]. See Figure 45(c). This buckling mode, not reported for steel girders, is a transverse column type buckling mode with no post-buckling strength. Theoretically this buckling mode can occur in long girders without intermediate stiffeners loaded with distributed load or closely spaced concentrated loads.

For none of the tests $V_{\exp}/V_R$ is less than 0.98 and the scatter is small, the coefficient of variation being 0.10

5.9. Design of stiffeners

Rigid end post

The rigid end post act as a bearing stiffener resisting the reaction at the girder support, and as a short beam resisting the longitudinal membrane stresses in the plane of the web. See Figure 36(a). It is assumed that the reaction force is acting on the inner stiffeners as shown in Figure 47(a).

A rigid end post may comprise of one stiffener at the girder end, and an inner, double-sided transverse stiffener, together forming the flanges of a short beam of length h_f, see Figure 47(a). The strip of web plate between the stiffeners forms the web of the short beam. Alternatively, an end post may be in the form of an inserted section, connected to the end of the web plate.

The rigid end post shall resist the tensile force N_h according to eqn. (3.5.7). Based on bending of the end post, a requirement for the cross section area of the extra stiffener can be derived. This requirement is simplified to the following rule of thumb:

$$A_e \geq 4h_f t_w^2 / e \text{ and } e > 0.1h_f \qquad (3.5.26)$$

where e is the distance between the stiffeners. See Figure 47(a).

The inner stiffener is designed for the reaction force R_{Ed} as a load-bearing stiffener.

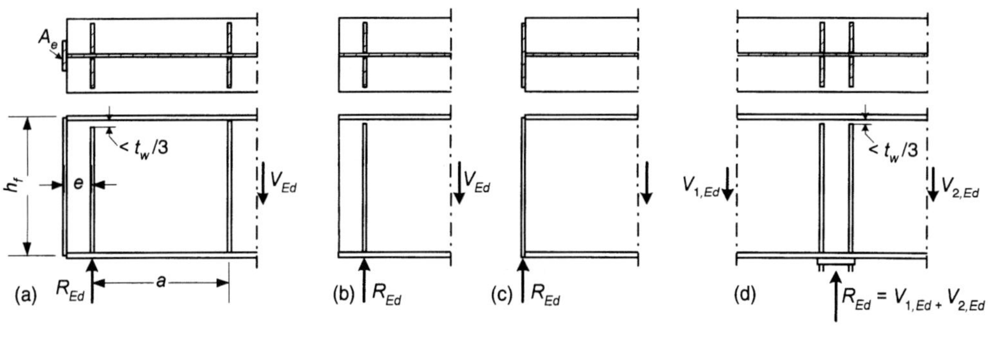

Figure 47 (a) Rigid end post. (b) and (c) Non-rigid end post. (d) Inner support of continuous girder

If an end post is the only means of providing resistance against twist at the end of a girder, the second moment of area of the end-post section about the center-line of the web (I_{ep}) should satisfy:

$$I_{ep} \geq b_w^3 t_f R_{Ed} / 250 W_{Ed} \quad (3.5.27)$$

where:

- t_f is the maximum value of flange thickness along the girder
- R_{Ed} is the reaction at the end of the girder under design loading
- W_{Ed} is the total design loading on the adjacent span.

A gap, smaller than one third of the web thickness, can be left between the top of the stiffener and the upper flange as shown in Figure 47(a) and (b), unless there is a load on the top flange.

Non-rigid end post

A non-rigid end post may be a single pair of stiffeners as shown in Figure 47(b) or a single stiffener at the girder end as in Figure 47(c). It is assumed to act as a bearing stiffener resisting the reaction at the girder support. If the end post is the only means of providing resistance against twist at the end of a girder, the requirement in (27) shall be satisfied.

Bearing stiffeners at a inner support of an continuous girder

Bearing stiffeners of an inner support can consist of two pair of stiffeners as in Figure 47(d) or a single pair of stiffeners. It is assumed to act as a bearing stiffener resisting the reaction at the girder support. In both cases it is assumed to act as a rigid end-post when calculating the shear strength $V_{1,Rd}$ and $V_{2,Rd}$ for the adjacent web panels.

Also in this case a gap, smaller than one third of the web thickness, can be left between the top of the stiffeners and the upper flange as shown in Figure 47(d), unless there is a load on the top flange.

Intermediate transverse stiffeners

Intermediate stiffeners, acting as ***rigid supports of interior panels*** of the web, should be checked for resistance and stiffness.

Other intermediate transverse stiffeners may be considered as ***flexible stiffeners***, their stiffness being considered in the calculation of k_τ in expression (3.5.16b).

The stiffeners can consist of a stiffener on one side of the web only, or pairs of stiffeners, see Figure 40(b).

Intermediate transverse stiffeners acting as rigid supports for the web panel should have a second moment of area fulfilling the following:

$$I_{st} \geq 1.5 h_w^3 t_w^3 / a^2 \qquad \text{if } a / h_w < \sqrt{2}: \quad (3.5.28)$$

$$I_{st} \geq 0.75 h_w t_w^3 \qquad \text{if } a / h_w \geq \sqrt{2}: \quad (3.5.29)$$

The second moment of area I_{st} shall be calculated for a cross section consisting of the stiffener itself and strips of the web, width $15 t_w$, on both side of the stiffener. See Figure 40(b).

The strength of intermediate rigid stiffeners should be checked for an axial force equal to V_{Ed} minus $\rho_v b_w t_w f_{ow}$ of the web with the considered stiffener removed. If the result is negative there is no special requirement for strength, only requirements for stiffness given above.

Longitudinal stiffeners

Longitudinal stiffeners may be either rigid or flexible. In both cases their stiffness should be taken into account when determining the buckling coefficient k_τ in expression (3.5.19) used in the slenderness parameter λ_w in expression (3.5.16b). This means that if a sub-panel governs the value of λ_w then the stiffener may be considered as rigid.

The strength should be checked for axial stresses if the stiffeners are taken into account for resisting such stresses due to bending or axial load on the girder.

Corrugated or closely stiffened webs

In plate girders with transverse reinforcement in the form of corrugations or closely spaced transverse stiffeners ($a / b_w < 0{,}3$) the flat parts between stiffeners can buckle locally, Figure 45(k), and the transverse reinforcement may deform with the web in an overall buckling mode, Figure 45(l).

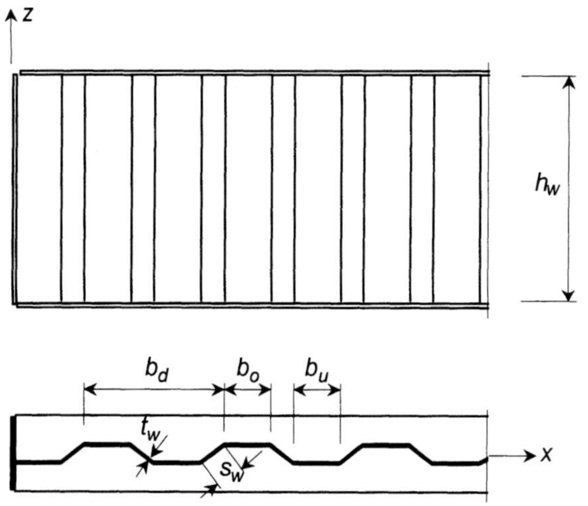

Figure 48: Corrugated web

The shear force resistance with respect to ***local shear buckling*** of flat parts is

$$V_{w,Rd} = 0.7\, \rho_v\, t_w\, h_w\, f_{ow} / \gamma_{M1} \tag{3.5.30}$$

where ρ_v is found in Table 5, non-rigid end post, for the slenderness parameter

$$\lambda_w = 0.35 \frac{b_m}{t_w} \sqrt{\frac{f_{ow}}{E}} \tag{3.5.31}$$

The width b_m is equal to the greatest width of the corrugated web plate panels b_o, b_u or s_w. In a closely stiffened flat web b is the greatest width between the stiffeners.

The factor 0.7 has been added to fit the test results. The need for this factor may be due to reduced strength in the heat affected zones from welding, Edlund et al.(2001).

The shear force resistance with respect to ***global shear buckling*** is determined according to

$$V_{o,Rd} = \chi_o h_w t_w \frac{f_{ow}}{\gamma_{M1}} \tag{3.5.32}$$

where:

$$\chi_o = \frac{0.60}{0.8 + \lambda_{ow}^2} \quad \text{but not more than } 0.6 \tag{3.5.33}$$

$$\lambda_{ow} = \sqrt{\frac{h_w t_w f_{ow}}{V_{o,cr}}} \tag{3.5.34}$$

$$V_{o,cr} = \frac{60\,E}{h_w} \sqrt[4]{I_z I_x^3} \tag{3.5.35}$$

$$I_z = \frac{b_d}{b_u + b_o + 2s_w} \frac{t_w^3}{10.9} \tag{3.5.36}$$

I_x is the second moment of area of the corrugated web per unit width, see Figure 48. For a closely stiffened flat web $I_z = t_w^3 / 10.9$.

6. PATCH LOADING RESISTANCE

6.1. Beam webs without stiffeners

The resistance of an unstiffened web to transverse forces applied through a flange, is governed by one of the following modes of failure

- crushing of the web close to the flange, accompanied by plastic deformation of the flange
- crippling of the web in form of localized buckling and crushing of the web close to the f-lange, accompanied by plastic deformation of the flange
- buckling of the web over most of the *depth* of the member
- overall buckling of the web over a large part of the *length* of the member. This failure mode is most likely to take place when there are a number of transverse forces or a distributed load along the member.

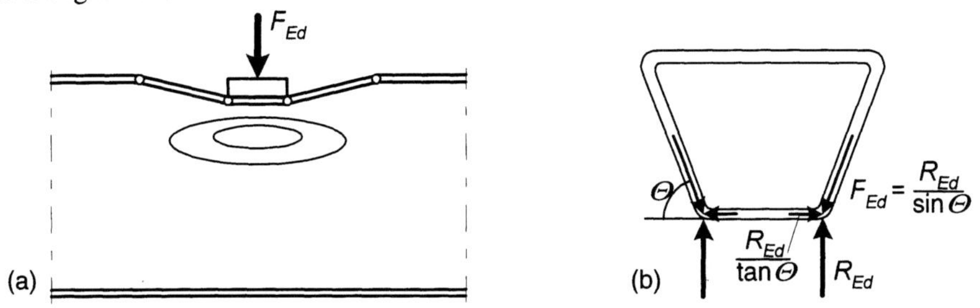

Figure 49: (a) Buckling of the web and plastic hinges in the flanges under a concentrated load, (b) Forces in inclined webs

A distinction is made between three types of load application, as follows:

a) Forces applied through one flange and resisted by shear forces in the web, *patch loading*, see Figure 51(a).
b) Forces applied to one flange and transferred through the web directly to the other flange, *opposite patch loading*, see Figure 51(b).
c) Forces applied through one flange close to an unstiffened end, *end patch loading*, see Figure 51(c).

For box girders with inclined webs the resistance should be checked for both web and flange. The load effects are the components of the external load in the plane of the web and flange respectively. See Figure 49(b). The effect of the transverse force on the moment resistance of the member should be considered.

The resistance of a longitudinally stiffened web is increased due to the presence of the stiffeners but no information is given in Eurocode 9.

The design approach in Eurocode 9 follows the same procedure as for other buckling cases. The resistance F_R is given as a reduction factor ρ times the yield resistance F_y, where the reduction factor is a function of the slenderness parameter $\lambda = \sqrt{F_R / F_{cr}}$ where F_{cr} is the elastic buckling load. The simple expression $\rho = 0.6/\lambda$ but not more than 1.0 is used for ρ, see Figure 50(a). This gives conservative results, Lagerqvist (1994) and Tryland (1999). Then

$$F_R = \rho F_y = 0{,}6\sqrt{F_{cr} F_y} \quad \text{but not more than } F_y \tag{3.6.1}$$

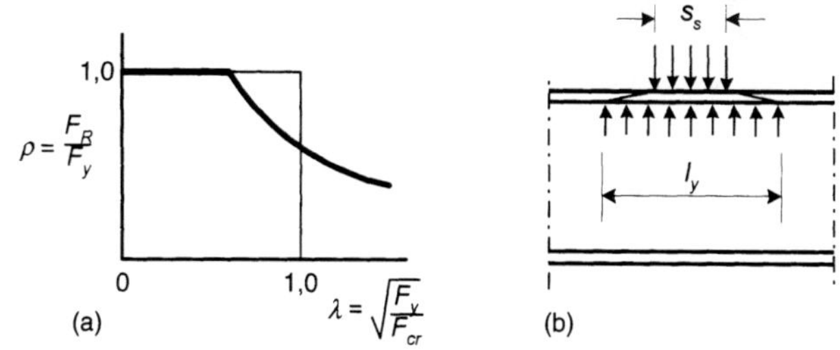

Figure 50: (a) Reduction factor ρ for (b) patch loading

The elastic buckling load is

$$F_{cr} = k_F \frac{\pi^2 E}{12(1-\nu^2)} \frac{t_w^3}{b_w} \tag{3.6.2}$$

where k_F is a coefficient depending on loading condition.

The yield load for patch loading can be expressed as

$$F_y = l_y t_w f_{ow} \tag{3.6.3}$$

where l_y is an effective loading length, depending on actual loading length s_s and stiffness of loaded flange. See Figure 50(b). f_{ow} is the 0.2 proof strength of the web material.

Inserting expression (3.6.2) and (3.6.3) into (3.6.1) and introducing the resistance factor γ_{M1} the design resistance F_{Rd} for a transverse force is

$$F_{Rd} = 0{,}57\, t_w^2 \sqrt{\frac{k_F\, l_y f_{ow} E}{b_w}} \frac{1}{\gamma_{M1}} \quad \text{but not more than} \quad t_w l_y \frac{f_{ow}}{\gamma_{M1}} \tag{3.6.4}$$

where k_F is given in Figure 51.

There is an interaction between concentrated loading and bending moment. See e.g. Tryland (1999). However, no formula is given in the ENV version of Eurocode 9.

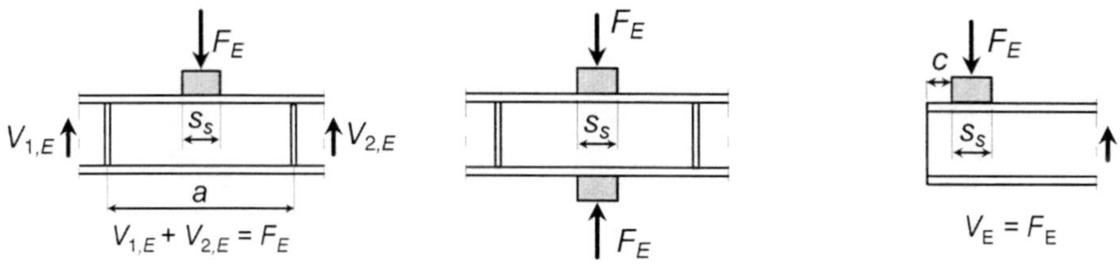

application a) application b) application c)

$$k_F = 6 + 2\left(\frac{b_w}{a}\right)^2 \qquad k_F = 3{,}5 + 2\left(\frac{b_w}{a}\right)^2 \qquad k_F = 2 + 6\,\frac{s_s + c}{b_w} \le 6$$

Figure 51: Load applications and buckling coefficients

The length of stiff bearing, s_s, on the flange is the distance over which the applied force is effectively distributed and it may be determined by dispersion of load through solid material at a slope of 1:1, see Figure 52. s_s should not be taken as more than b_w.

If several concentrated loads are closely spaced, the resistance should be checked for each individual load as well as for the total load. In the latter case s_s, should be taken as the centre distance between the outer loads.

The effective loaded length l_y, is calculated using the dimensionless parameters

$$m_1 = \frac{f_{of}\, b_f}{f_{ow}\, t_w} \tag{3.6.5}$$

$$m_2 = 0{,}02\left(\frac{b_w}{t_f}\right)^2 \quad \text{if} \quad \frac{(s_s + 4t_f)\, b_w\, f_{ow}}{k_F\, E\, t_w^2} > 0{,}2 \quad \text{else} \quad m_2 = 0 \tag{3.6.6}$$

For box girders, b_f in expression (3.6.5) is limited to $25t_f$ on each side of the web.

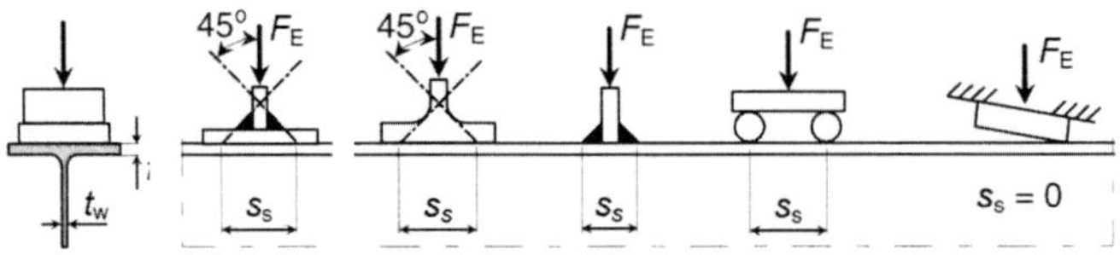

Figure 52. Length of stiff bearing

For load application (a) and (b) in Figure 51, l_y is given by

$$l_y = s_s + 2t_f\left(1 + \sqrt{m_1 + m_2}\right) \tag{3.6.7}$$

For load application (c) in Figure 50, l_y is given by the smaller of expressions (3.6.7), (3.6.9) and (3.6.10). Note that $s_s = 0$ if the loading device does not follow the change in the slope of the girder end.

$$l_{ef} = \frac{k_F\, E\, t_w^2}{2\, f_{ow}\, b_w} \le s_s + c \tag{3.6.8}$$

$$l_y = l_{ef} + t_f\sqrt{\frac{m_1}{2} + \left(\frac{l_{ef}}{t_f}\right)^2 + m_2} \tag{3.6.9}$$

$$l_y = l_{ef} + t_f\sqrt{m_1 + m_2} \tag{3.6.10}$$

6.2. Beam webs with stiffeners

Transverse stiffeners at supports and under concentrated loads are normally placed in pairs on both sides of the web, and are attached to the loaded flange.

The stiffener is designed as a compressed member (column) with a cross section composed of two stiffeners and a strip of the web having a width of $25t_w$ at interior stiffeners and $12t_w$ at the ends of the beam.

If both flanges are supported laterally, the stiffener member is supposed to have a buckling length equal to $0{,}8b_w$. If only one flange is supported laterally, the buckling length of the bar will depend on the connecting structure, but with a minimum value of $1{,}6b_w$. Consideration must also be taken to any lateral reaction from the unsupported flange.

7. Flange induced buckling

To prevent the possibility of the compression flange buckling in the plane of the web due to the curvature of the beam, the ratio b_w/t_w of the web shall satisfy the following expression

$$\frac{b_w}{t_w} \le \frac{kE}{f_{of}} \sqrt{\frac{A_w}{A_{fc}}} \tag{3.7.1}$$

where A_w is the area of the web and A_{fc} is the area of the compression flange. The value of k should be taken as follows:

Plastic rotation utilized	0.3
Plastic moment resistance utilized	0.4
Elastic moment resistance utilized	0.55

When the girder is curved in elevation, with the compression flange on the concave face, the following expression should be checked as well

$$\frac{b_w}{t_w} \le \frac{kE}{f_{of}} \sqrt{\frac{A_w}{A_{fc}}} \frac{1}{\sqrt{1 + \dfrac{b_w E}{3rf_{of}}}} \tag{3.7.2}$$

where r is the radius of curvature of the compression flange. When the girder has transverse web stiffeners, the limiting value of b_w/t_w may be increased by a factor $1 + (b_w / a)^2$.

References

Basler, K.(3.6.1961). Strength of plate girders under combined bending and shear. *ASCE Journal*, St 7, October 1961.

Benson, P.G.(1992). Shear buckling and overall web buckling of welded aluminium girders. *Royal Institute of Technology*, Division of Steel Structures, Stockholm, PhD thesis.

Brown, K.E.P.(1990). The post-buckling and collapse behaviour of aluminium girders. *University of Wales* College of Cardiff, PhD thesis.

Bulson, P.S., (1970). *The Stability of Flat Plates*. Chatto & Windus, London

Burt, C.A.(1987). The ultimate strength of aluminium plate girders. *University of Wales* College of Cardiff, PhD thesis.

Crate, H. and Lo, H.(1948). *NACA Tech. Note* No. 1589.

Edlund, S., Jansson, R. and Höglund, T.(2001). Shear buckling of Welded Aluminium Girders. *9th Nordic Steel Construction Conference*, Helsinki.

Eurocode 9, (1998) ENV 1999-1-1. *Eurocode 9: Design of Aluminium Structures – Part 1-1: General rules – General rules and rules for buildings*. CEN (European Committee for Standardization).

Evans, H.R. and Lee, A.Y.N.(1984). An appraisal, by comparison with experimental data, of new design procedures for aluminium plate girders. *Proc. Inst. Civ. Eng.* Structures & Buildings, Feb. 1984.

Evans, H.R. and Hamoodi, M.J. (1987). The collapse of welded aluminium plate girders - an experimental study. *Thin-Walled Structures* 5.

Evans, H.R. and Burt, C.(1990). Ultimate load determination for welded aluminium plate girders. Aluminium Structures: advances, design and construction. *Elsevier Applied Science*, London and New York.

Galambos, Th.V. (ed) (1988). *Guide to Stability Design Criteria for Metal Structures*, 4th edition. John Wiley & Sons.

Hamoodi, M.J.(1983). The behaviour of reinforced aluminium web plates in a shear loading. *University of Wales* College of Cardiff, MSc thesis.

Höglund, T.(1972). Design of thin plate I girders in shear and bending with special reference to web buckling. *Royal Inst of Technology*, Dept. of Building Statics and Structural Engineering, Stockholm.

Höglund, T.(1995). Shear buckling of Steel and Aluminium Plate Girders. *Royal Inst of Technology*, Dept. of Structural Engineering, Technical Report 1995:4, Stockholm

Klöppel, E.K. and Sheer, I.(1960). *Beulwerte ausgestreifter Rechteckplatten*. Wilhelm Ernst & Sohn, Berlin.

Lagerqvist, O. (1994). Patch loading. Resistance of Steel Girders Subjected to Concentated Forces. Ph.D. thesis, *Luleå University of Technology*, Division of Steel Structures, Luleå, Sweden.

Rockey, K. and Škaloud, M.(1968). Influence of flange stiffeners upon the load-carrying capacity of webs in shear. *IABSE Final Report, Eighth Congress* New York.

Rockey, K.C. and Evans, H.R.(1970). An experimental study of the ultimate load capacity of welded aluminium plate girders loaded in shear. Research Report, *University of Wales* College of Cardiff.

Seah, M.H.(1984). The behaviour of welded aluminium alloy plate girders reinforced with carbon fibre reinforced plastic. *University of Wales* College of Cardiff, MSc thesis.

Škaloud, M.(1971). Ultimate load and failure mechanism of webs in shear. *IABSE Colloquium*, London.

Timoshenko, S.P., and Gere, M.G. (1961). *Theory of elastic stability*. McGraw-Hill Book Company, New York, Toronto, London

Tryland, T. (1999). Aluminium and Steel Beams under Concentrated Loading. Dr.Ing. Thesis. *Norwegian University of Science and Technology*, Trondheim, Norway.

Chapter 4

DESIGN OF CONNECTIONS

F. Soetens, B.W.E.M. van Hove

Department of Structural Design, Eindhoven University of Technolgy, The Netherlands
Department of Building Structures, TNO Building and Construction Research, The Netherlands

PREFACE

Although a lot of research on aluminium alloys has been carried out in the past, relatively little attention has been given to their structural behaviour. Therefore, in the past most design rules for aluminium alloys were based on design rules for steel.
Applying similar design rules is permissible since aluminium and steel structures do have very similar structural behaviour. This means in particular the application of limit state design methods which allow a better description of the real, non linear behaviour of a structure than allowable stress methods are capable of.
However, the use of limit state design methods requires knowledge of the structural behaviour of members and connections. In the past decades quite a number of studies have been carried out to further enlarge the above mentioned knowledge. In particular the joint industry research initiated by Technical Committee 2 "Aluminium Alloy Structures" of the European Convention for Constructional Steelwork should be mentioned here.
Main subjects of research have been: stability (global and local stability); connections (bolted, welded, adhesive bonded connections); and last but not least fatigue (in particular fatigue of welded connections).
The above research effort has resulted in an adequate knowledge of the structural behaviour of aluminium members and connections today. Moreover design rules in national codes have been up-dated, but most important a (draft) European code "Eurocode 9: Design of aluminium structures" has been edited in 1998, which contains up-to-date design rules for aluminium structures.
In the following chapters attention is given to the state of the art concerning connections. In chapter 1 joining technology is described while in chapter 2 the design of joints is dealt with, in particular the design of welded, bolted, and adhesive bonded joints. In chapter 3 a number of structural aluminium applications is dealt with an emphasis on connections. At last in chapter 4 final remarks are given.

1. JOINING TECHNOLOGY

1.1. General

Well designed joints are essential to answer the satisfactory performances of a structure. In aluminium frameworks with riveted or bolted gusset plates it has been estimated that the weight of the joints is about 10% of the weight of the structure; in cost terms the ratio is much larger. A significant weight advantage results from the use of welding which reduces the above ratio to about 4%. Welding may also be preferred for general engineering purposes because it simplifies fabrication and assembly, which reduces cost. However where site assembly is required joints with mechanical fasteners – bolts or rivets – may be necessary. Furthermore, such joints provide useful system damping which is virtually absent in continuous welded structures.

When designing connections it is distinguished between:

- joints in primary structures, and
- joints in thin-walled structures.

In primary structures typical element thicknesses are larger than 3 mm. Joining techniques applied are:

- welding;
- mechanical fastening i.e. bolts and rivets;
- adhesive bonding;
- hybrid connections.

Sometimes special connections are used such as profile to profile joints, snap joints etc.

In addition to the above joining techniques in primary structures for thin-walled structures specific joining techniques are used such a screws, (blind) rivets, and spot welding.

1.2. Joints in primary structures

1.2.1. Welding

Welding is defined as the joining of materials by the use of heat – sometimes force or combining heat and force – with or without a filler metal. The welding of aluminium is widely established and has been developed into an important method of joining.

Inert gas shielded are welding processes (TIG and MIG) have considerably extended the possibilities for welding aluminium, and are even used in ordinary workshops.

Advantages of welded connections is saving of work and material, absence of drilling and overlap, tight joints, and no crevice corrosion in case of butt welds, see also fig. 1. By the extrusion technique groove preparation and backing can be integrated in the profile, strength

reduction in heat affected zones can be compensated by locally increasing the thickness, and difference in thickness can be levelled out. Butt-welded joints are preferable in most cases.
It gives a favourable state of stress in members in bending and tension. More details of welded connections are given in chapter 2.

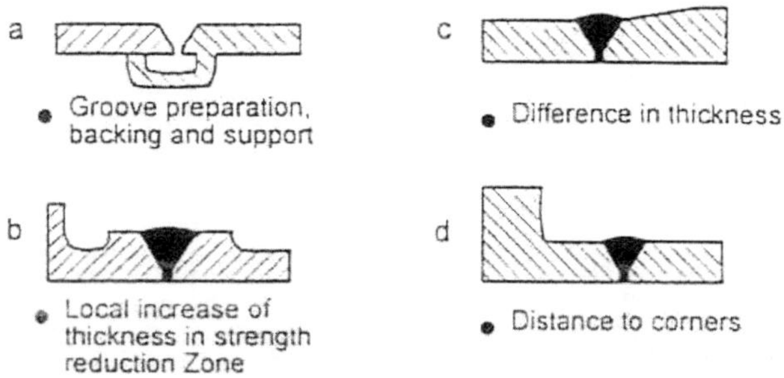

Figure 1: Welded connections

The aluminium welding processes commonly used in workshop practice are:
- Gas welding (autogenous welding);
- Metallic arc welding (flux-coated electrodes);
- Inert gas shielded arc welding with a non-consumable tungsten electrode (TIG), or with a consumable metal electrode (MIG);
- Laser welding;
- Friction welding.

Gas welding as well as metallic arc welding commonly used in steel structures, are not widely used in aluminium structures because of low efficiency and bad quality in particular susceptibility for defects.
Inert gas shielded arc welding (TIC and MIG) are the most wide-spread welding techniques that are used for structural applications. In fig. 2 the principle of TIG welding is shown.

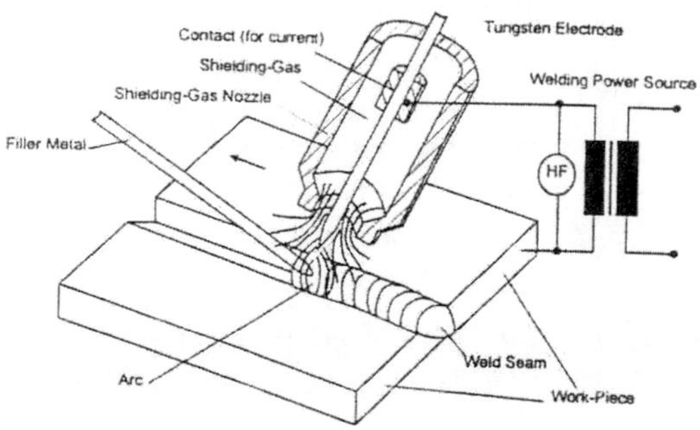

Figure 2: Principle of TIG Welding

During TIG Welding an arc is maintained between a tungsten electrode and the work-piece. The molten metal and the electrode are shielded by an inert gas flow (argon, helium, or a mixture). The filler metal is usually introduced manually.
A major difference between the welding of steel and the TIG (or MIG) welding of aluminium is the adhering oxide film on the aluminium surface which has to be removed and the shielding of the molten metal against oxygen from the atmosphere.
In fig. 3 the principle of MIG welding is shown. Analogous to TIG welding, MIG welding is conducted using inert gases. The arc is drawn between a melting wire electrode and the work piece.

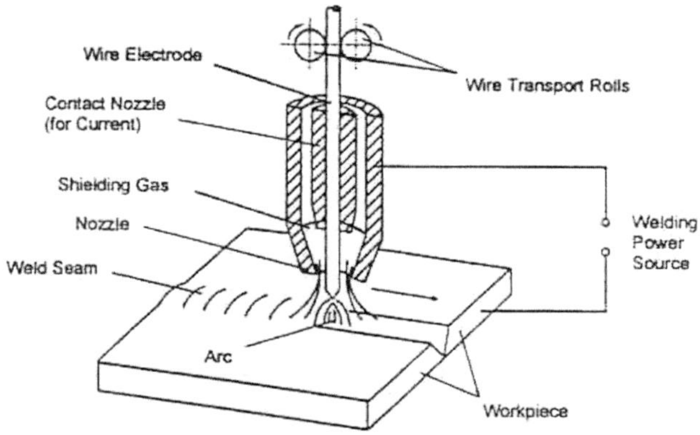

Figure 3: Principle of MIG Welding

The current used with MIG welding is much higher than with TIG welding which results in a better penetration (thicker material can be welded) and a higher welding speed. MIG welding is the more efficient process compared to TIG welding and it is also easier to mechanize.
Unalloyed aluminium and most aluminium alloys are entirely suitable for welding. Increases in strength by work hardening or age hardening are partially or fully lost under the action of the welding heat. As with any type of welding, undesirable distortions of shape or weld cracks can arise. Assuming that the component to be welded has been correctly designed for this purpose, these risks can be met, but only appropriate handling and welding procedures, the selection of a suitable filler material, and by a welding method suited to the job. In this connection, some of the properties of the material itself are significant for welding technology: its thermal expansion, tendency of shrinkage, modulus of elasticity, and its melting point.
Although the following processes are much less used compared to TIG and MIG welding some of them have specific applications.
With laser welding of aluminium two types of lasers are used i.e. solid-state lasers (Nd-YAG) and gas-discharge lasers (CO2). The first ones operate at lower energy and are most suited for

thinner material while the gas-discharge lasers operate at higher energy and are used for thicker material.

With friction welding usually rotational symmetrical parts are rotated while being pressed together, thereby generating heat of friction which causes the parts to weld. This process also enables to weld aluminium to steel.

A more recently developed process is friction stir welding (FSW). In fig. 4 the principle of friction stir welding is shown. A specially shaped rotation pin moves between the abutting faces of the joint which generates frictional heat that creates a plasticised region around the tool. The contact of the shouldered region of the tool with the workpieces also generates significant frictional heat as well as preventing plasticised material from being expelled. The tool is moved along the joint line, forcing the plasticised material to coalesce behind the tool to form a solid-phase joint.

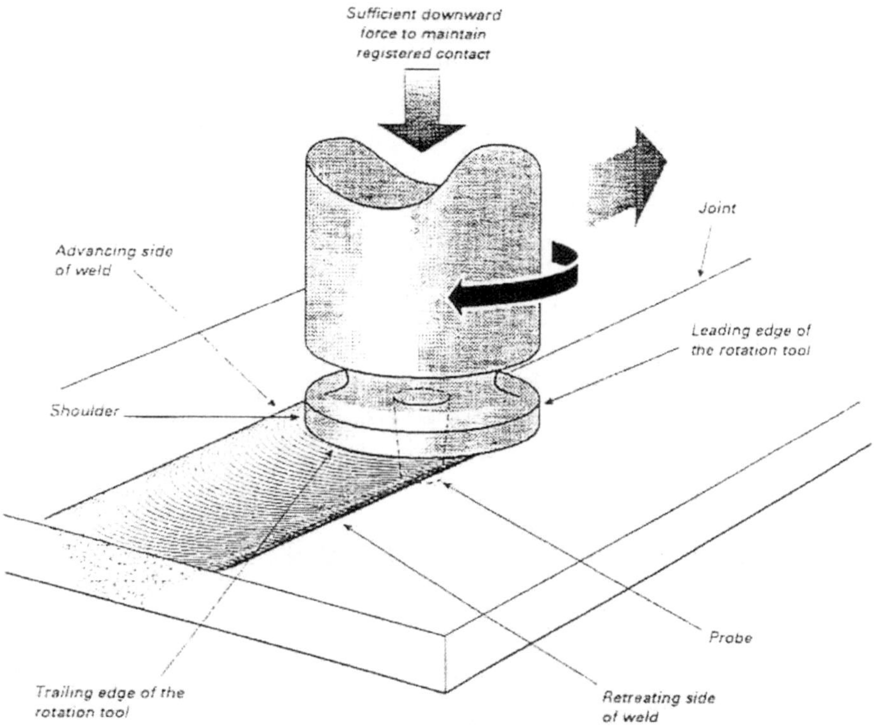

Figure 4: Friction stir welding with a rotating tool

Friction stir welding can be regarded as an autogenous keyhole joining technique, where consolidated welds are solid-phase in nature and do not show fusion welding defects. No consumable filler material or edge preparation is normally necessary. The distortion is significantly less than that caused by arc fusion welding techniques.

In addition to laser welding and friction welding solid state welding processes to be mentioned, are:

- explosion welding;
- ultrasonic welding;
- diffusion welding;
- cold or hot pressure welding.

Except for welding aluminium these processes enable aluminium to be welded to a wide range of other metals.

1.2.2. Mechanical fastening

In primary structures most commonly used mechanical fasteners are bolts, either aluminium or steel bolts. In some cases also rivets are used but they are considered to be rather outdated and uneconomical. At the contrary for thin-walled structures special rivets have been developed with a wide field of application, see chapter 1.3.

Compared to welded joints mechanical fasteners have the advantage that there is no softening of the materials due to the influence of heat. Furthermore, mechanical fasteners can be used on site contrarily to welding which is an in-shop method. So for the assembly on site preferably bolting is the joining method.

Aluminium bolts offer the advantage that changes in the tightness of the joints due to thermal expansion, which can occur when using steel bolts, are avoided.

The aluminium bolt material is chosen to match the material forming the components to be joined. Further information is given in chapter 2.

The following general points should be born in mind with bolts:

- Excessively high pressure on the surface of the aluminium when the fastener is tightened, can be avoided by fitting hard aluminium washers under de head of the bolt and the nut.
- When bolts are loosened and done up again frequently, the thread in the aluminium component or on the bolt itself, can quickly become worn. In such cases it is recommended to use inserts, see also fig. 5.
- For joints exposed to moisture the aluminium bolts should be sealed.

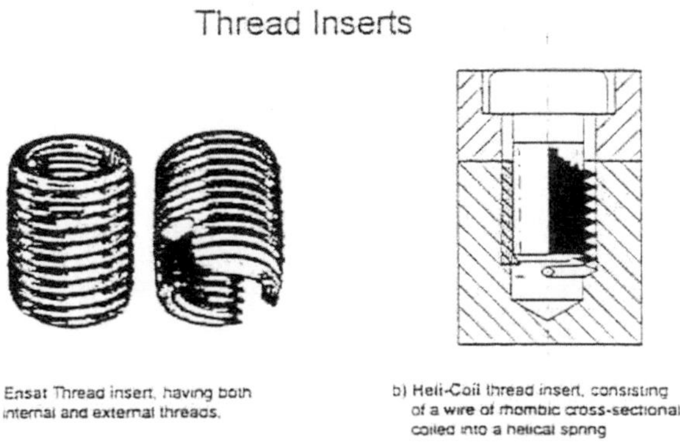

Figure 5: Thread inserts in aluminium components

Steel bolts used in aluminium structures exposed to weathering or other corrosive environments must be protected against corrosion; for example the steel bolts may be galvanised. However, it is becoming increasingly common to use bolts made of stainless steel. To avoid excessive surface pressure steel bolts are also fitted under the bolt head and nut with galvanised washers to arrive at a more evenly distributed load.

Combined steel bolts and female threads in materials of low shear stress is avoided by using threat inserts. The latter enable a higher load bearing capacity of the joint; in case of frequently loosening and tightening up of the bolt the thread of the aluminium component is not worn.

When used in primary structures solid aluminium rivets (see fig. 6) are driven while cold. Contrarily to hot-driven rivets, cold-driven rivets do not shrink and therefore do not press the sheets together. That means that cold-driven rivets are loaded in a similar way as bolts (see also chapter 2).

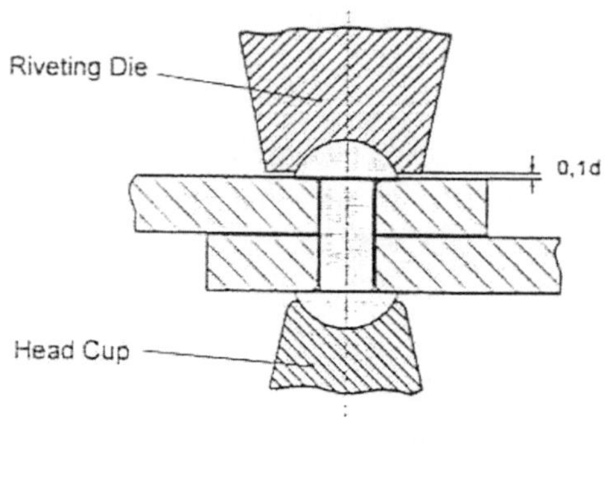

Figure 6: Aluminium solid rivets

Sometimes special mechanical joints are used such as profile to profile joints, snap joints, and corner joints. In all these cases use is made of extruded connections like the tongue and groove connection shown in fig. 7. Many times these types of connections are combined with other types of fasteners.
In the example of fig. 7 locking of the profiles in longitudinal direction is possible by the use of screws.

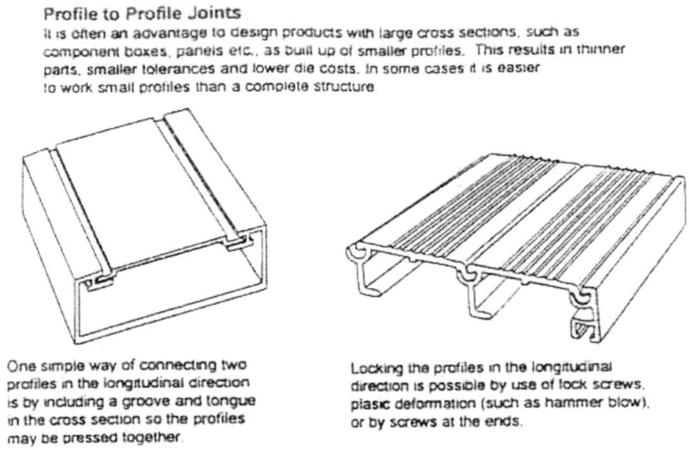

Figure 7: Special mechanical joint

1.2.3. Adhesive bonding

Although not yet widely used in structural applications, adhesive bonding of aluminium will become more and more important in particular because of the geometrical possibilities enabled by extrusion.
Since adhesive bonding is not very familiar outside the aerospace industry in the following the most important aspects with respect to bonding technology will be reviewed.
Adhesive bonding is defined as the process of joining parts using a non-metallic substance (adhesive) which undergoes a physical or chemical hardening reaction causing the parts to join together through surface adherence (adhesion) and internal strength of the adhesive (cohesion).

Adhesive Bonding of Aluminium

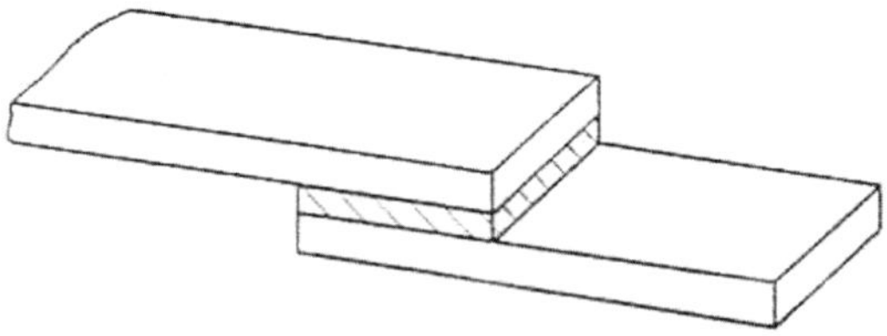

Figure 8: Adhesive bonding of aluminium

Adhesive bonding has a number of advantages compared to mechanical fastening and welding. But at the same time a number of disadvantages as well .

The advantages are:

- continuous joints, no holes (fasteners), no heat input (welding) which affects parent metal strength and also results in deformations;
- more uniform stress and strain distributions; stress concentrations at a lower level than with mechanical fasteners; in particular for cyclic loaded structures (fatigue conditions) this results in a longer life time of the structure.
- joining of aluminium with other materials;
- more flexibility in designing;
- very thin materials and small components can be adhesively bonded;
- tight joints; liquid as well as gas tight joints;
- isolating or conductive (electricity, heat);
- decreasing vibrations.

The disadvantages are:

- specific joint design; lower stress levels, shear forces, large surface areas needed;
- surface treatment;
- curing time adhesives;
- adhesion difficult to control; NDT methods hardly available;
- demounting/repair difficult;
- restricted structural behaviour at high temperature.

Figure 9: Adhesive joint design. Left figure not suitable. Right figure typical adhesive joint design.

Adhesive joint are composite systems whose strength depends on both the geometrical design and loading type as well as on the individual strengths of the different members of the joint as illustrated in fig. 10.

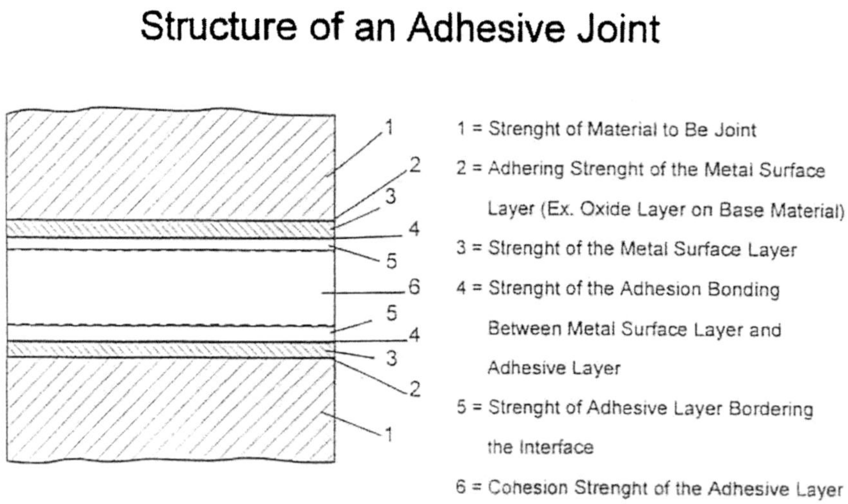

Figure 10: Different members of an adhesive bonded joint

Adhesion is defined as the force acting between the adhesive and the surface of the material. This force is the result of the mechanical interlocking between the adhesive and the material surface roughness as well as the physical and/or chemical interaction between the adhesive and the material.

Cohesion is the strength of the adhesive itself. This is a result of the mechanical entangling and interlocking of the adhesive molecules and their physical and/or chemical affinity for each other.

The separation of adhesive bonded joints occurs due to the failure of adhesion or cohesion or a mixed adhesion and cohesion failure, see figure 11.

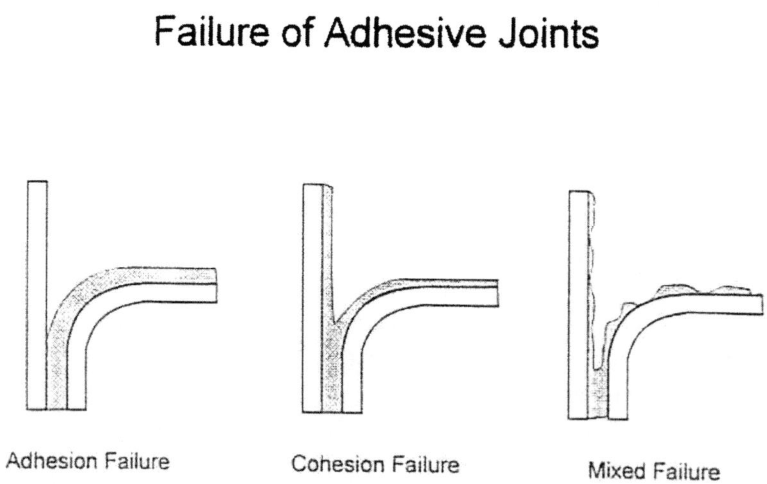

Figure 11: Failure of adhesive joints

When designing bonded aluminium structures, one of the primary decisions to be made in selection of the right adhesive. Many factors have to be evaluated such as:
- function of the adhesive: structural or non structural;
- curing temperature: cold-curing, heat-curing;
- specific curing conditions: humidity needed, absence of oxygen;
- number of components: one component, two components;
- application form: liquid, paste, solid, tape;
- chemical composition: epoxy, acrylic, polyurethane;
- physical behaviour: thermosetting, thermoplastic;

The above described factors have to do with the variety of adhesives available. On the other side one has to bare in mind the application.

Factors involved with the application are:
- structural aspects: strength, durability, flexibility;
- materials to be connected: aluminium- aluminium or aluminium – other material;
- surface treatment: higher strength but more particular higher durability requires more rigorous surface treatments;
- curing: time needed, temperature, pressure, in-shop or on-site;
- application: apparatus, "open" time adhesive;
- quality assurance: testing, destructive, non-destructive.

Although far from complete in the following some guidance will be given how to arrive at selecting the right adhesive.

As far as structural applications are concerned three types of adhesives are used: phenolic adhesive systems; epoxides; and acrylics.

For semi structural applications: polyurethanes; hot melt systems; and silicones.

For non-structural applications: cyanoacrylates; contact adhesives.

The adhesives mentioned above are suited to bond aluminium to aluminium but also aluminium to steel, or stainless steel, or zinc. Adhesion depends to a large extent on the surface treatment applied.

Surface treatment prior to bonding results in:

- higher initial adhesion of the adhesive and the aluminium surface;
- higher durability of the joint.

The better the interaction between adhesive and the aluminium surface the higher the strength of the interface between them and also the less susceptible the interface is for ageing i.e. decrease of adhesion due to environmental attack.

Preparing the surface prior to bonding the first step should be cleaning and degreasing. Further steps can be: mechanical pretreatment such as grit blasting, brushing, or chemical pretreatment like etching, anodising, chromate conversion of the surface. For aluminium anodising is a well-known surface treatment which results in a very stable oxide layer on the bare metal with a high resistance against environmental attack.

A final step can be the application of a primer which can further enhance the adhesion to the surface as well as decrease the ageing susceptibility of the interface. A primer is sometimes used to bridge the time span between preparing the surface and the actual bonding of the parts.

Whether or not one has to carry out a more sophisticated surface treatment depends on the application. However, almost all applications need at least the first step mentioned i.e. cleaning and degreasing.

To illustrate the effects of ageing in fig. 12 the results are given of accelerated tests (high temperature, high humidity) for phenolic resp. epoxy adhesives without mechanical stressing (fig. 12a) and with mechanical stressing (fig. 12b).

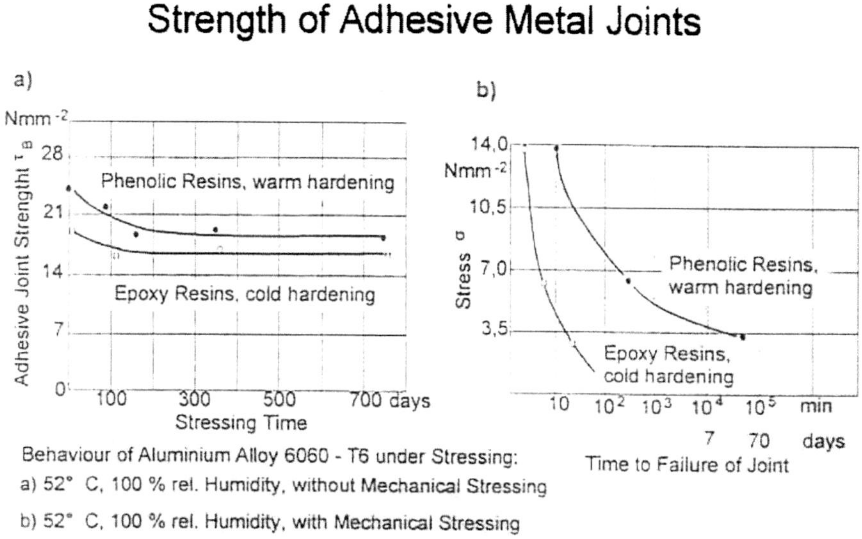

Figure 12: Accelerated tests on adhesive bonded joints.

Applying adhesive bonded joints it is essential to design the joints for bonding and not simply substitute it for other means of joining. Care should therefore be taken that stress concentrations be avoided and the loads be carried over as large an area as possible.

Wherever possible, the adhesive should be loaded in shear so that peel and cleavage stresses are avoided.

Most widely used are lap joints as shown in fig. 13.

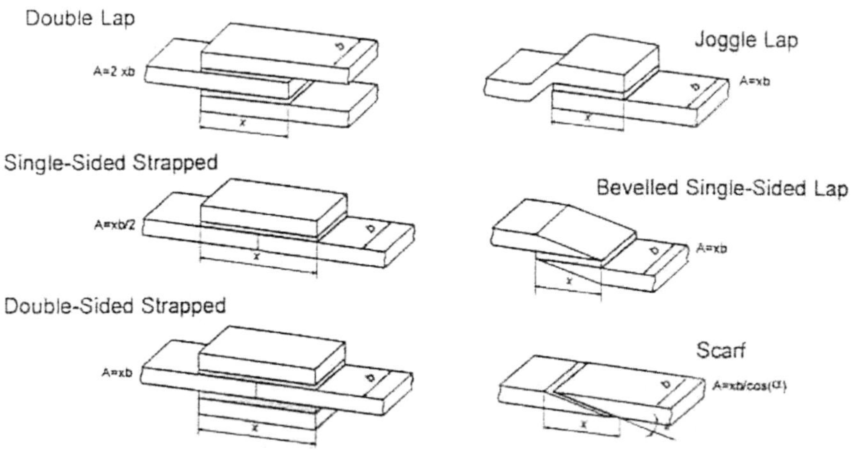

Figure 13: Adhesive bonded lap joints

Other examples for joint design which have proven to be successful are given in figs. 14 and 15.

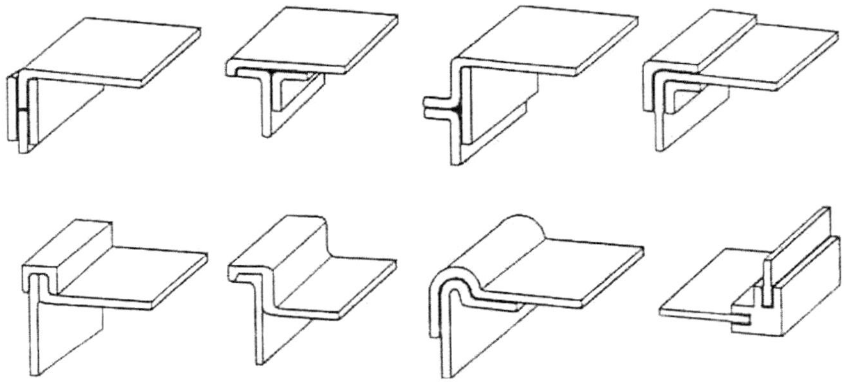

Figure 14: Examples of adhesive bonded corner joints

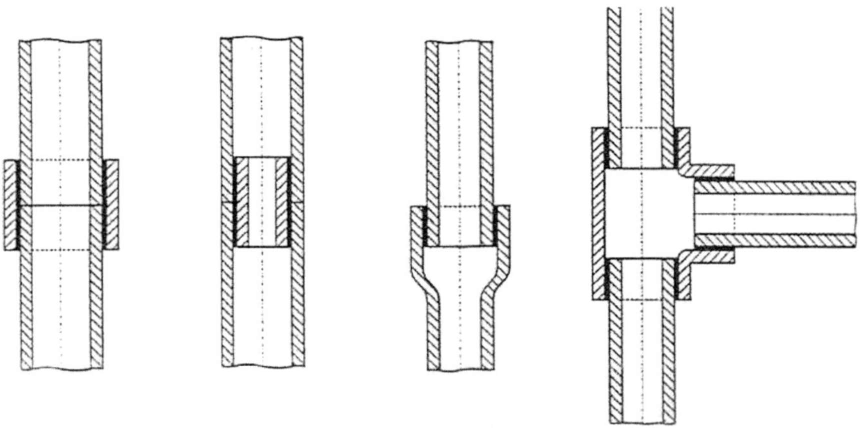

Figure 15: Examples of adhesive bonded tube joints.

Other successful examples of adhesive bonded structural applications to be mentioned are:
- strengthening / stiffening of structural components;
- bonded sandwich structures;
- bonded, built-up profiles.

1.2.4. Hybrid joints

In some applications different joining methods are combined. Well-known combinations in primary structures are welding and bolting as shown in fig. 16.

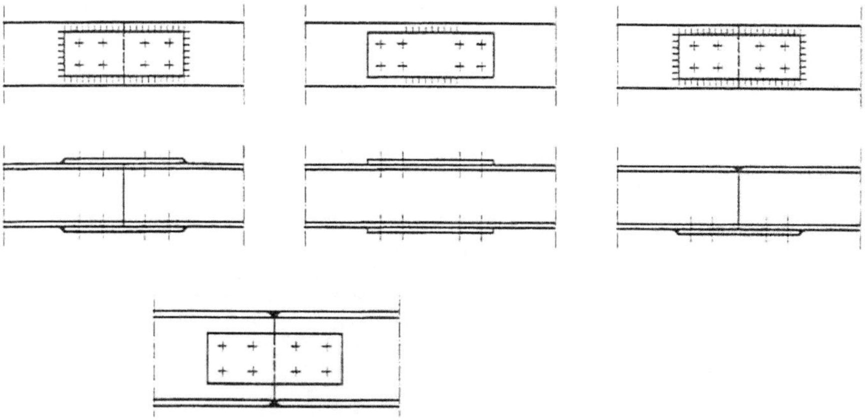

Figure 16: Hybrid joints

1.3. Joints in thin-walled structures

The joining methods as described for primary structures, can also be used in thin-walled structures. In addition to that specific joining methods for thin-walled structures have been developed. The latter methods will be shortly reviewed in the following.

1.3.1. Spot welding

Two main groups of spot welding procedures are available:
- arc spot welding, and
- resistance spot welding.

The main advantage of arc spot welding is the ability to weld from one side, a much lower equipment cost, and the portability of the gun, all compared to resistance welding. For those reasons in particular MIG spot welding has been quite widely applied but nowadays resistance spot welding is by far the most widely used method.
Resistance spot welding is the general name for a group of processes which rely on the resistance of a metal to the flow of electrical current to produce the heat needed for coalescence. Because of aluminium's high coefficient of electrical conductivity current levels for welding aluminium must be much higher than for a low conductivity metal like steel.

Consequently, while aluminium can be welded by all the usual resistance welding methods, special care is needed to achieve the desired results.
Resistance spot welding produces a local weld "spot" by clamping two (or sometimes more) sheets of metal between two electrodes for a brief interval with the metal under pressure, see fig. 17. The heat required for coalescence is generated by the bulk electrical resistance of the metal and also by the interface resistance between the metal sheets. It is a fusion welding process because melting must occur at the interface between the joint members to cause coalescence, form a cost nugget and joint the member together.

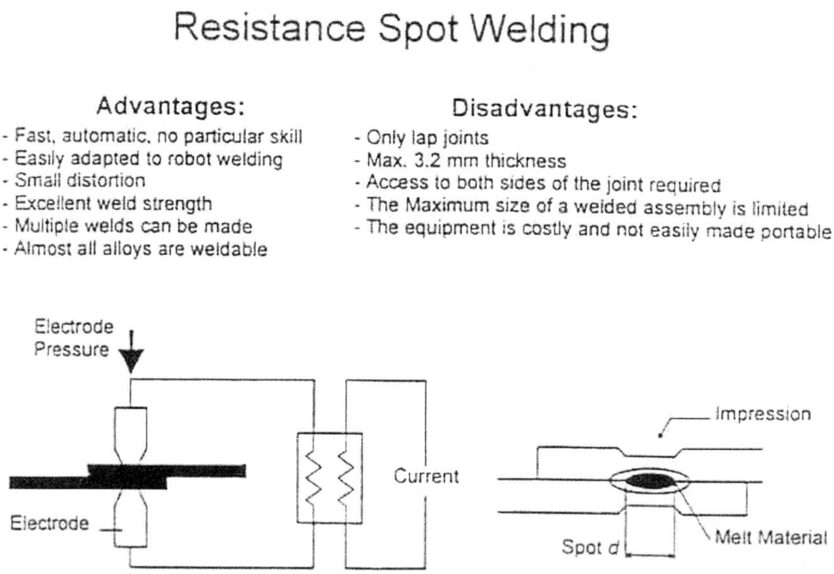

Figure 17: Resistance spot welding

Resistance spot welding is a high production joining method for fabricating sheet structures ranging from aircraft to cooking intensils. It has a number of advantages over other methods:

- Welds are usually completed in a fraction of a second and weld-to-weld times are often less than one second;
- The process is automatic and requires no particular skills.
- It is easily adapted to robotic welding.

It has also some disadvantages such as:

- Only lap joints can be welded;
- The thickness is limited to about 3 mm;
- Both sides of the joint must be accessible.

Resistance roll spot welding is similar to resistance spot welding, except that the conventional electrodes are replaced by rotating wheel electrodes. Welds are made repetitively, usually at uniform spacing. When the welds are spaced, it is termed intermittent seam welding. When the welds overlap each other, it is termed seam welding. The latter is often used to make gas or liquid tight joints.

1.3.2. Mechanical fastening

With thin-walled structures special mechanical fasteners such as screws and (blind) rivets are used.
The most usual application of screws is the fastening of thin to thin or thin to thick material.
Thread forming or thread cutting screws are installed in predrilled or punched holes or screw grooves, which can be easily extruded.
Self drilling screws are provided with a drilling cutter or a sharp bore bit to drill the hole, see fig. 18.

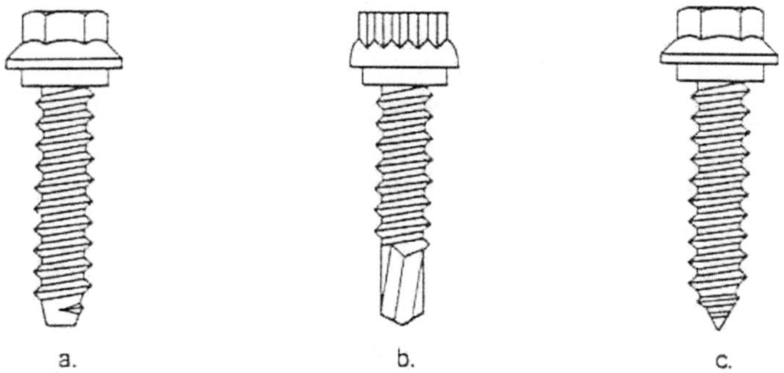

a. a) Thread forming screw

b. b) Self drilling (and thread forming) screw with drilling cutter

c. c) Self drilling (and thread forming) screw with sharp bore bit

Figure 18: Thread forming and selfdrilling screws

When components have to be riveted together in situations where the rivet is not accessible from both sides, blind riveting systems provide a solution to the problem. Using a suitable tool the rivets are inserted from one side and a closing head is formed. Blind rivets are always hollow, but may be provided with filler pins. The various systems available have proved themselves over a long period of time and are nowadays often used in stead of solid rivets.
Probably most well-known is the so called pop rivet, see fig. 19.

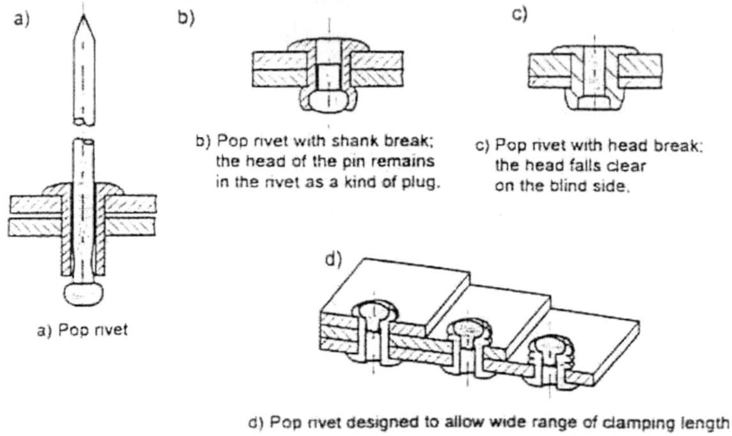

Figure 19: Pop-rivets

Pop rivets consist of a riveting pin and a rivet sleeve with a flat or countersunk head. In the riveting process the tool first pulls the sheets tightly together and then forms the closing head. Finally, the pin breaks at a predetermined point. In some types the pin breaks at a point within the rivet shank and its head then remains as a sort of sealing plug. In other types the head breaks off and falls clear on the blind side.

Huckbolts are not blind rivets in the strict sense of the term, since the bolts are introduced from the rear, while the closing operation is performed on the working side. The particular advantage of this method is that the rivet bolt itself is not deformed. The bolts are made of steel or aluminium (AlMg5, 5056A). The joint can withstand very high shear and tensile stresses. Huckbolts are fasteners consisting of two parts, the rivet bolt itself, with a mushroom or countersunk head, and a closing collet, see figure 20. The bolt has a smooth, cylindrical shank to withstand shear stresses.

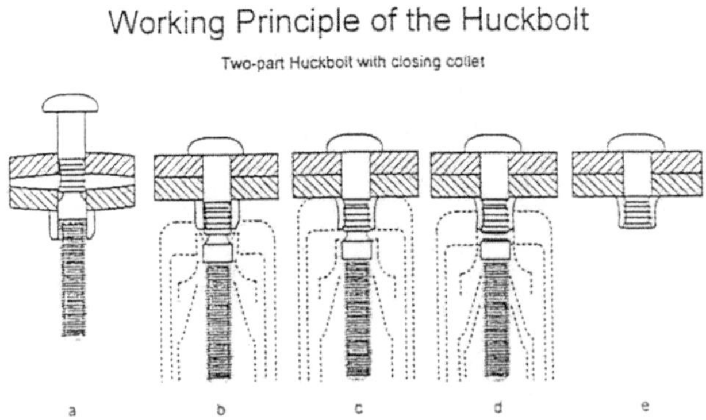

Figure 20: Huckbolts

1.3.3. Adhesive bonding

Adhesive bonding is used both in primary structures as well as in thin-walled structures. Similar aspects as described in 1.2.3, have to be dealt with when applying bonding technology in thin-walled structures. Very nice examples of thin-walled adhesive bonded structures are sandwich panels, consisting of thin outer skins of aluminium sheet adhesive bonded to a (isolating) core material, forming an element with excellent mechanical properties.

1.3.4. Hybrid joints

With thin-walled structures different joining methods can be combined. Widely used combinations are: spot-welding and adhesive bonding; riveting and adhesive bonding; screws and adhesive bonding.
In some applications the two joining methods are complementary for example spot-welding and adhesive bonding; in others the fasteners are meant to hold the parts in position while the adhesives are curing and finally joint strength is arrived at.

2. DESIGN OF JOINTS

2.1. Principles of design

Connections are an important part of every structure not only from the point of view of structural behaviour, but also in relation to the cost of production. As mentioned in chapter 2 a variety of joining methods is available for aluminium structures. Correct selection is governed by a large number of factors, structural and non-structural.
This chapter focuses on structural requirements.
For the design of a connection two quantities have to be considered:

F_{sd}= the force in the connection caused by the design load;

F_{rd}= the design strength (resistance) of a connection (see also fig. 21).

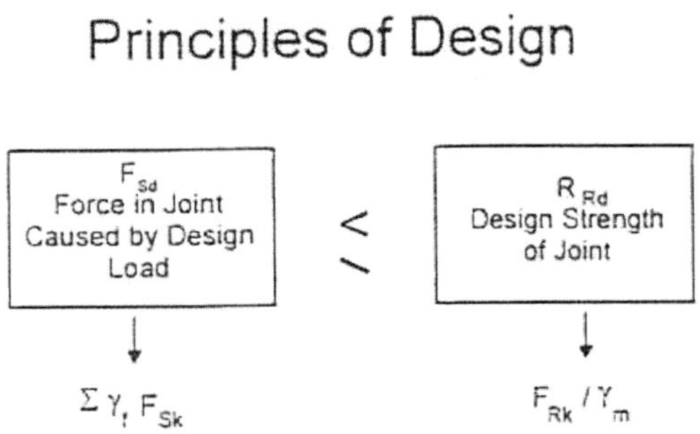

where F_{Sk} = force in connection caused by characteristic load
F_{Rk} = characteristic strength of connection
γ_f = appropriate load factor
γ_m = appropriate material factors

Figure 21: Principles of design

The forces in connections are dependent on the:

- loads on the jointed elements;
- stiffness of the jointed elements;
- stiffness and deformation capacity of the connection.

Stiffness of a connection is important because it determines the stiffness of the whole structure or of its components. Moreover, stiffness can also influence the forces in a connection.

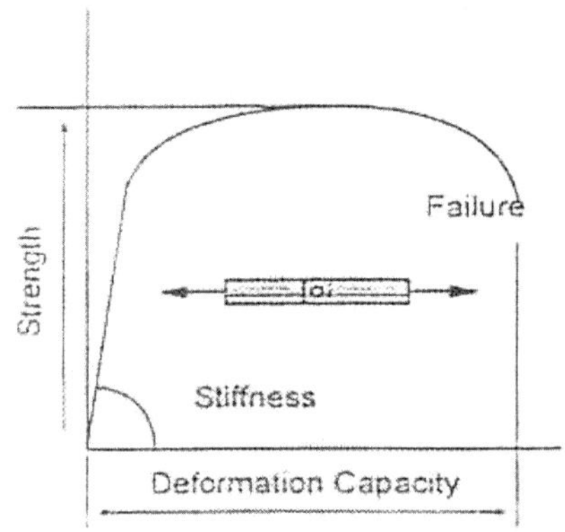

Figure 22: Strength, stiffness and deformation capacity

The deformation capacity of a connection is also important. A connection with no deformation capacity can cause a brittle fracture of a structure or element. This primarily applies to a continuous construction, where such influences as settlings and fluctuating temperatures are normally not included in a design calculation. Local overloading can be eliminated if the connection can deform sufficiently. In the case of simply supported structures, the deformation capacity of the separate fasteners can be important if more fasteners are used.

2.2. Design of welded connections

2.2.1. General

In the design of welded joints consideration should be given both to the strength of the welds and to the strength of the heat-affected zone. The design guidance given here which complies with Eurocode 9, applies to:

- Strength members under predominantly static loads.
- The welding process MIG for all material thicknesses and TIG only for thicknesses up to 6 mm and for repair.
- The welder and welding procedure are approved in accordance with qualification requirements as specified resulting in a normal quality level.
- Combinations of parent and filler metal as given in fig. 23.

If, in case of strength members, the above conditions are not fulfilled representative test pieces have to be welded and tested. For partial strength or non-strength members a lower quality can be specified by the designer resulting in lower design strength values. The latter can be obtained by using a partial safety factor γ_M = 1,6 instead of γ_M = 1,25 as usual for strength members.

In order to ensure the welding quality qualification specimens have to be welded according to a written welding procedure specification. This enables to approve the welder and the welding method as well as to determine the welding parameters and other relevant data, which can be added to the welding procedure specification. And, if necessary, these qualifications specimens can be subjected to mechanical testing to prove the design and the procedures.

Parent metal combinations *)						
Parent metal	7020	6082	6061	6060	5454	5083
7020	5356 5183 4043					
6082	5356 5183 4043	5356 5183 4043				
6061	5356 5183 4043	5356 5183 4043	5356 5183 4043			
6060	5356 4043	5356 4043	5356 4043	5356 4043		
5454	5356 5183	5356 5183	5356 5183	5356	5356 5183	
5083	5356 5183	5356 5183	5356 5183	5356	5356 5183	5356 5183
*) Filler metals for parent metal combinations are shown in one box, which is located at the intersection of the relevant parent metal row and column.						

Figure 23: Selection of filler metal

2.2.2. Welded connections

Contrarily to structural steels with aluminium due to the heat-input by welding a heat-affected zone (HAZ) has to be taken into account for the following classes of alloys:

- heat-treatable alloys in any heat-treated condition above T4 (6xxx and 7xxx series).
- Non-heat-treatable alloys in any work-hardened condition (3xxx and 5xxx series).

The severity of the HAZ is given in fig. 24 by means of a softening factor ρ_{HAZ} . Applying this factor to the ultimate strength (f_u) of the alloy results in the limiting strength of the HAZ.

Alloy series	Condition	ρ_{HAZ} (MIG)	ρ_{HAZ} (TIG)
6xxxx	T5	0,65	0,60
	T6	0,65	0,50
7xxx	T6	0,80	0,60
5xxx	H22	0,86	0,86
	H24	0,80	0,80
3xxx	H14, 16, 18	0,60	0,60

Figure 24: HAZ softening factor ρ_{HAZ}

The HAZ is assumed to extend a distance b_{HAZ} in any direction from a weld as shown in fig. 25. The values of b_{HAZ} depend on the welding process and the material thickness, see fig. 26.

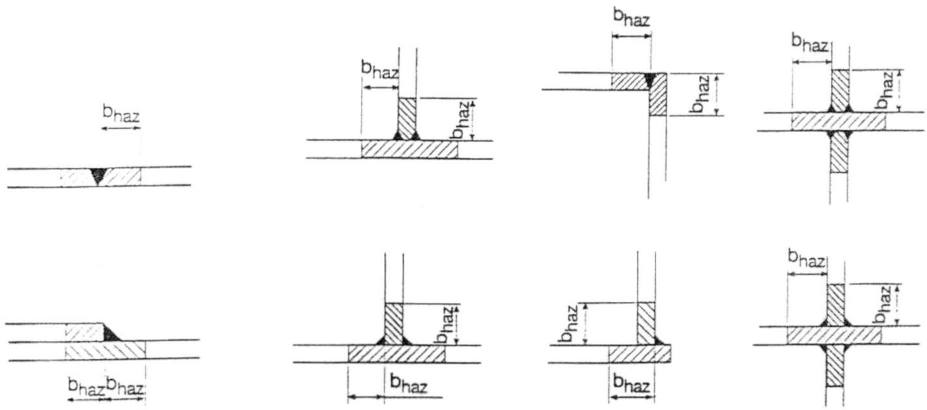

Figure 25: Extent of heat-affected zones (HAZ)

	b_{HAZ} in mm	
Thickness *t* in mm	MIG	TIG
$0 < t \leq 6$	20	30
$6 < t \leq 12$	30	-
$12 < t \leq 25$	35	-
$t > 25$	40	-

Figure 26: Distance b_{HAZ} in mm

In case of temperature build-up higher than 60°C when multi-pass welds are laid down, higher values of b_{HAZ} have to applied.

It is noted that both severity and extent differ for TIG and MIG welding. For TIG welding a higher extent (larger HAZ area) and more severe softening occurs due to the higher heat-input. For the limiting strength of weld metal (f_w) the values according to fig. 27 may be used, provided that the combinations of parent metal and filler metal as given in fig. 23, are applied. For the limiting strength of weld metal it is distinguished according to the filler metal used. The choice of filler metal can have a significant influence on the strength of the weld metal. With the design of welded connections in aluminium structures it is noted that – except for the strength reduction in the HAZ – also the strength of the weld metal usually appears to be lower than the strength of the parent metal. It is recommended to take that into account with the design of members as well.

Limiting strength	Filler metal	Alloy								
		3103	5052	5083	5454	6060	6005A	6061	6082	7020
f_w [N/mm^2]	5356		170	240	220	160	180	190	210	260
	4043A	95	-	-	-	150	160	170	190	210

For alloy 5754 the values of alloy 5454 and for alloy 6063 the values of alloy 6060 can be used. If filler metals 5056A, 5556A or 5183 are used then the values for 5356 have to be applied. If filler metals 4047A or 3103 are used then the values of 4043A have to be applied. For different combinations of alloys the lowest limiting strength of the weld metal has to be used.

Figure 27: Limiting strength values of weld metal f_w

2.2.3. Design of welds

Using butt welds in aluminium structures for strength members full penetration butt welds shall be applied. In that case the effective thickness of the butt weld equals the thickness of the connected members. With different member thickness the smallest one shall be taken into account.

Reinforcement or undercut of the weld within the limits as specified should be neglected for the design.

The effective length of the weld shall be taken as equal to the weld length when run-on and run-off procedures are sues. Otherwise the total length shall be reduced by twice the thickness *t*.

Partial penetration butt welds shall only be used for strength members when verified by testing that no serious weld defects are apparent. In other cases partial penetration butt welds shall be only applied with a higher γ_M value (see also 2.2.1) because of the high susceptibility for weld defects of partial penetration butt welds.

For partial penetration butt welds effective lengths is referred to full penetration butt welds. As far as the effective throat section t_e is considered fig. 28 is governing the design.

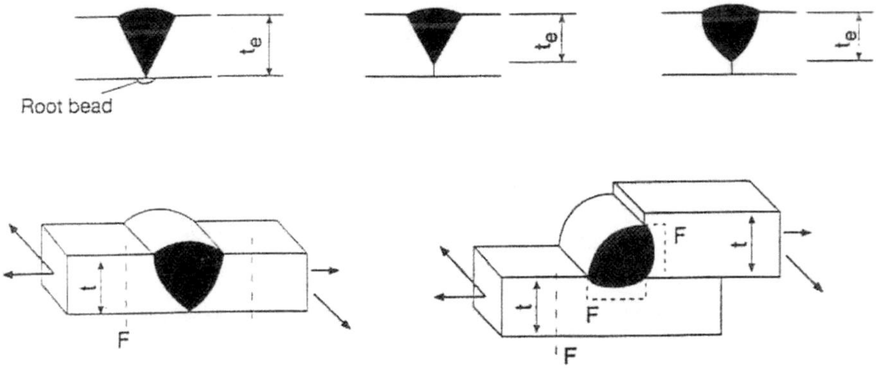

Figure 28: Effective throat thickness t_e (butt welds)
Failure planes (F) adjacent to a weld

As far as the design of butt-welded joints is concerned, the following holds: the stresses occurring in the weld should be related to the design strength of the weld metal. For example a tensile stress, perpendicular to the weld axis, has to satisfy:

$\sigma_\perp \leq \dfrac{f_w}{\gamma_M}$ where f_w = limiting strength weld metal (see fig. 27)

γ_M = partial safety factor; usually $\gamma_M = 1{,}25$

For fillet welds the throat section (see fig. 29) shall be taken as the governing section, since the actual strength of a fillet weld is well approximated by considering the throat section and the forces acting on that section. The throat section is determined by the effective length and the effective throat thickness of a fillet weld.

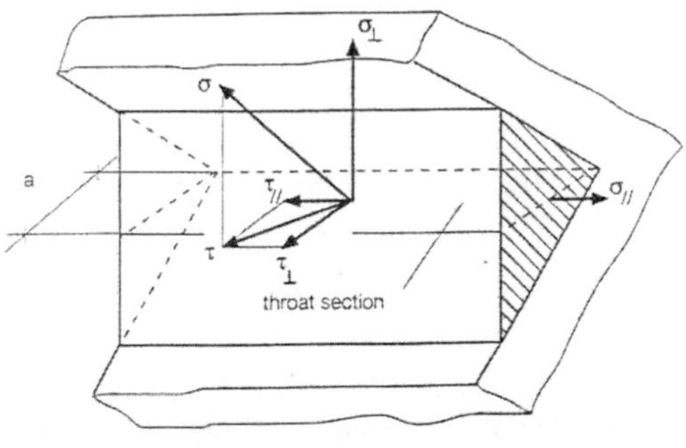

Figure 29: Stresses $\sigma_\perp, \tau_\perp and \tau_{//}$ acting on the throat section of a fillet weld

The effective length shall be taken as the total length of a fillet weld when:

- Run-on/run-off procedures are used.
- The length of the fillet weld is at least 8 times the throat thickness.
- The length of the fillet weld does not exceed 100 times the throat thickness with non-uniform stress distributions along the length of the weld. With uniform stress distributions a length restriction does not apply.

If the above requirements are not fulfilled an effective length has to be taken into account, see Eurocode 9. With non-uniform stress distributions and thin long welds the deformation capacity at the ends of the welds may be exhausted before the middle part of the weld yields; thus the connection fails by a kind of zipper-effect.

For the effective throat section of a fillet weld the height a of the largest triangle which can be inscribed within the weld, has to be applied, see fig. 30. When the qualification specimens show a consistent, positive root penetration, the throat thickness may be increased by 20 % with a maximum of 2 mm.

With deep penetration fillet welds the additional throat thickness may be taken into account provided that consistent penetration has been proved by test.

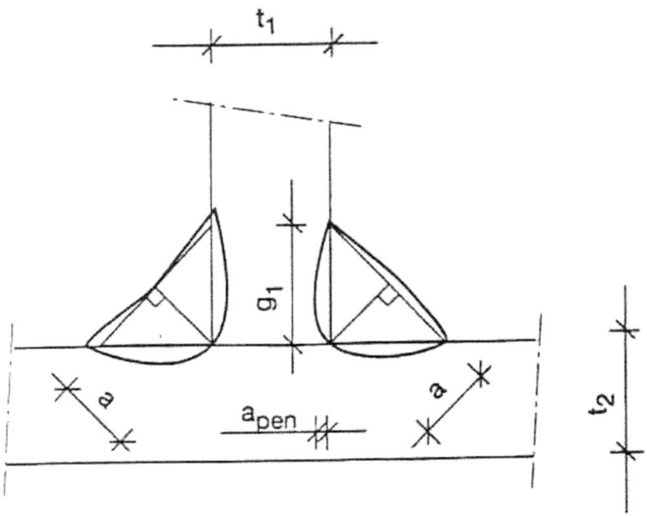

Figure 30: Effective throat thickness a; positive root penetration a_{pen}.

As far as the design of fillet welds is concerned, the forces acting on fillet welded joints, should be resolved into stress components with respect to the throat section, see fig. 29.

This may result in a normal stress $\sigma_\perp$ perpendicular to the throat section, a shear stress $\tau_\perp$ acting on the throat section perpendicular to the weld axis, and a shear stress $\tau_{//}$ acting on the throat section parallel to the weld axis. A normal stress $\sigma_{//}$ acting along the weld axis does not have to be considered.

For the design the above stress components shall be combined to a comparison stress σ_c as follows:

$$\sigma_c = \beta\sqrt{\sigma_\perp^2 + 3(\tau_\perp^2 + \tau_{//}^2)}$$ where β = 1,0 for the alloys of fig. 27.

For the design the following shall be applied:

$$\sigma_c \leq \frac{f_w}{\gamma_M}$$ where f_w = limiting strength weld metal (fig. 27)

γ_M = partial safety factor; usually γ_M = 1,25

$$\sigma_\perp \leq \frac{f_w}{\gamma_M}$$

For the design of fillet welds the above β-formula can be used similarly to steel structures. Contrarily to steel where also simplified formulas are applied, for aluminium the β-formula is preferred since it allows for:

- A better approximation of the actual strength of a fillet weld which is more critical than with steel structures, where the weld strength exceeds the parent metal strength.
- A consistent approach towards the design of a fillet weld since the β-formula accounts for the direction of the forces acting on the weld. The difference in strength between transverse and longitudinal fillet welds is taken into account automatically.
- Simple calculations with simple joints. Often only one or two stress components govern the design and subsequently the β-formula reduces to a simple design formula.

2.2.4. Design strength HAZ

As stated earlier in 2.2.1. consideration should be given to the strength of the heat-affected zone (HAZ). Similarly to the design of welds the following applies:

- The forces acting on the failure plane F (see fig. 28) result in respective stress components.
- The design strength of the HAZ which is arrived at as follows:

$$\sigma_{HAZ} \leq \frac{f_{a,HAZ}}{\gamma_M}$$ for forces perpendicular to the failure plane

$$\tau_{HAZ} \leq \frac{f_{v,HAZ}}{\gamma_M}$$ for shear forces in the failure plane.

where:

$f_{a,HAZ} = \rho_{HAZ} f_u$

$$f_{v,HAZ} = \rho_{HAZ} \frac{f_u}{\sqrt{3}}$$

ρ_{HAZ} = softening factor (see 3.2.2)

γ_M = partial safety factor (see 3.2.1)

f_u = ultimate strength parent metal

It is noted that the above design guidance about the heat-affected zones is dealing with welded connections as such. With the design of members the effect of heat-input by welding shall be taken into account in a proper way.

2.2.5. Design of connections with combined welds

The design of connections with combined welds can be reduced to the design of the constituent welds by using one of the two following methods:

1. The loads acting on the joint are distributed to the respective welds, which are most suited to carry them (plastic design approach).
2. The welds are designed for the stresses occurring in the parent metal of the different parts of the joint (linear-elastic design approach).

With method 1 it has to be checked whether the welds possess sufficient deformation capacity to allow for such a simplified load distribution. Besides, the assumed loads in the welds should not give rise to overloading of the connected members.

With method 2 the above problem do not exist, but may be difficult to determine the stresses in the parent metal of the different parts of the joint.

Assuming a simplified load distribution (method 1) is the most commonly applied method. Since the actual distribution of loads between the welds is highly indeterminate, such assumptions have been found to be an acceptable and satisfactory design practice. However, these assumptions rely on the demonstrated ability of welds to redistribute loads by yielding.

Residual stresses and other stresses not participating in the transfer of loads need not be considered for the design.

2.3. Design of bolted and riveted connections

2.3.1. General

The design guidance given in this chapter applies to:

- Strength members under predominantly static loads;
- Mechanical fasteners in primary structures i.e. bolts and solid rivets, see also chapter 1.2.2.
- Positioning of holes as given in chapter 2.3.2.
- Elastic or plastic design of the connection.
- Non preloaded and preloaded bolts, the latter in slip resistant connections.

2.3.2. Positioning of holes

The positioning of holes for bolts and rivets shall be in conformity with the limits of validity of the rules used to determine the design resistances of the bolts and the rivets. The positioning shall also be such as to prevent corrosion and local buckling as well as to facilitate the installation of the bolts and the nuts.

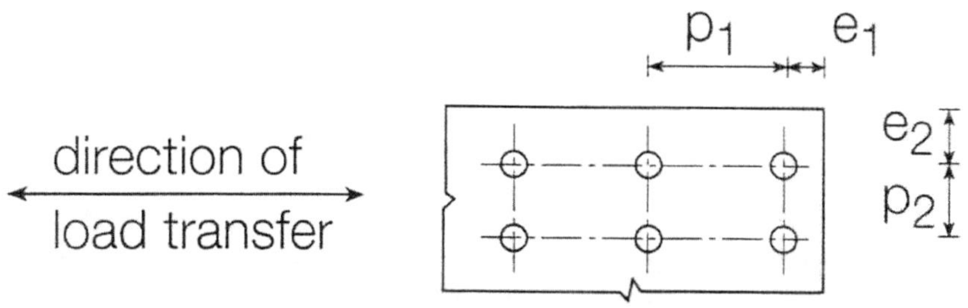

Figure 31: Positioning of holes

To fulfil the above requirements the following distances shall be applied (see also fig. 31):

* end distance e_1	minimum	: 1,2 d
* edge distance e_2	maximum	: 4 t + 40 mm (corrosive environment)
		12 t or 150 mm (non-corrosive environment)
* spacing p_1	minimum	: 2,2 d
	maximum	: 14 t or 200 mm
* spacing p_2	minimum	: 2,4 d
	maximum	: 14 t or 200 mm

2.3.3. Deductions for fasteners holes

The net area of a member (A_{net}) is arrived at by deducting the maximum sum of the sectional areas of the holes in any cross section perpendicular to the member axis, provided the fastener holes are not staggered.

For staggered holes: $A_{net} = A - (2\,dt - s^2t/4p)$
(see fig. 32)

or $A_{net} = A - dt$

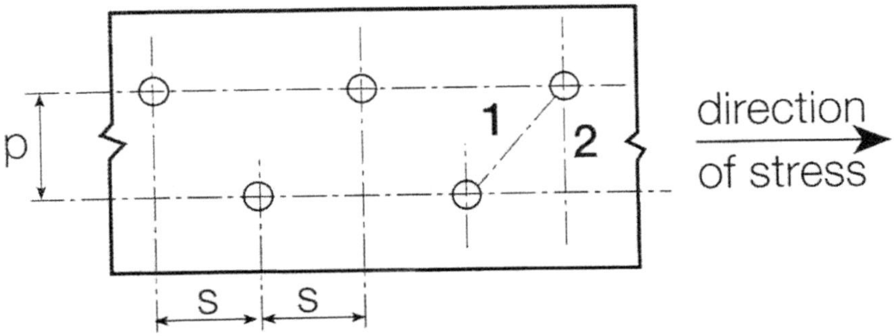

Figure 32: Tension member, staggered holes

The above deductions hold for tension members; for compression members no deductions for fastener holes have to be applied. In case of holes in tension flanges as well as webs in beams specific rules apply. In case of shear usually A_{net} shall be taken into account. At the end of a member block shear should be verified.
In the case of unsymmetrical connected members such as angles specific guidance is given in Eurocode 9.

2.3.4. Categories of bolted connections

Two types of bolted connections can be distinguished:

- shear connections;
- tension connections.

Within shear connections three categories have been defined:

- Category A: bearing type.

In this category (galvanised) steel, stainless steel, or aluminium bolts (or solid rivets) may be used.. The design ultimate shear load shall not exceed the design shear resistance nor the design bearing resistance of the bolt, see also 2.3.6.

- Category B: slip resistant at serviceability limit state.
  In this category preloaded high strength steel bolts with controlled tightening shall be used. The design serviceability shear load should not exceed the design slip resistance (see 2.3.7), while the design ultimate shear load shall not exceed the design shear resistance nor the design bearing resistance, see 2.3.6.

- Category C: slip resistant at ultimate limit state.
  In this category preloaded high strength steel bolts with controlled tightening shall be used. The design ultimate shear load shall not exceed the design slip resistance (see 2.3.7), nor the design bearing resistance obtained from 2.3.6.

In addition at the ultimate limit state the design resistance of the net section at bolt holes shall be taken as:

$$N_{net,Rd} = \frac{A_{net} \cdot f_{0,2}}{\gamma_{M1}}$$

Within tension connections two categories have been defined:

- Category D: non-preloaded bolts.
  In this category low carbon or high strength (galvanised) steel, stainless steel, or aluminium bolts shall be used without preloading. The design ultimate tensile load shall not exceed the design tensile resistance of the bolt, see 2.3.6.

- Category E: preloaded high strength steel bolts.
  In this category preloaded high strength steel bolts with controlled tightening shall be used. The design ultimate tensile load shall not exceed the tensile resistance of the bolt, see 2.3.6. Such connections are often used to improve fatigue resistance; in that case the design tensile load shall should not disrupt the contact of the surfaces of the members connected.

2.3.5. Distribution of forces between fasteners

In the following cases the distribution of forces between fasteners at the ultimate limit state due to the bending moment shall be proportional to the distance from the centre of rotation, and due to the shear load shall be equally divided (see fig. 33a):

- category C slip-resistant connections;
- other shear connections where the design shear resistance of a fastener ($F_{v,Rd}$) is less than the design bearing resistance ($F_{b,Rd}$).

In all other cases the distribution of forces between fasteners due to the bending moment and the shear load at the ultimate limit state may be assumed plastic as for example indicated in fig. 33b.

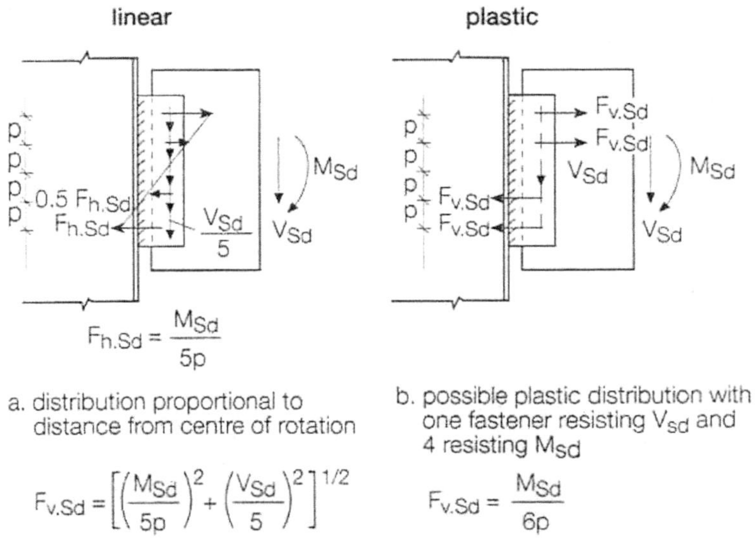

Figure 33: Distribution of forces between fasteners.

In a lap joint the distribution of forces between fasteners may be assumed equal to a maximum length of L = 15d, where d is the nominal diameter of the bolt of rivet. For long joints (L > 15 d) a reduction of the force per fastener shall be applied as follows: the design shear resistance $F_{v,Rd}$ of all the fasteners as specified in 3.3.6 shall be multiplied by a reduction factor β_{Lf}, where:

$$\beta_{Lf} = 1 - \frac{L_j - 15d}{200} \quad \text{but } 0{,}75 \leq \beta_{Lf} \leq 1{,}0$$

L_j : see fig. 34.

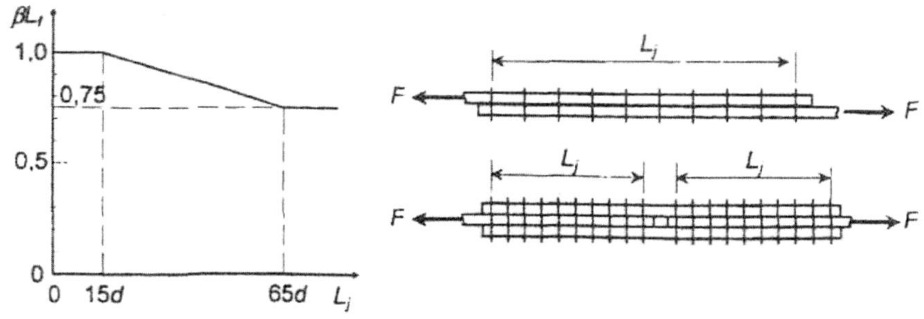

Figure 34: Long lap joints

2.3.6. Design resistance of bolts and rivets

The design resistances given here apply to standard manufactured steel bolts of strength grades 4.6, 5.6, 8.8 and 10.9, or aluminium bolts, or stainless steel bolts. Bolts including nuts and washers shall conform with relevant standards.

At the ultimate limit state the design shear force $F_{v,Ed}$ on a bolt shall not exceed the lesser of:

- the design shear resistance of the bolt $F_{v,Rd}$;
- the design bearing resistance of the bolt $F_{b,Rd}$.

At the ultimate limit state the design tensile force $F_{t,Ed}$ on a bolt, including any force due to prying action, shall not exceed the design tension resistance of the bolt $F_{t,Rd}$.

Bolts subjected in the ultimate limit state to both shear and tensile forces shall in addition satisfy:

$$\frac{F_{v,Ed}}{F_{v,Rd}} + \frac{F_{t,Ed}}{1{,}4F_{t,Rd}} \leq 1{,}0$$

For the design resistance of the bolts the following holds:

- Shear resistance per shear plane:

$$F_{v,Rd} = \frac{0{,}6 f_{ub} A}{\gamma_{Mb}}$$ for strength grades lower than 10.9.

$$F_{v,Rd} = \frac{0{,}5 f_{ub} A}{\gamma_{Mb}}$$ for strength grade 10.9, stainless steel and aluminium bolts.

- Bearing resistance:

$$F_{b,Rd} = \frac{2{,}5\alpha f_u d\, t}{\gamma_{Mb}}$$

where α is the smallest of $\frac{e_1}{3d_o}; \frac{P_1}{3d_o} - \frac{1}{4}; \frac{f_{ub}}{f_u}$ *or* 1,0

- Tension resistance:

$$F_{t,Rd} = \frac{0{,}9 f_{ub} A_s}{\gamma_{Mb}}$$

where A_s is the stress area of the bolt.

For rivets similar formula apply. However, tension in aluminium rivets is not recommended. Note that the bearing resistance is dependent on the edge distance as well as the spacing of the fasteners.

2.3.7. High strength bolts in slip-resistant connections

Design may be based on calculations for joints where the proof strength of the connected parts is higher than 200 N/mm^2. Otherwise, the strength of joints using high strength steel bolts should be proved by testing.
For the design of slip-resistant connections the following has to be distinguished.
In the ultimate limit state slip is not allowed (see also 2.3.4, Category C). Thus, the design ultimate shear load shall not exceed:

- the design slip resistance $F_{s,Rd}$;
- the design bearing resistance $F_{b,Rd}$ (see 2.3.6).

In the serviceability limit state slip is not allowed (see also 2.3.4, Category B). Thus, the design serviceability shear load shall not exceed the design slip resistance $F_{s,Rd}$; and the design ultimate shear load shall not exceed the design shear resistance $F_{v,Rd}$ nor the design bearing resistance $F_{b,Rd}$ (see 2.3.6).
In addition to the above at the ultimate limit state the design resistance of the net section at bolt holes shall be taken as:

$$N_{net,Rd} = \frac{A_{net} \cdot f_{0,2}}{\gamma_{M1}}$$

The design slip resistance of a preloaded high-strength bolt shall be taken as:

$$F_{s,Rd} = \frac{n\mu}{\gamma_{Ms}} F_{p,cd}$$ for shear connections

$$F_{s,Rd} = \frac{\mu}{\gamma_{Ms}} (F_{p,cd} - 0{,}8\, F_{t,Ed})$$ for shear connections where the bolts are subjected

to an additional tensile force $F_{t,Ed}$.

Where:
$F_{p,Cd}$ is the design preloading force.
μ is the slip factor.
n is the number of friction interfaces
γ_{Ms} is the partial safety factor for slip resistance and shall be taken as:
γ_{Ms} = 1,25 for the ultimate limit state.
γ_{Ms} = 1,10 for the serviceability limit state.

Notice that slotted or oversized holes shall not be used. The usual hole clearances for bolts apply.
The design preloading force $F_{p,Cd}$ for high strength bolts with controlled tightening shall be taken as:

$$F_{p,Cd} = 0{,}7\, f_{ub}\, A_s$$

Tightening procedures shall be such that the level of the design preloading force is arrived at.
The design value of the slip factor μ is dependent on the surface treatment applied. The value of μ for lightly shot blasted surfaces without surface protection treatments shall be taken as
$\mu = 0{,}3$.
In case of other surface treatments or when higher values of μ are needed, a sufficient number of tests on representative test specimens has to be carried out.

2.4. Design of adhesive bonded connections

2.4.1. General

The design guidance given in this chapter applies under the condition that:
- Strength members are subjected to predominantly static loads;

- Adhesive bonding shall not be used for main structural joints unless considerable testing has established its validity.
- The joint design is such that only shear forces have to be transmitted (see 2.4.2).
- Appropriate adhesives are used (see 2.4.3)
- The surface preparation procedures before bonding do meet the specifications as required by the application (see 2.4.3).

2.4.2. Joint design

With the design of adhesive bonded joints shear forces should be looked after; tensile forces (in particular peeling or other forces tending to open the joint) should be avoided or should be transmitted by complementary structural means, see fig. 35.

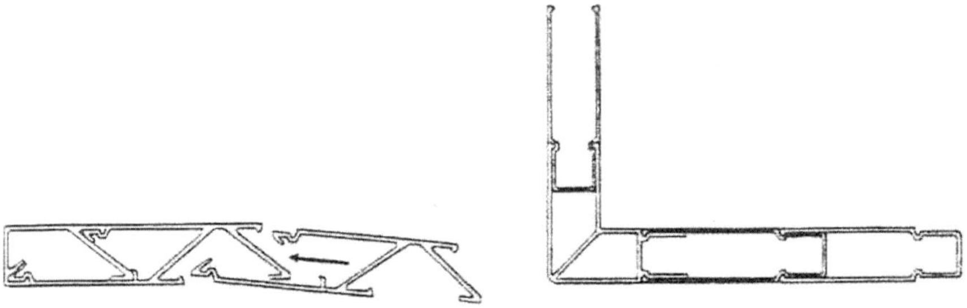

Figure 35: Extruded members; in-plane tensile forces transmitted by snapping parts; out of plane shear loading transmitted by adhesive bonding.

Loads should be carried over as large an area as possible. Increasing the width of joints usually increases the strength pro rata. Increasing the length is beneficial only for short overlaps. Longer overlaps do result in more severe stress concentrations in particular at the ends of the laps, see fig. 36.

Furthermore, uniform distribution of stresses and sufficient deformation capacity to enable a ductile type of failure of the component are to be strived for. In case the design strength of the joint exceeds the yield strength of the connected member sufficient deformation capacity is ensured.

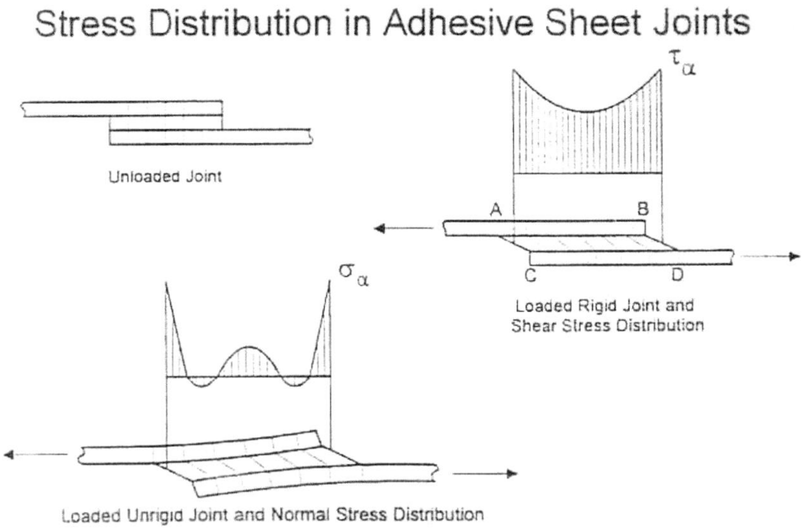

Figure 36: Stress distribution in adhesive sheet joints.

2.4.3. Limiting strength adhesives

As far as the mechanical properties are concerned high strength adhesives should be used for structural applications, see fig. 37. However, also the toughness should be sufficient to overcome stress/strain concentrations and to enable a ductile type of failure.

The influence of the adhesives E-modulus on the strength and stiffness of the joint is not significant. But, low E-modulus adhesives are more sensitive to creep. Concerning other adhesives properties it is noted that in the temperature range –20° C up to +60° C the adhesive properties do not vary too much as long as the glass transition temperature is not exceeded.

Pre-treatments of the surfaces to be bonded have to be chosen such that the bonded joint meets the design requirements during service life of the structure. An overview of pre-treatments from simple degreasing up to very sophisticated chemical pre-treatments is given in 1.2.3.

For structural applications the limiting shear strength values of adhesives $f_{v,adh}$ as given in fig. 37, may be used.

Adhesive types	$f_{v,adh}$ N/mm^2
1-component, heat cured, modified epoxide	35
2-components, cold cured, modified epoxide	25
2-components, cold cured, modified acrylic	20

Figure 37: Limited shear strength values of adhesives

The adhesive types as mentioned in fig. 37 may be used under the conditions as given earlier in 2.4.1 and 2.4.2. The values given are based on results of extensive research. However, it is allowed to use higher shear strength values than the ones given in fig. 37 provided that adequate tests are carried out, see 2.4.5.

2.4.4. Design shear strength adhesives

The design shear strength shall be taken as:

$$\tau \leq \frac{f_{v,adh}}{\gamma_{M,adh}}$$

where:

τ is shear stress in the adhesive layer
$f_{v,adh}$ is limiting shear strength value adhesive, see fig. 37.
$\gamma_{M,adh}$ is the material factor for adhesive bonded and shall be taken as: $\gamma_{M,adh} = 3.0$

The above high value of $\gamma_{M,adh}$ has to be used since:
- the design of the joint is based on ultimate shear strength of the adhesive;
- the scatter in adhesive strength can be considerable;
- the experience with adhesive bonded joints is small;
- the mechanical properties of adhesive bonded joints suffer from ageing.

2.4.5. Design assisted by testing

Higher limiting shear strength values of adhesives than given in fig. 37 may be used when thick-adherend shear tests are carried out, see fig. 38. The results of these tests have to be evaluated as to arrive at reliable strength value of the applied adhesive.

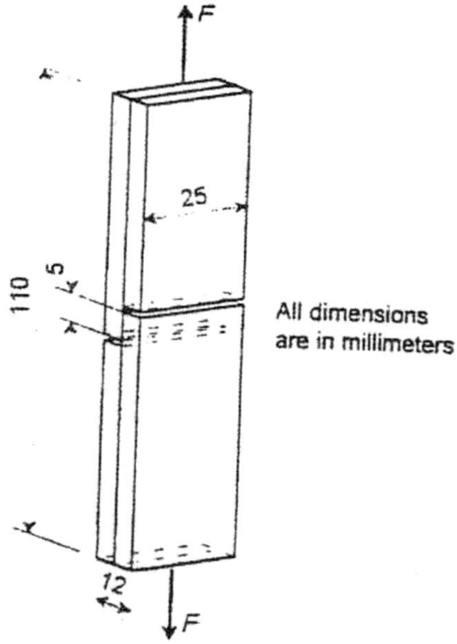

Figure 38: Thick adherend shear test specimen

The strength of adhesive bonded joints or members may also be determined by testing. Sample joints should be made at full scale, using the same manufacturing procedure as for production joints. These sample joints should be tested with similar joint construction and loading to that occurring in the actual structure.

2.5. Hybrid connections

When different forms of fasteners are used to carry a shear load or when welding and fasteners are used in combination (see also 1.2.4 and 1.3.4) the designer shall verify that they act together.

In general the degree of collaboration may be evaluated through a consideration of the load-displacement curves of the particular connection with individual kind of joining, or also by adequate tests of the complete hybrid connection.

In particular normal bolts with hole clearance shall not collaborate with welding.

Preloaded high-strength bolts in connections designed as slip-resistant at the ultimate limit state (Category C) may be assumed to share load with welds, provided that the final tightening of the bolts is carried out after the welding is completed. The total design load should be given by the appropriate design load of each fastener with its corresponding γ_M value.

If not demonstrated that different fasteners act together only one type of fastener may be accounted for with the design of the joint.

3. DESIGN EXAMPLES OF STRUCTURAL APPLICATIONS

In the current practice still many concepts for aluminium structures are based on steel solutions. However, most of the time this is not the best solution for an aluminium structure.
In order to take advantage of the beneficial properties of aluminium it is important to 'think in aluminium' from the beginning of the design process. This attitude will result in aluminium structures better optimised with respect to structural design, material selection, fabrication, transport, erection and maintenance. Of course the connections play an important role in this process. When the connections are in mind from the beginning of the design process, the structure will be better optimised.
In this chapter the design of some typical aluminium structures is described, where the 'thinking in aluminium' has proven to work out very well in practice. The design of the structure is described generally, while the specific design of the connections is described in more detail.

3.1. Traffic gantries

The Dutch Ministry of Transport and Water Management has initiated the development of traffic gantries focused on durable materials, low maintenance costs and aesthetic aspects. Alcoa Masten, a subsidiary of Alcoa USA, has taken up this opportunity in designing and manufacturing aluminium traffic gantries. Four aluminium gantries have already been realised in the south of the Netherlands (see figure 39).

Figure 39: Aluminium traffic gantry

3.1.1. Design concept

The traffic gantries are constructed by using a triangular space frame supported by two X-shaped column frames. Slenderness and transparency were objectives in the architectonic design.

The gantries are designed according to the Dutch code for aluminium structures NEN 6710, which is similar to Eurocode 9. The predominant load is the wind load. This is reflected in the overall size of the space frame structure: the horizontal dimension (depth) of the triangular cross-section is larger than its vertical dimension (height).

Apart from static requirements the traffic gantries are also evaluated for wind induced fatigue. This has had consequences for the design of the cross sections as well as the design of the connections.

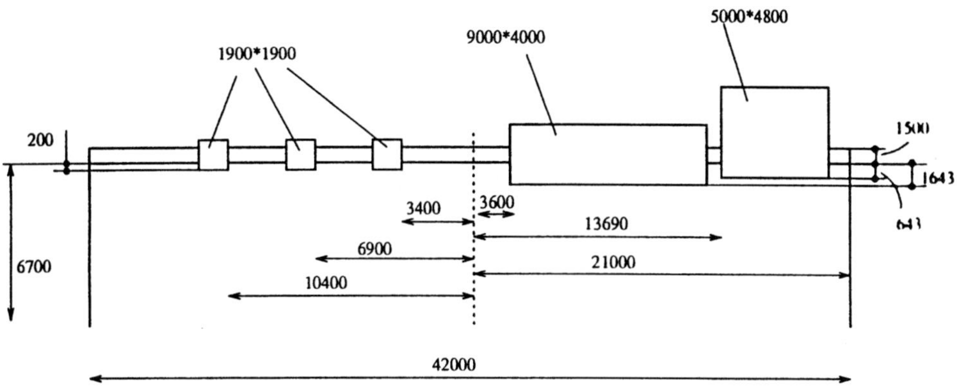

Figure 40: Schematisation and positions of boarding

3.1.2. Triangular space frame

The triangular space frame has two upper chords and one base chord, which are interconnected by braces. Both chords and braces are fabricated by means of extrusion, allowing geometry and cross sections to be optimised with respect to load transfer and connections. The joints between the braces and chords are established by welding.

The upper chords and the base chord both have two flat surfaces to facilitate an easy welding procedure for these joints. The small external bulges and the flat surfaces of the braces are designed for an easy positioning, which also benefits the welding of the braces to the chords.

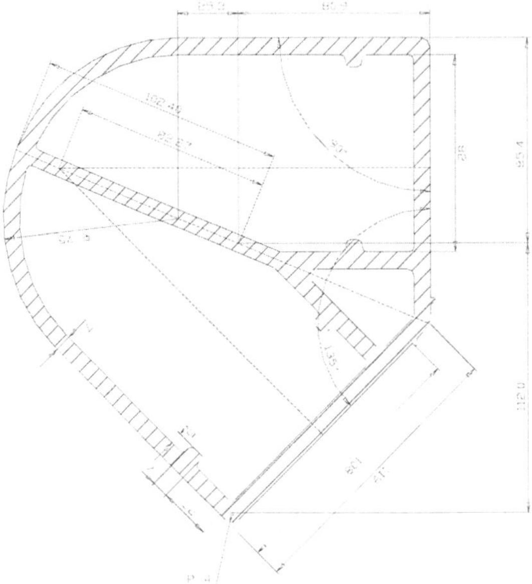

Figure 41: Cross section upper chords

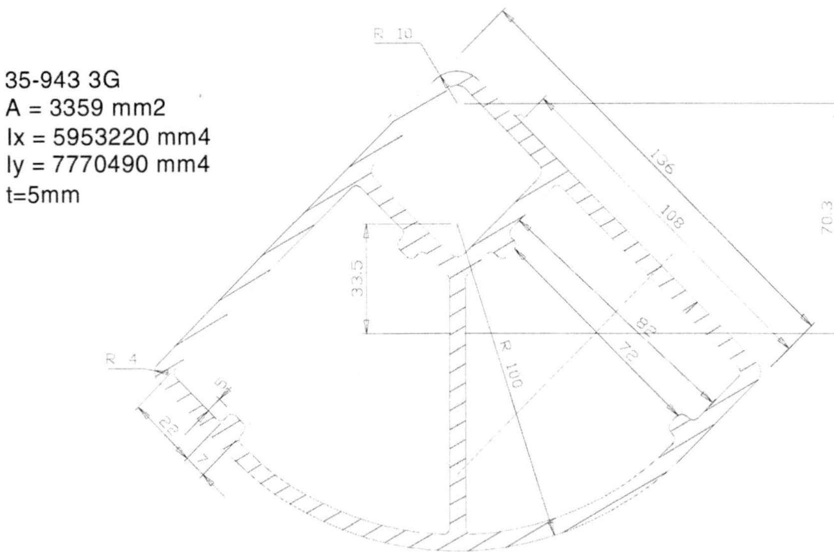

Figure 42: Cross section base chord

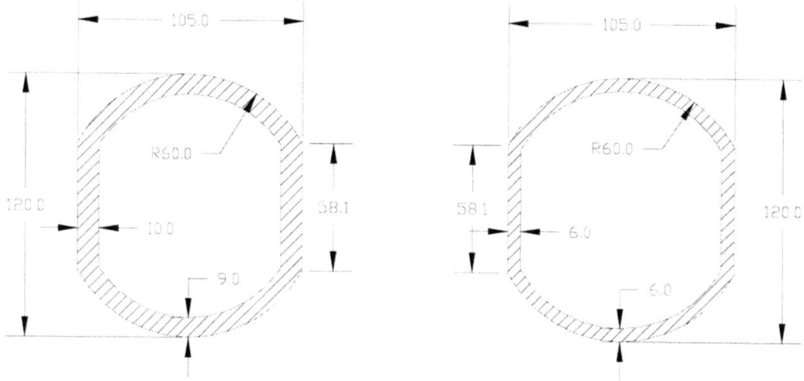

Figure 43: Cross section braces

The internal bridges in the cross-section of the chord beams are designed for a reduction of the stress concentrations along the perimeter of the welded connection. Relatively low 'hot spot' stresses are important, especially when fatigue is governing for the design, which is the case for the traffic gantries. A detailed FEM analysis on the stress concentration of the multiplanar joint has shown that the internal bridge has indeed an important reducing influence on the stress concentrations. Chord sections without these internal bridges have shown up to 8 times higher SCF-values than the same sections with internal bridges.

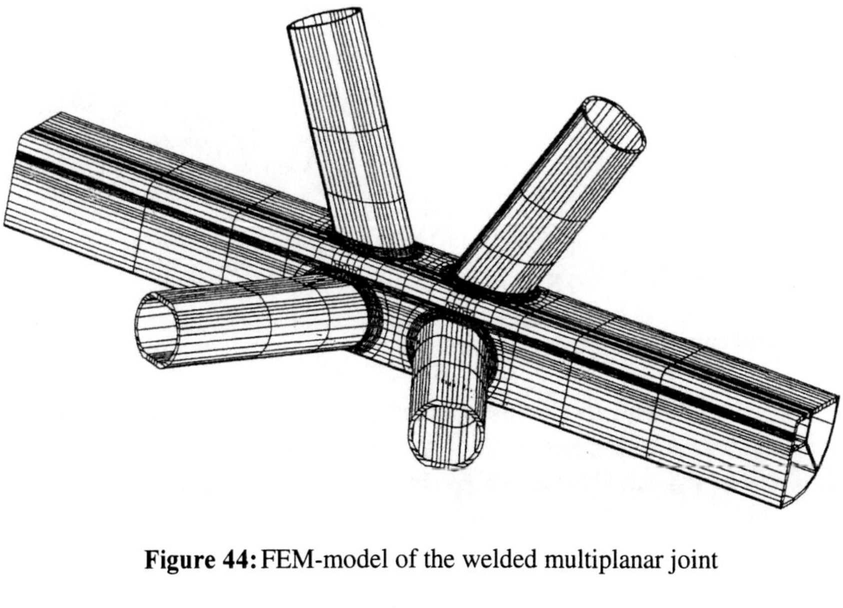

Figure 44: FEM-model of the welded multiplanar joint

The girder is prepared at the factory in sections up to 12 meters. The maximum span for which the gantry has been designed is 42 meters. So, depending on the actual length to be used, the sections are connected at the factory to the appropriate length and transported to the site. The internal legs in the aluminium chord sections are designed for the easy manufacturing of this section connection, which is realised by using two internal stainless steel strips for each connection between two chords. The stainless steel strips (width 80 mm, thickness 20 mm for the upper chords and 12 mm for the lower chord) are connected to the aluminium sections using M16 bolts. For the bolt holes a practical diameter of 18 mm is used, which is necessary to avoid problems during the assembly.

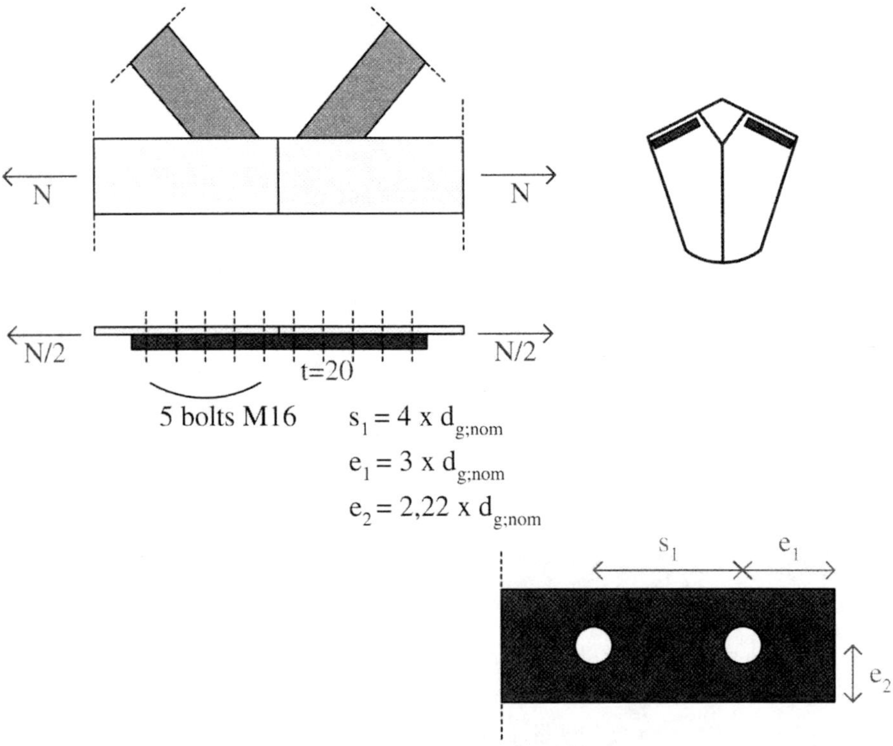

Figure 45: Connection between girder sections

3.1.3. X-shaped columns

In the first designs the cross-section of the columns consisted of four parts (of which two and two are matching), to be welded to one profile in the factory (see figure 46). The columns of the four gantries already realised (see figure 39) have been constructed in this way.

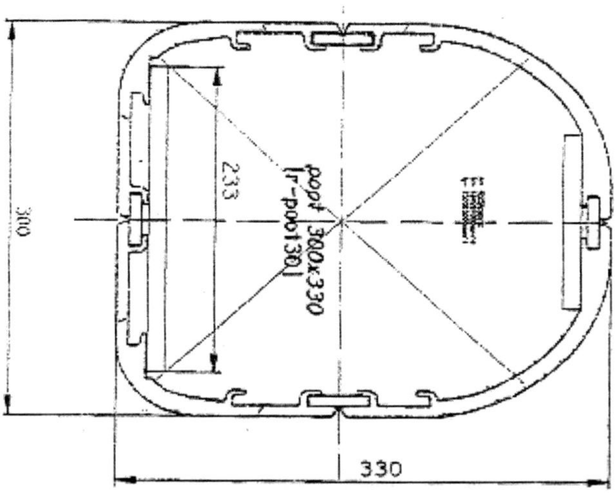

Figure 46: Cross section of columns assembled by four parts

However, the fabrication of this column type seems to be very laborous and expensive, such that the aluminium gantry is not in good competition with the current steel gantries. In newer designs the column section contains only one extrusion, where the cross section for the column profile above the 'waist' is different from the column profile below the 'waist'.

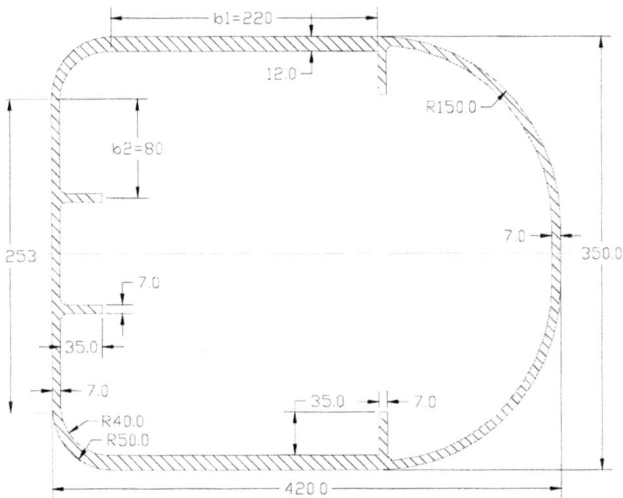

Figure 47: Cross-section column profiles below the waist

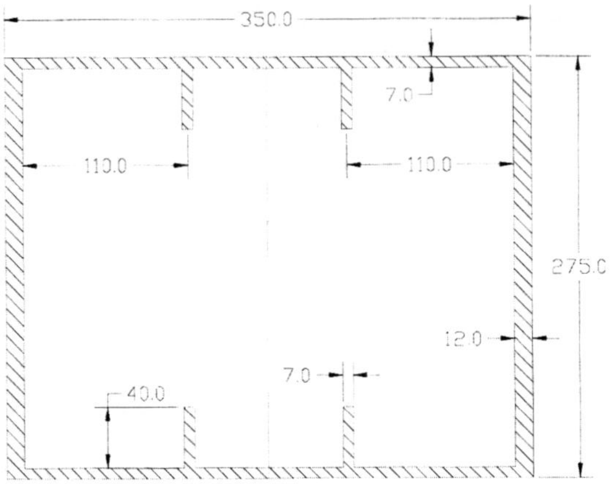

Figure 48: Cross-section column profiles above the waist

For the assembly of the X-shape a special connection has been designed, which is able to transfer the high bending moments, horizontal shear forces and vertical axial forces. The 'waist' that forms the X-shape is realised by a preloaded slip resistant bolted connection between the two upper column sections and the two lower column sections. The ends of these profiles are provided with bolt holes previously. Stainless steel plates, which also have the X-shape, are placed inside the four column profiles to form the connection. These stainless steel plates are also provided with bolt holes as well as with pre-welded stainless steel nuts. In the factory one single connection is fixed by screwing, such that the X-shaped column can be transported to site. The actual connection is realised by preloaded friction grip bolts.

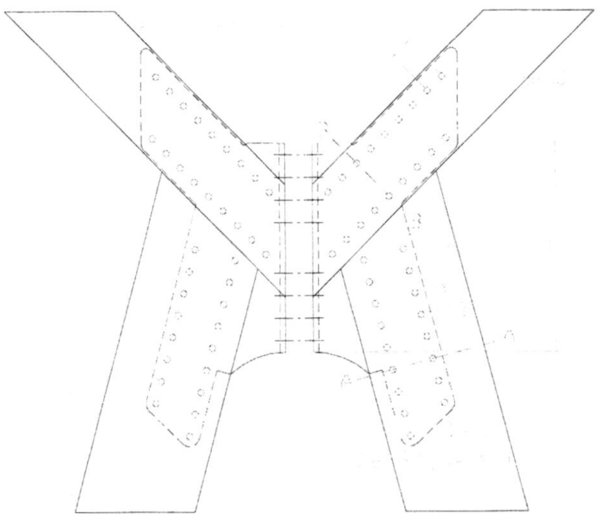

Figure 49: Preloaded slip-resistant connection at the waist

The connection at the column foot is a standard moment connection using foot plates and M39 bolts for each connection.

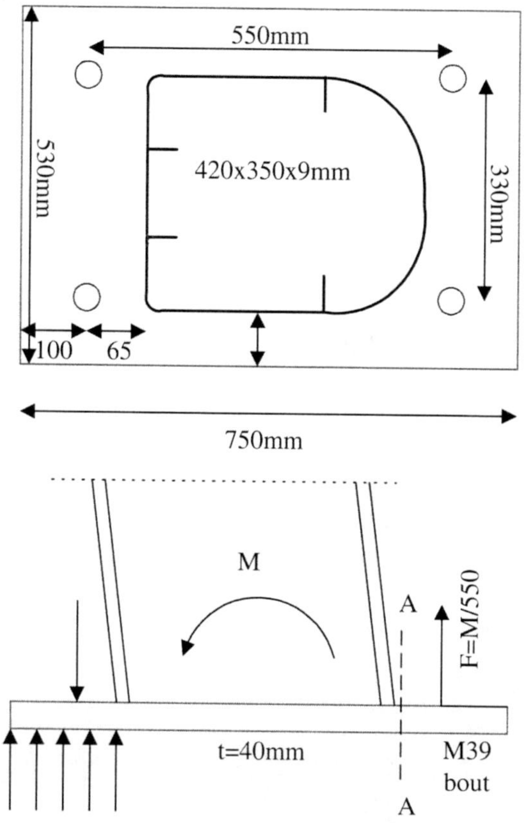

Figure 50: Connection at column foot

When the girder is attached to the columns the traffic flow has to be stopped for some time. Special provisions, i.e. centre cams for the easy positioning of the girder, have been taken to reduce this conjugation of the traffic. In practice the assembly of girder and columns has taken place in a time span of only 10 minutes.

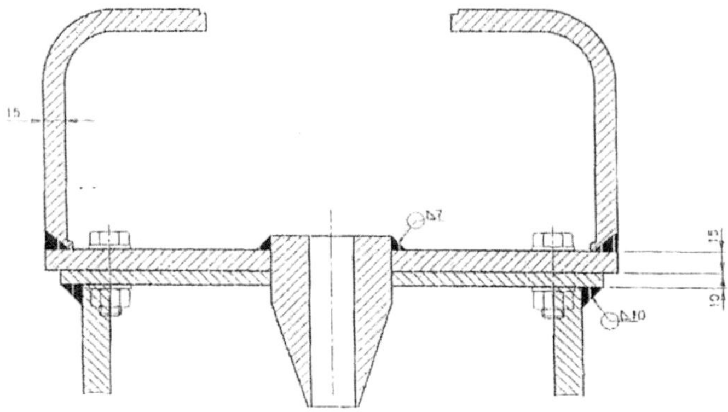

Figure 51: Centre cams

3.2. Office building

The new office of the Dutch Aluminium Centre is settled in an all aluminium office building, which has recently been built in Houten in the Netherlands. The design concept of this building is rather exceptional: it is a one-storey building of about 1000 m^2 supported by more than 350 slender aluminium columns. The architect Micha de Haas was inspired by the rural landscape of the environment of Houten and called his design the 'aluminium forest' (see figure 52).

Figure 52: View of the 'aluminium forest'

3.2.1. Structural system of the building

The office layer has a plan of 31.2 meter by 36.0 meter and a height of 5 meter. The supporting columns have a length of 6 meter and range in diameter from 90 mm up to 210 mm, increasing in steps of 30 mm. The position of the columns is rather irregular, but such that criteria for strength, stability and serviceability are satisfied. This is realised mainly by:

- rigid connection of the larger columns (diameter 210 mm) at the top as well as at the bottom;
- slight inclination (1:10) of these larger columns, which simulated a bracing system without destroying the 'aluminium forest' concept;
- concentration of the larger columns in three clusters, such that horizontal and rotational stability is satisfied.

The floor plan is constructed by primary beams at distances of 2400 mm. To include the supporting columns that are not located at this 2400 mm raster secundary beams are used lying under the primary beams. The aluminium beam section is similar to a steel HE160 beam. Where the span is too large two aluminium beams are placed above each other. These beams are connected by bolts which results in one assembled beam.

Figure 53: Structural floor plan

Figure 54: Floor beams and supporting columns during erection

Figure 55: View of the bolted beam-column connection

For the roof structure the beams are assembled using two standard aluminium Π-profiles (150 mm width and 100 mm height) in combination with RHS profiles with varying heights. This results in assembled girders of 220 mm, 250 mm or 350 mm height, or in a lattice girder of 500 mm height (see figure 56). The assembly of the Π and RHS profiles is realised by welds in case of the girders and by bolts in case of the lattice girders. Where the span or the loading is too large for one lattice girder two of these girders are placed and coupled next to each other.

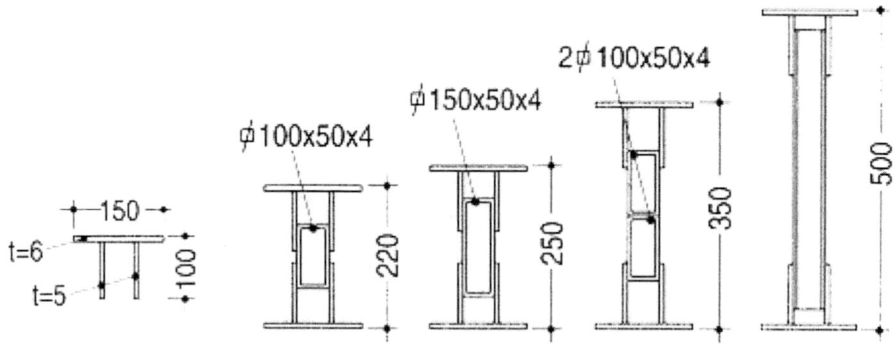

Figure 56: Cross sections of the assembled beams in the roof structure

Figure 57: View of single and double lattice girders

3.2.2. Bracing system

The bracing system, which is necessary for horizontal stability, is realised by slight inclination of the larger 210 mm columns. These columns have rigid connections at the bottom and at the top, which results in a satisfying horizontal deformation and a satisfying natural frequency.
It was required that the connection between the two inclined columns of the bracing system is located above the ceiling. Therefore the gusset plates, which are welded to the column top,

have 'ears' to realise a bolted connection at the top of the inclined columns. The connection beneath the ceiling is realised by an in-welded bush.

Figure 58: View of the simulated bracing system

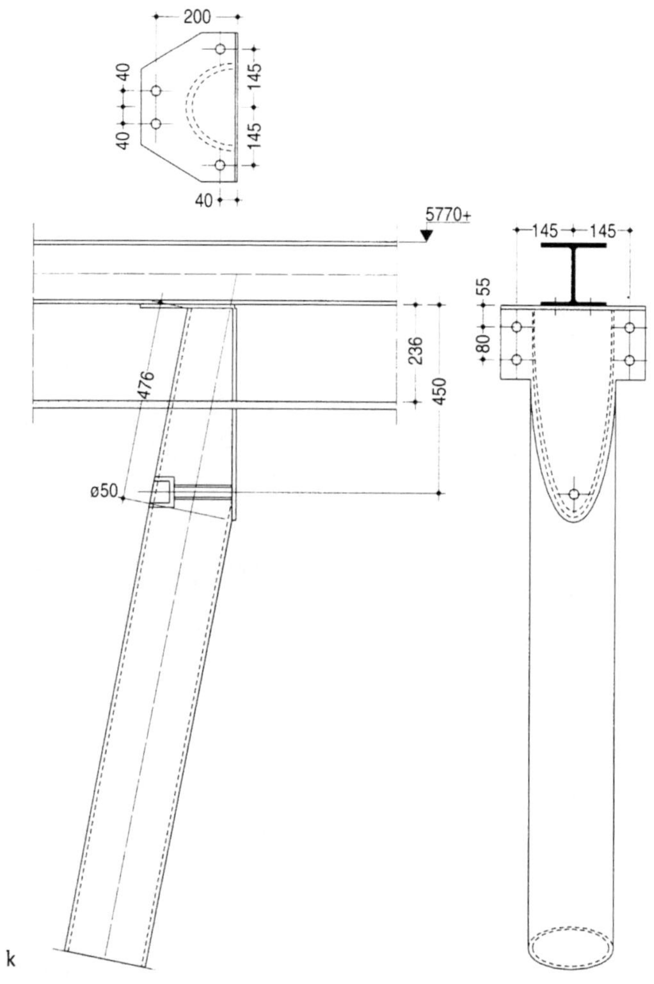

Figure 59: Beam to column connection in case of inclined columns

3.2.3. Exposition room

The exposition room is settled inside the office building and has a floor plan of 14.4 meter by 12.0 meter. The roof structure of this exposition room consists of two whole and two half triangular aluminium truss girders spanning 14.4 meter. This aluminium structure can be considered as very innovative because of the special requirements and development of:

- an esthetical 'streamline' design for the connection between elements and nodes,
- an industrial method for the production of the girders,
- an industrial method for the production of variable aluminium casted node elements in low serie numbers,

- an adhesive bonded connection suitable for the primary load transfer between nodes and elements.

The requirement for a 'streamline' detail has been governing for the truss design. The lower part of the trusses is only 3.5 meter above floor level, which involves that elements and details have to be designed for architectonic qualities as well as structural properties. An adhesive bonded connection between nodes and elements fulfils both requirements, because then no visible connectors have to be applied.

Figure 60: Triangular truss girders

An industrial production of the structure is met by using single truss girders as a basic structure. These single truss girders are located at the sides of the roof structure. The triangular girders, which are situated in the middle part of the roof structure, are constructed by two of these single truss girders. Together with the roof girders, which are located 50 mm above the nodes, the truss girders create a stable triangular system. The two bottom chords of the triangular truss are horizontally connected at the nodes. For esthetical reasons the upper chords are placed along a curved line.

Figure 61: Shop fabrication of the triangular truss girder

For the casted nodes a production method with 'lost foam' has been developed. At first a simple polystyrene basic part, representing the connection of one element to the node, has been made. This has been done by producing one light aluminium expansion mould with loose screw thread inserts. This has resulted in polystyrene basic parts with left and right screw threads.

After this the polystyrene nodes have been assembled by manually cutting two or three or four polystyrene basic parts under the right angle and connecting these parts by adhesive bonding. Then the nodes are provided with a coating and bedded in sand by vibration.

After this the nodes are casted by fluid aluminium. During the casting the polystyrene mould, which contains 99% air, is melting and the residue disappears through the coating into the sand. The casted product has the typical polystyrene surface, but the inside is fully aluminium. Finally, the nodes were heat-treated for improvement of the strength.

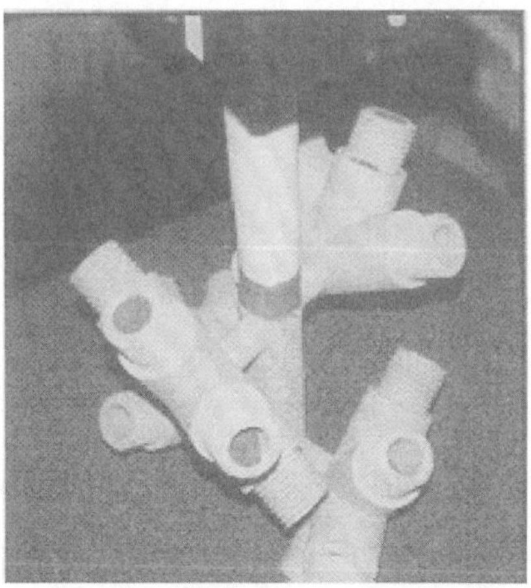

Figure 62: Fabrication of nodes

The connection between elements and nodes is realised by adhesive bonding. However, a direct connection between nodes and elements is not possible due to differing tolerances. Adhesive bonded connections require very strict tolerances, and the product tolerances of the nodes are not small enough due to the manual assembly of the polystyrene moulds. This problem is tackled by taking single sleeves which are adhesively bonded in the tubes. At the ends these sleeves have right and left screw threads which fit to the screw threads of the nodes. The adhesive chosen for this application (two component epoxy) has already proven its quality in airplanes. Before the adhesive is applied a special cleaning treatment is used.

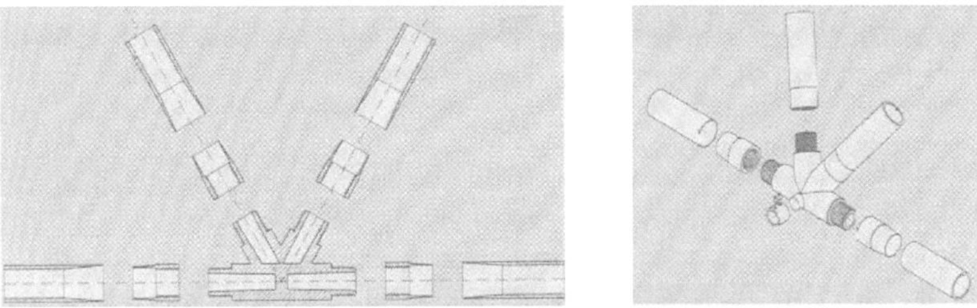

Figure 63: Assembly of nodes and elements

The assembly of the truss girders, including the transverse roof girders, took only one day.

Figure 64: One day erection at location

3.3. Bridges

Since 1995 world wide initiatives have been taken in the design of aluminium bridges. In Europe, particularly in Norway and Sweden, approximately 30 bridges have been built over the

last couple of years. This revival of aluminium bridges is mainly due to the market-pull to lightweight and sustainable bridges. A proper design of the connections in aluminium bridges is important, because it may simplify the assembly and erection, which benefits the initial costs of aluminium bridges.
Also in the Netherlands several new initiatives have been taken in the design and construction of aluminium bridges. Three of these initiatives are described in this subchapter.

3.3.1. Leidschenveen bridges

During 1997 and 1998 an international contest has been held for the building of 58 pedestrian and traffic bridges in the planned area of Leidschenveen, near The Hague. In the tradition of Dutch city planning the area is criss-crossed with canals. To accommodate traffic circulation, a bridge system has been designed for 15 different bridge types, consisting of 45 pedestrian/cycling bridges and 13 traffic bridges.

Figure 65: Leidschenveen traffic bridge

The initial concept of the aluminium bridge is a lens-shaped bridge, build up out of longitudinal beams of extruded elements. The lens shape accommodates both architectural and constructive requirements, providing a slender 'wing' to cross the water and adapting itself to

the change in bending moment in the bridge girders. In addition the concept of longitudinal beams provides a modular system that can be used for any required bridge width.
The cross-section of the longitudinal beam is assembled by two identical extruded flanges separated by a flat web plate. The bridge height is varied by means of variation the heigth of the web plates. The triangular shape of the extrusions provides a torsion-stiff profile, is able to withstand concentrated loads and prevents local buckling. The extruded shapes are designed to accommodate adhesive bonded connections, which simplifies the assembly of the bridge girders.

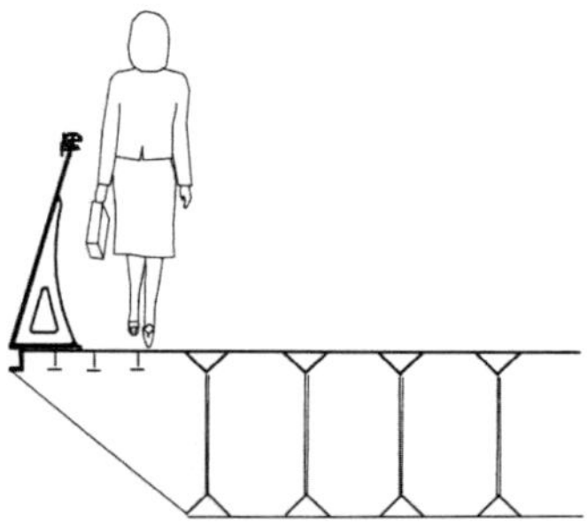

Figure 66: Initial design

In the final concept of the aluminium bridge the design differed highly from the initial design. Aluminium bridge deck sections in transverse direction replace the triangular extruded flanges (in longitudinal direction) of the initial design. This kind of bridge decks have shown to be able to withstand concentrated wheel loads.

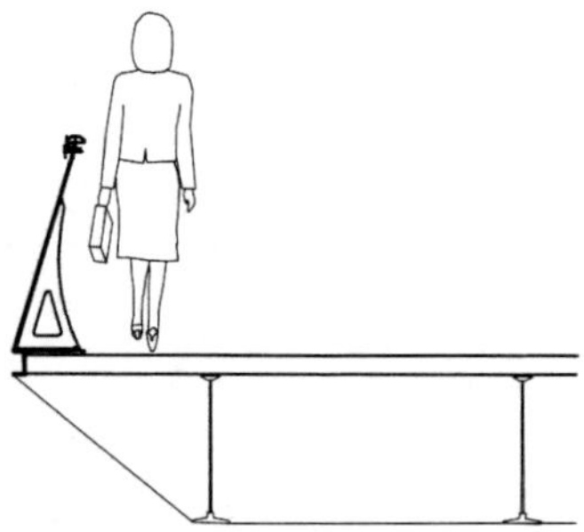

Figure 67: Detailed design

The deck elements can also be applied in longitudinal direction. In this case the stiffness of the bridge decks reduces the number of beams to be applied. The typical top flange extrusion and the larger bottom flange extrusion have been designed for this situation.

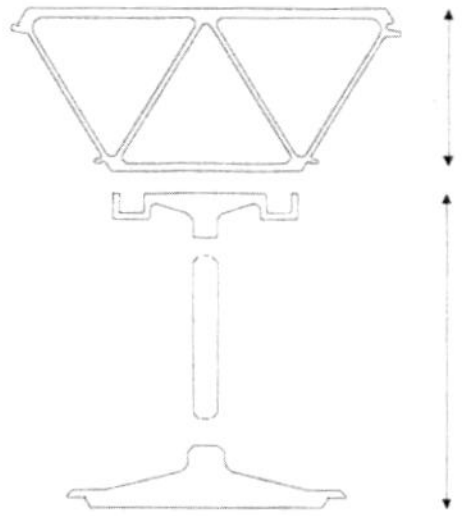

Figure 68: Extrusions

Connections between extrusions, web plates, etcetera are established by welding. Some concept details are shown in figure 69.

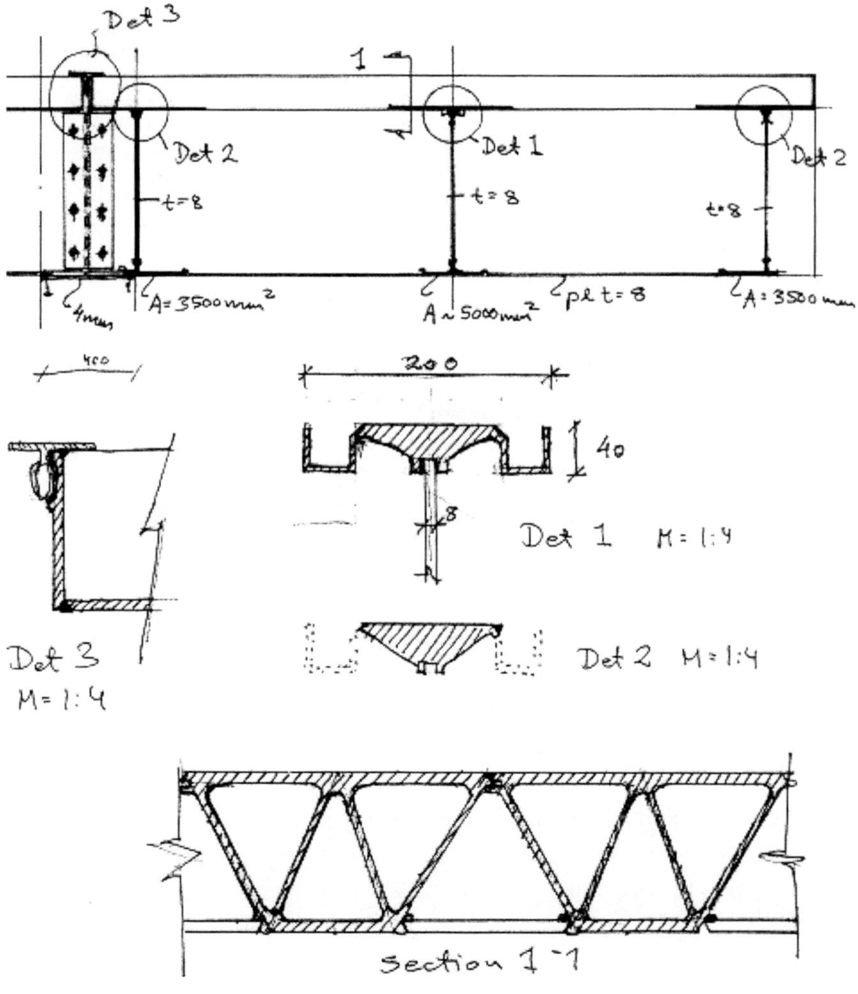

Figure 69: Details

3.3.2. Vinex bridges

In 1999 a three year joint industry project has been started which aims at further stimulating the use of aluminium for bridges. One of the four subprojects that is dealt with is called 'Vinex bridges'. Vinex bridges are pedestrian, cycling and traffic bridges in new-built residential areas, like the Leidschenveen bridges described before.
In this subproject a modular system of bridges has been developed for spans from 8 to 16 meter. This modular system is based on three different standard sections of 3 meter width, which can be easily connected to a total bridge width of 12 meter when necessary.

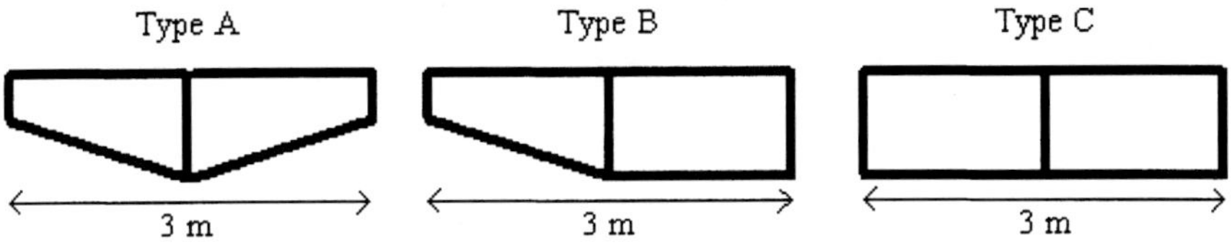

Figure 70: Vinex bridges: three types of standard sections

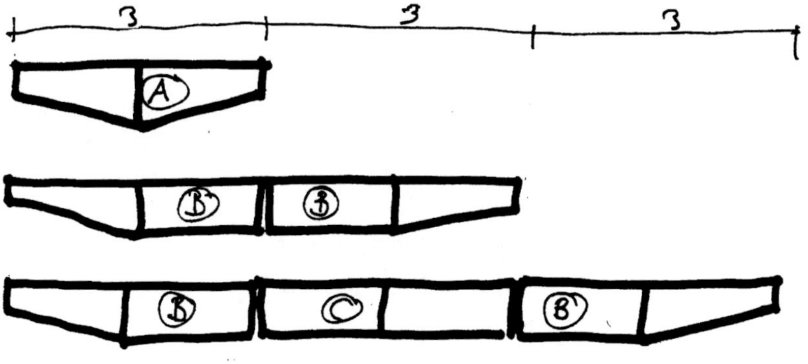

Figure 71: Combinations of standard sections

The deck in the Vinex bridges is a standard aluminium deck profile, for example the deck as shown in figure 72. The bridge deck is used in longitudinal direction of the bridge, which involves simple connections for the coupling between standard bridge sections. Besides this the deck profile is taken into account for the main bearing structure, i.e. the deck profile forms the top flange of a rectangular hollow structure.

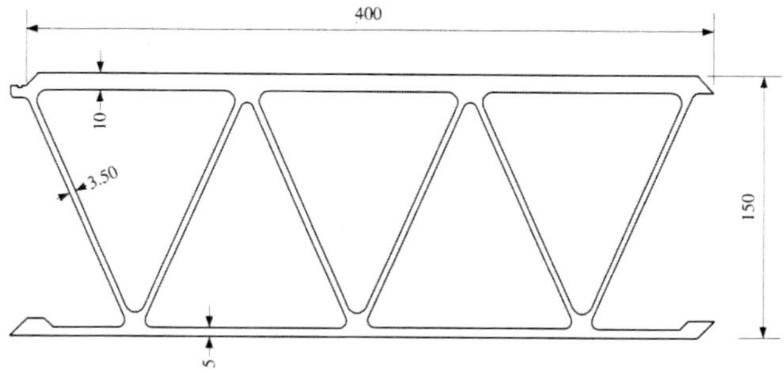

Figure 72: Bridge deck

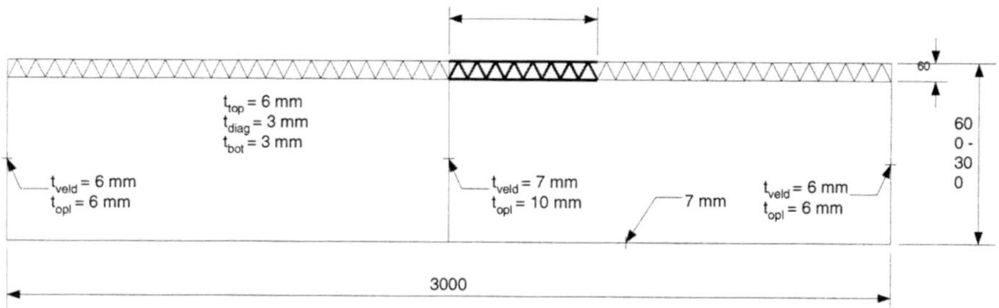

Figure 73: Example of a rectangular standard bridge section

The three standard sections have been worked out for different situations. In the next figures a possible cross section including details are given for bridge section A. For better welding of the thin-walled bridge deck profiles to the relatively thick web plates it is advised to use thick bearing strips, that are locally supported by transverse strips.

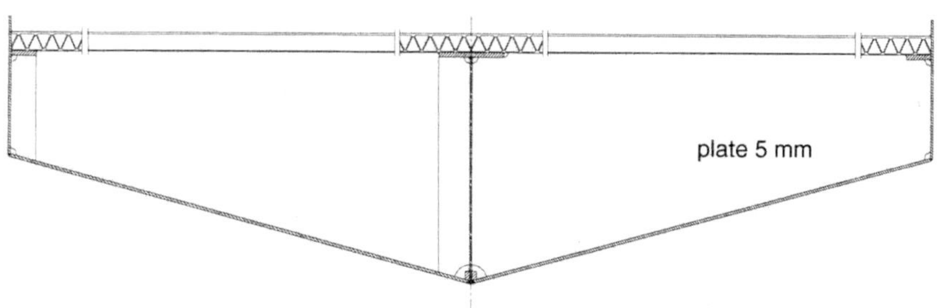

Figure 74: Bridge section A

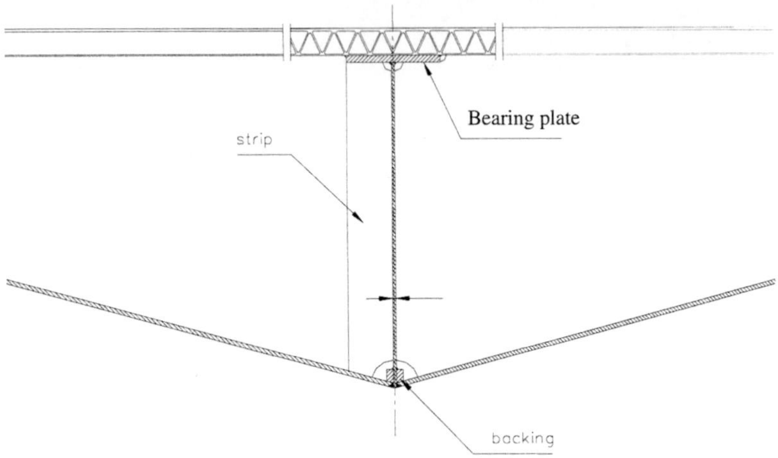

Figure 75: Detail of the connection between bridge deck and web

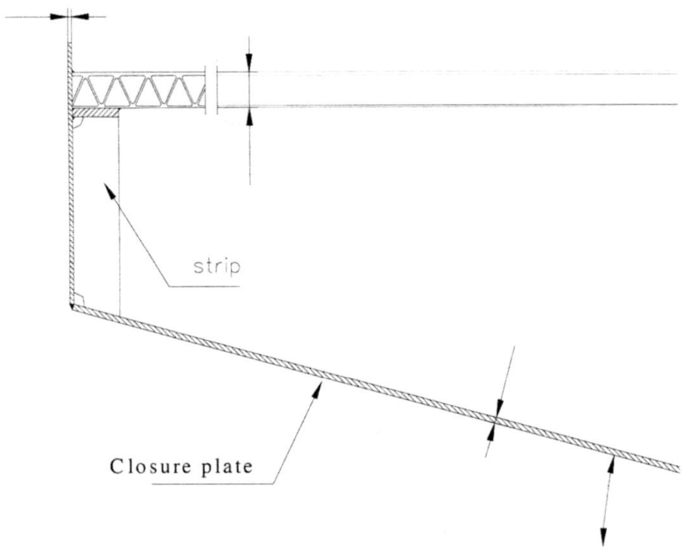

Figure 76: Detail for edges of bridge section A

3.3.3. Bridge decks

In 2000 a fundamental study on the optimal cross-section for aluminium bridge deck profiles has been started. For so far existing extruded bridge deck profiles, using a triangular core between a top plate and a bottom plate, seems to be a well designed cross-section, especially when concentrated wheel loads are governing for the structural design.
The geometry of such a bridge deck section has been worked out for different situations. The cross-section in figure 77 is the optimised design for a deck span of 2 meter. The detailed design meets various requirements, where fabrication method and connections are as important as other aspects (strength, stiffness, local buckling, fatigue, vibrations, et cetera).

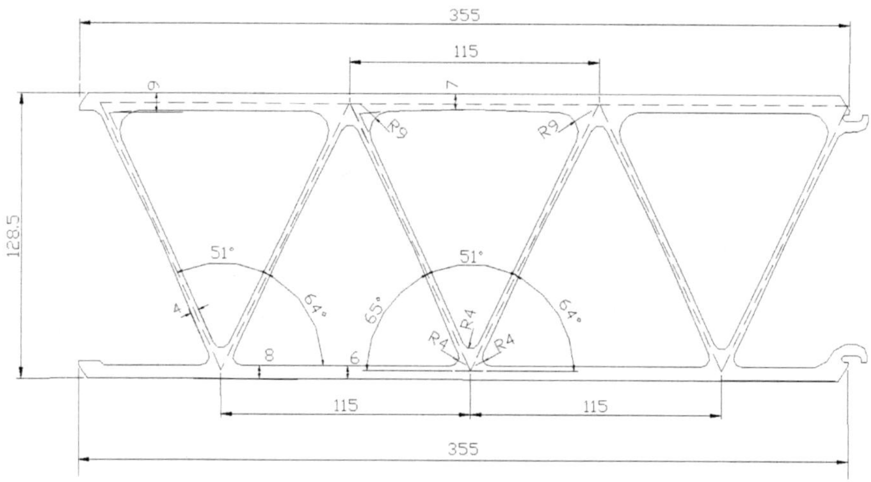

Figure 77: Optimised bridge deck profile

The bridge deck profiles are welded together to larger bridge deck sections in the factory. The section width, which is dependent on transport requirements, is limited to 3 meter. On site the bridge deck sections are connected by means of bolts. Two alternative bolted connections, which satisfy the strength and stiffness requirements, have been worked out.

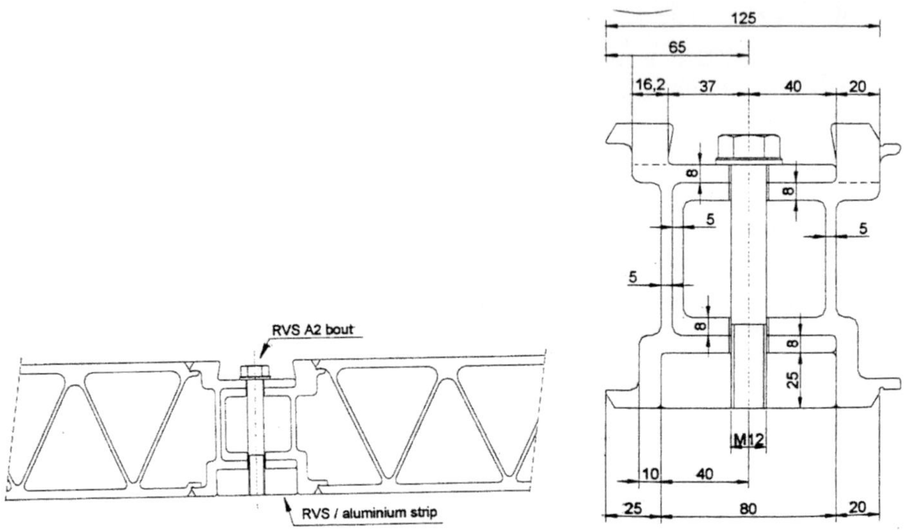

Figure 78: Connection between deck sections – type 1

For the first type of connection two different extrusions are used, which are welded to the deck profile, shoved into each other and connected by means of stainless steel bolts. The design of this connection is such that the upper side of the bolts is beneath the upper side of the deck profiles, which enables a fully flat upper side of the bridge deck. Further a solid stainless steel or aluminium strip is used at the bottom side of the connection. This strip has screw thread inserts, which enables the connection to be assembled from the upper side only. This may be necessary in case of bridge deck renovations, where it is often not possible to reach the under side of the bridge deck.

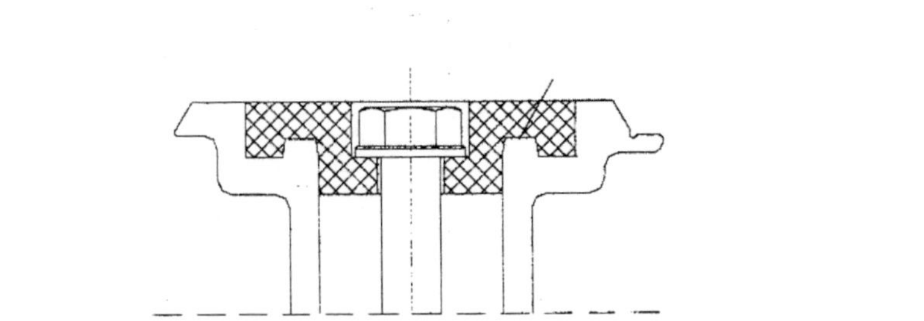

Figure 79: Connection between deck sections – type 2

For the second type of connection also two extrusions are used to be welded to the deck profiles. At site the connection is realised by strips, which keep the deck profiles on its place while turning on the bolts. The extrusions have conic parts, which benefits the accurate and easy positioning of the strips. As in type 1 the solid strip at the bottom side has screw thread inserts, and the upper side of the bolted connections is beneath the upper side of the deck profiles.

4. FINAL REMARKS

In the previous chapters joining of aluminium components is dealt with. After a short introduction of the subject attention is given to the joining technology both for primary as well as thin-walled structures. Subsequently welding, mechanical fastening, and adhesive bonding are reviewed.

Then, the design of connections is described following closely the design rules of Eurocode 9. The latter is regarded as the most up-to-date standard for structural aluminium applications.

Further, a number of recently designed – and some have been actually built – aluminium structures are reviewed with a focus on connections used with these structural designs.

The authors are of the opinion that the subject "connections in aluminium structures" is dealt with such that designers are aware of the main aspects to be looked after. For further reading a number of references is attached to this document.

References

[1] "TALAT, Training in Aluminium Application Technologies", ATP – Aluminium Training Partnership, EAA, Brussels, 1994.

[2] Mazzolani, F.M., "Aluminium Alloy Structures", 1st edition, Pitman, London, 1985.

[3] Soetens, F., "Welded connections in aluminium alloy structures", HERON, Vol. 32, 1987.

[4] Thomas, W.M., "Friction stir welding and related friction process characteristics", Proceedings INALCO, Cambridge, 1998.

[5] Adams, R.D., Wake, W.C., "Structural Adhesive Joints in Engineering", Elsevier, London, 1984.

[6] Eurocode 9 "Design of aluminium structures", European Committee for Standardisation (CEN), Brussels, 1998.

[7] Soetens, F., "Design of welded aluminium connections", Stahlbau, Ernst & Sohn, Berlin, 1998.

Chapter 5

FATIGUE DESIGN

B. Atzori, G. Meneghetti

University of Padua - Department of Mechanical Engineering

1. BASIC CONCEPTS ON FATIGUE

Fatigue is a phenomenon of progressive damage of materials due to the repetition of applied loads. If these loads were statically applied, then they would not lead to failure. In machine design the critical range where fatigue failures may occur is between 10^4 and 10^6 load cycles concerning constant amplitude fatigue, while it is much higher concerning variable amplitude fatigue. So machines and structures may undergo unexpected fatigue failures even after a long in-service life, when their reliability is thought to be proved.

Load repetitions can be due to the following causes:

1) a variation of the in-service loading conditions (for example cranes)
2) dynamic effects such as vibrations (for example wind action on cranes)

In the usual operating conditions the two causes are superimposed.

Since fatigue damage is not detectable by simple inspection, it is difficult to predict the number of cycles at which failure may occur. In fact fatigue failure is of brittle type, even for ductile materials and no visible phenomena occur before failure, such as the large plastic strains that precede static failure.

Concerning mechanical components without notch effect subjected to constant amplitude fatigue loads, the main parameters are:

- the number of load cycles n
- the applied stresses.

Usually stresses are calculated in the engineering way, but in some circumstances the true stresses should be adopted.

The characteristic parameters able to describe the fatigue load cycle can be (see fig. 1.1):

1) σ_{max}, σ_{min}
2) σ_m, σ_a
3) σ_m, $\Delta\sigma$
4) σ_a, load ratio $R=\sigma_{min}/\sigma_{max}$

In the field of classic fatigue analysis (components without defects) parameters listed at points 2 or 4 are adopted, while in presence of defects (fracture mechanics approach) parameters listed at point 3 are

adopted. Concerning the applied load cycles, in the field of strain controlled fatigue testing we consider half of a cycle while in the field of load controlled fatigue we consider the complete cycle.
Woehler in the mid eighties discovered the fatigue of steel components and performed some constant amplitude fatigue tests. He reported the test results on a diagram (σ_a, N), where the number of cycles to failure N is reported in abscissa. It should be noted that Woehler realized that fatigue life depends on the stress amplitude (see fig. 1.2). In the Woehler diagram the number of cycles to failure was indicated by N while the general number of cycles sustained without failure is indicated by n.
Typical values of the inverse slope of the Woehler curve are:
k=8÷10 for smooth specimens
k=3÷4 for notched specimens
Due to the previous values of k, notched specimens are characterised by a steeper fatigue curve (i.e. a lower value of k).
Now we analyse the factors that influence the fatigue behaviour. They can be divided into two groups, named internal parameters and external parameters.
Internal parameters affecting fatigue strength: they are already present in the mechanical component at the end of the manufacturing process:
I1) material
I2) absolute dimensions (Kd)
I3) surface roughness (Kl)
I4) shape (Kf)
I5) surface treatments

External parameters affecting fatigue strength: they depends by the in-service condition:
E1) mean stress different from zero
E2) type of loading
E3) environmental characteristics and temperature
E4) variations in the applied load cycles
E5) previous load history

The different coefficients K, which take into account these parameters, are to be applied to the material fatigue limit $\sigma_{a\infty}$. In absence of more precise information, the fatigue limit will be assumed in correspondence of a fatigue life $N_A=2\cdot 10^6$.
The actual load histories are not of constant amplitude type, but of variable amplitude type as shown in the right side of fig. 1.1. Then the load history must be processed and represented as a cumulative of loading shown in fig. 1.3, by means of some algorithms that will be briefly reviewed later. Standard shape of cumulative of loading can be defined and then the corresponding fatigue strength curves, similar to the Woehler curves (see again fig. 1.3), can be experimentally determined. They are named Gassner curves, introduced by Gassner in the 40's.
It should be noted that the experimental results have different usefulness and cost in the different zones of the Woehler curves (high cycle or low cycle fatigue), as depicted in fig. 1.4.
In order to clarify the method adopted to analyze variable amplitude loading histories, it is convenient to consider a loading history made of some loading blocks, as shown in fig. 1.5.
Each loading block characterised by a stress amplitude σ_{ai} and a number of cycle n_i, will produce a fatigue damage equal to $D_i=n_i/N_i$. The total damage is then given by:

$$D = \sum_i D_i = \sum_i \frac{n_i}{N_i}$$

If the loading history were made of one block only, fatigue failure will occur when n_i=N_i, that is D=D_i=1. The same idea can be extended to variable amplitude load histories (Miner hypothesis); then failure will occur when:

$$D = \sum_i D_i = \sum_i \frac{n_i}{N_i} = 1$$

Even if Miner hypothesis is often criticized, there is not any modification able to better interpret the experimental fatigue tests results.

By applying this hypothesis the improvement or worsening of the fatigue strength due to the application of load cycles respectively above or below the fatigue limit are completely neglected. In order to take into account the change in the fatigue limit position in the Woehler diagram due to the applications of load cycles above the fatigue limit, it is usual (and recommended also in some Fatigue Design Standards) to modify the Woehler slope by considering a new slope equal to 2·k-1 (being k the slope) beyond the constant amplitude fatigue limit. In practical application it is more convenient to extend the same slope k, since fatigue assessment will be more simple and to the safe–side. The Standards in force, which recommend the modification of the Woehler slope, define a new variable amplitude fatigue limit at a number of cycle usually equal to $10^{\cdot 8}$.

A load history such that reported in fig. 1.5 is often re-arranged in terms of an histogram, as shown in fig. 1.6. If we consider that the number of loading blocks is very high, then the histogram takes the shape of a cumulative spectrum (named also cumulative of loading).

Concerning the fatigue assessment, two different safety factors can be defined, a stress-based coefficient or a life-based coefficient respectively, as depicted in fig. 1.7 referring to a constant amplitude load history. The two coefficients are obviously correlated to each other and it should be noted that the life-based factor is much higher than the stress-based factor since $\gamma_N=\gamma_\sigma^{\,k}$

It is worth noting that γ_N is equal to the inverse of the damage. In fact if we consider a variable amplitude load history, it will be obtained:

$$\gamma_N = \frac{1}{D} = \frac{1}{\sum_i \frac{n_i}{N_i}}$$

Since usually the material fatigue strength data are available with reference to a constant amplitude load history, it could be useful to consider a constant amplitude load which is equivalent to the given variable amplitude load history. That means the actual and the equivalent load histories must produce the same fatigue damage, as fig. 1.8 and 1.9 illustrate. This procedure is adopted by the Design Standards, which in particular propose two different possibilities to define the equivalent constant amplitude load history, as shown in fig. 1.10. Actually one could choose any of the equivalent loads represented by a line parallel to the Woehler slope (see fig. 1.8). In general, the substitution of a variable amplitude load history with a constant amplitude one is useful in order to accelerate the experimental fatigue tests.

On the other hand the most correct way to predict the actual in-service fatigue behaviour is to perform fatigue tests by applying the variable amplitude load history that the mechanical component will experience in service. This technique was adopted initially by prof. Gassner in the 40's and is based on different types of load spectra as represented in fig. 1.11. For any given load spectrum a fatigue curve similar to the Woehler curve named Gassner curve can be experimentally determined. The Gassner curve reports the total number of cycles to failure in the abscissa as a function of the maximum stress amplitude, as shown in fig. 1.12. The experimental determination of the Gassner curve is expensive and time consuming, but it can be derived also analitically for a given cumulative type and a Woehler curve by imposing that for different maximum stress amplitudes failure is reached, i.e. by imposing the condition:

$$\sum_i \frac{n_i}{N_i} = 1$$

Finally the method to derive a load spectrum from a given or measured load history, like the one represented in fig. 1.13, will be briefly reviewed.
The most widespread method is the "rain flow", which does not require that the load history is periodic or has a constant mean value. If the load history starts and finishes at the highest peak stress values, then such a method is coincident with the reservoir method, as considered in the Fatigue Design Standards. We will apply the rain flow method to a load history supposed to be periodic and such that the start and end point coincide with the highest stress peaks, as represented in fig. 1.14. The method consider the peak or the trough value of a stress history and the three previous points (peaks or troughs). If the stress values of mid points are in between the first and the last values then one cycle is counted (and eliminated from the load history) with a range $\Delta\sigma=\sigma_2-\sigma_3$ and mean value $\sigma_m=(\sigma_2+\sigma_3)/2$. If the values of the mid points are not in between the first and the last ones, nothing is modified and then the following point together with its three previous ones is checked for further cycle counting. In the case of a non periodic stress history we may obtain at the end of the counting procedure a residual stress history that cannot be counted. For this residual history simply the ranges $\Delta\sigma$ between a peak and the following trough are separately counted.
Further readings on the matter treated in this chapter can be found in [Atzori 00-a, Atzori 00-b, ASTM 96, Suresh 91, Fuchs 80 and Frost 74].

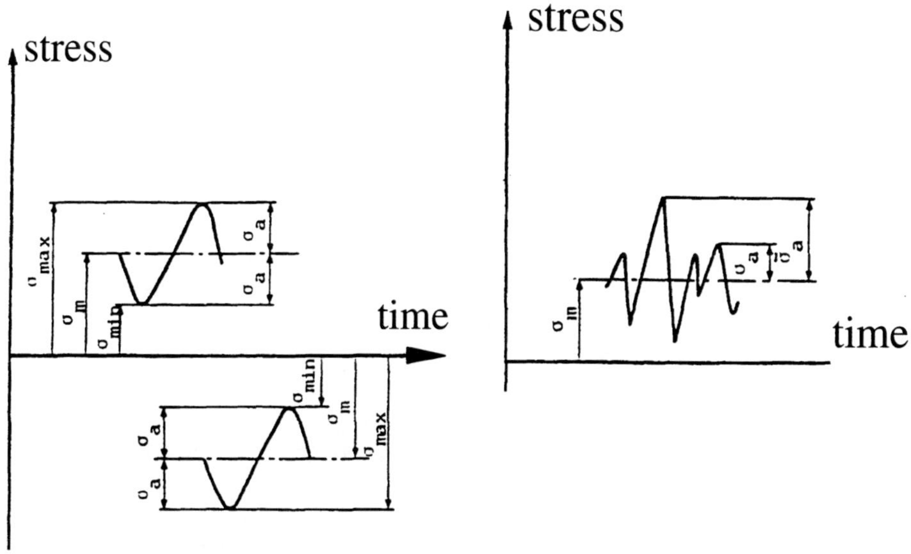

Figure 1.1: Constant and variable amplitude fatigue loads

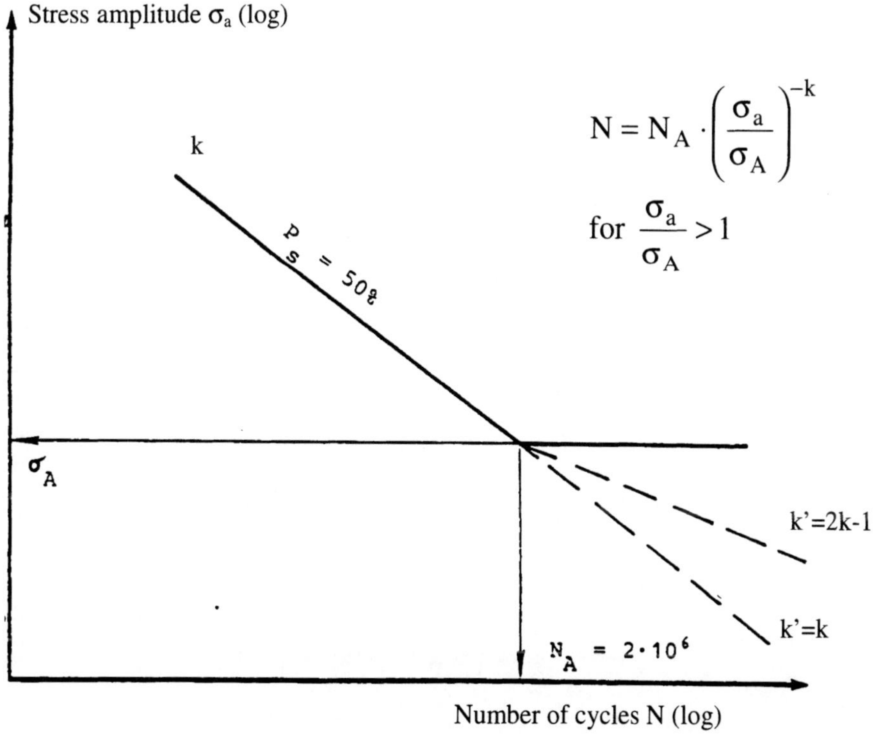

Figure 1.2: Modified Woehler curve

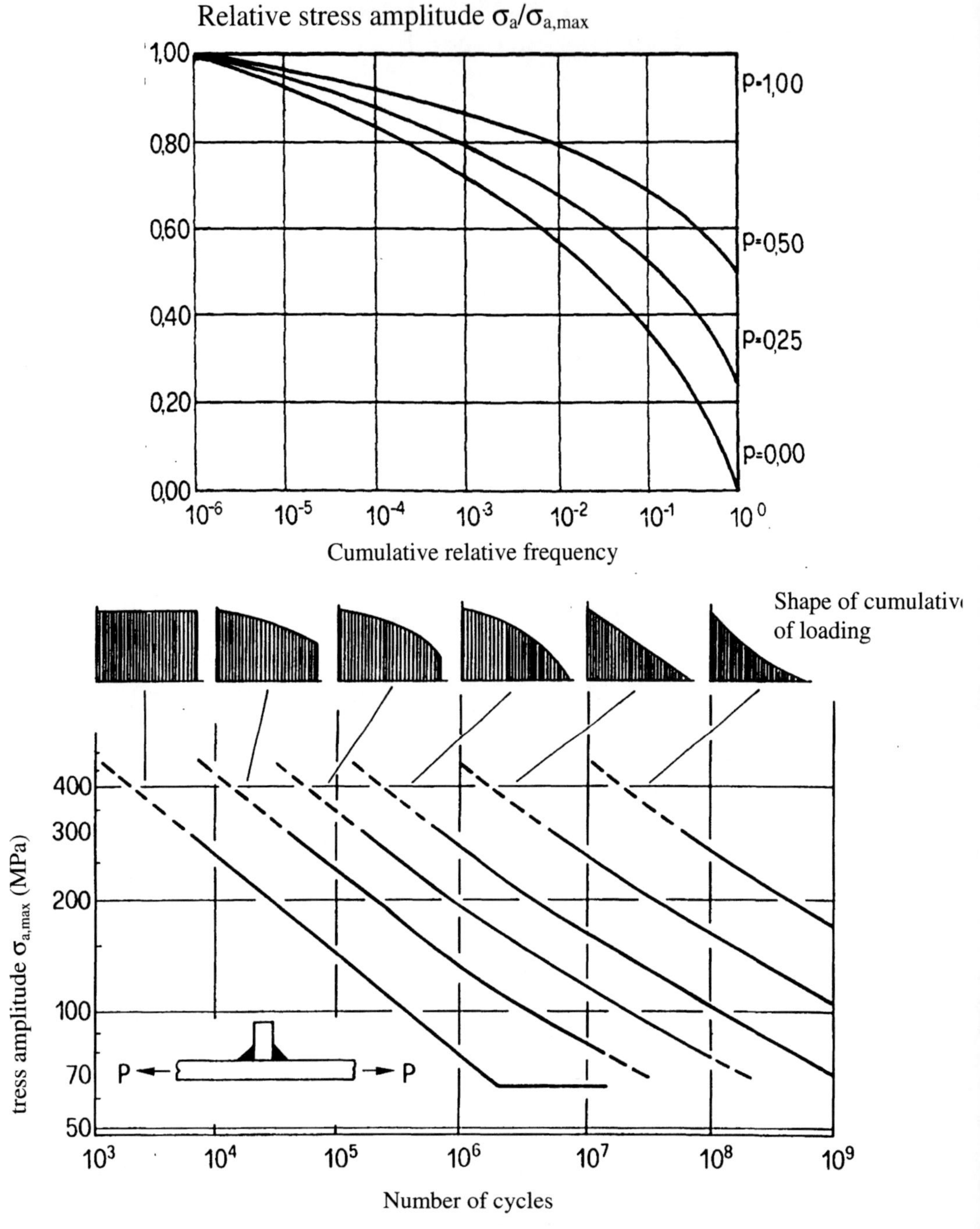

Figure 1.3: Cumulative of loading and Gassner curves

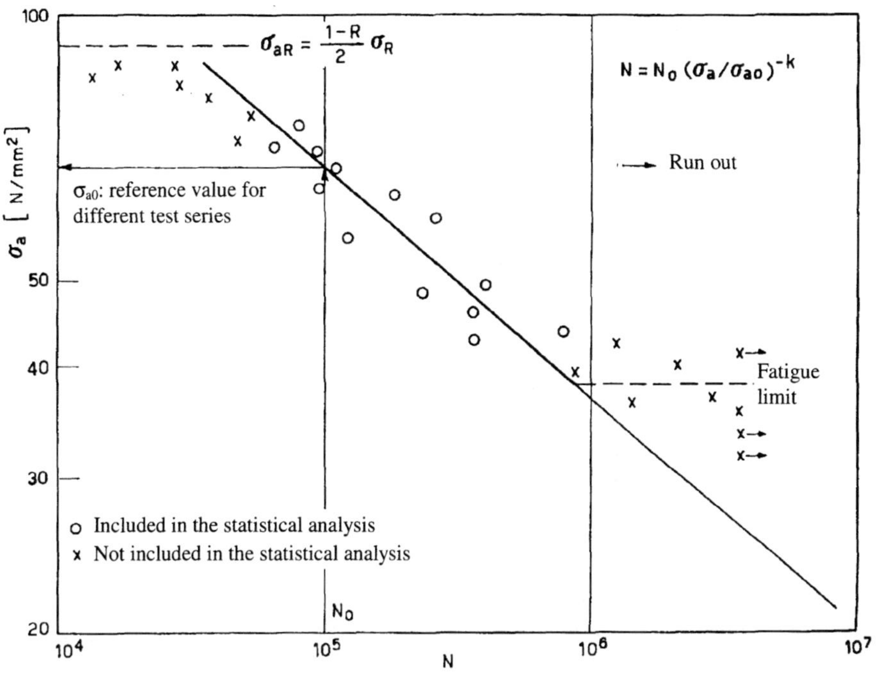

Figure 1.4: Different kinds of fatigue tests results

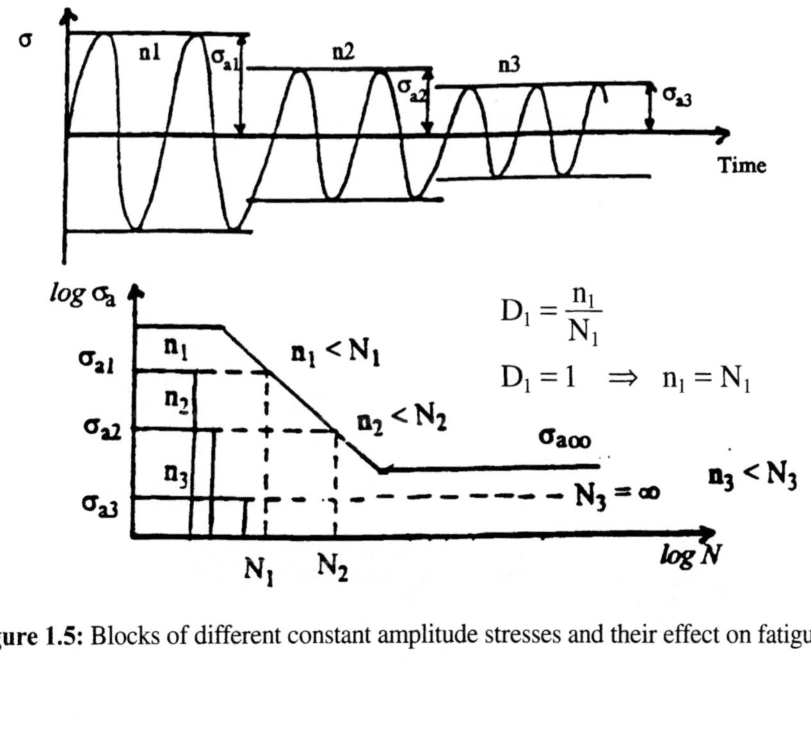

Figure 1.5: Blocks of different constant amplitude stresses and their effect on fatigue life

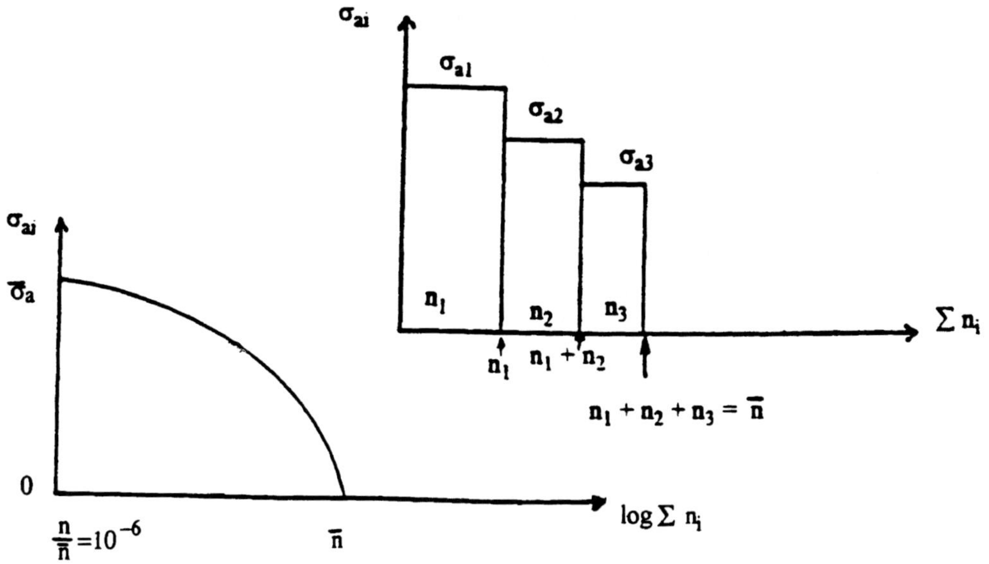

Figure 1.6: Representation of the loading history as a cumulative histogram or spectrum

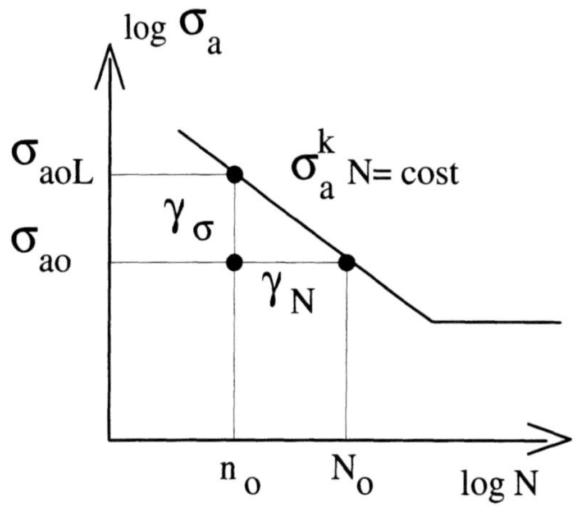

$$\gamma_\sigma = \frac{\sigma_{aoL}}{\sigma_{ao}} \qquad \gamma_N = \frac{N_o}{n_o}$$

since :

$$n_o \cdot \sigma_{aoL}^k = N_o \cdot \sigma_{ao}^k$$

then :

$$\frac{N_o}{n_o} = \left(\frac{\sigma_{ao_L}}{\sigma_{ao}}\right)^k \Rightarrow \gamma_N = \gamma_\sigma^k$$

N.B.: $\gamma_N = \frac{N_o}{n_o} = \frac{1}{D}$

Figure 1.7: Evaluation of stress (or life) safety factor for a constant amplitude loading

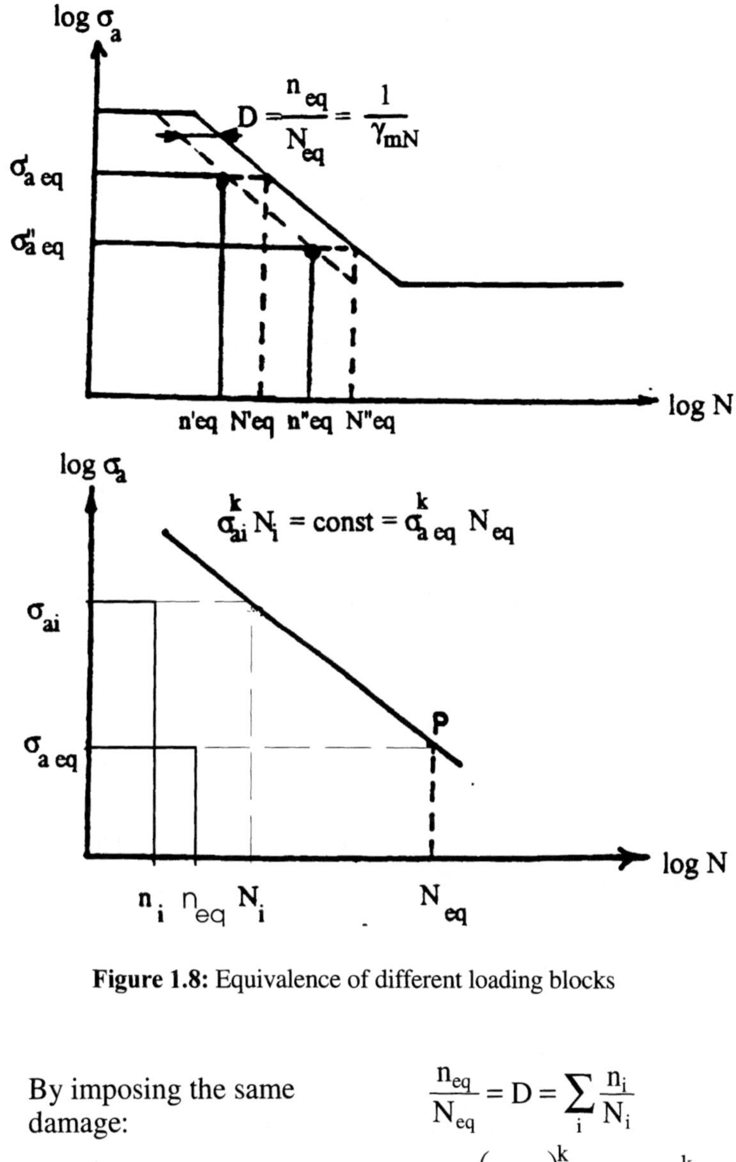

Figure 1.8: Equivalence of different loading blocks

By imposing the same damage: $$\frac{n_{eq}}{N_{eq}} = D = \sum_i \frac{n_i}{N_i}$$

then: $$n_{eq} = \sum_i n_i \cdot \left(\frac{\sigma_{ai}}{\sigma_{a_{eq}}} \right)^k = \sum_i \frac{n_i \cdot \sigma_{ai}^k}{\sigma_{a_{eq}}^k}$$

furthermore: $$n_{eq} \cdot \sigma_{a_{eq}}^k = \sum_i n_i \cdot \sigma_{ai}^k = \text{cost}$$

Figure 1.9: calculation of the equivalent loading history

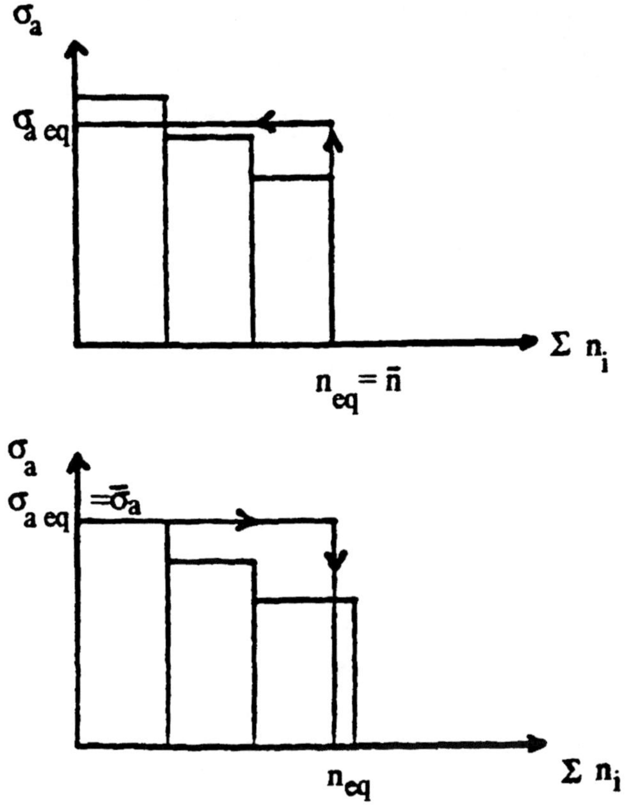

Figure 1.10: Different possibilities of substitution of a variable amplitude fatigue history with a constant amplitude one

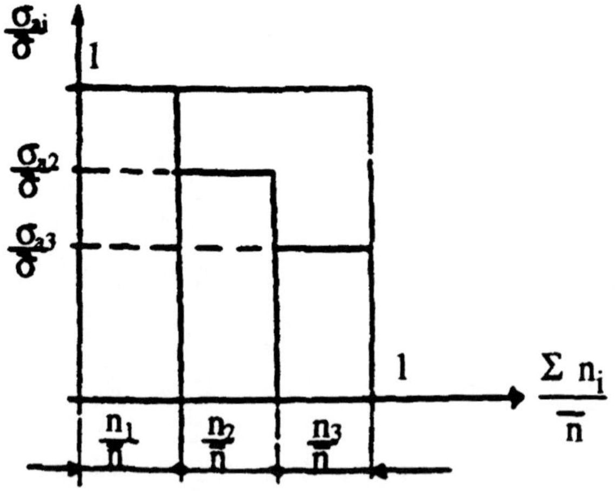

Figure 1.11: Definition of a cumulative type

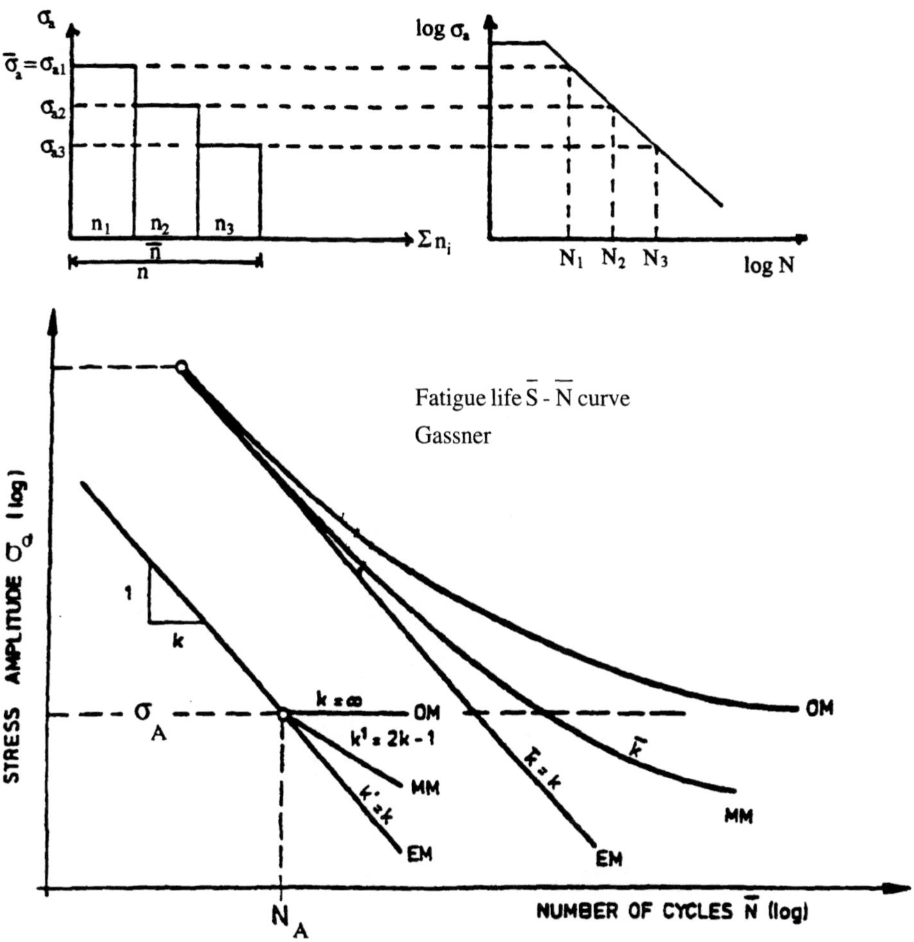

Figure 1.12: Definition of a cumulative type and theoretical evaluation of the corresponding Gassner curve

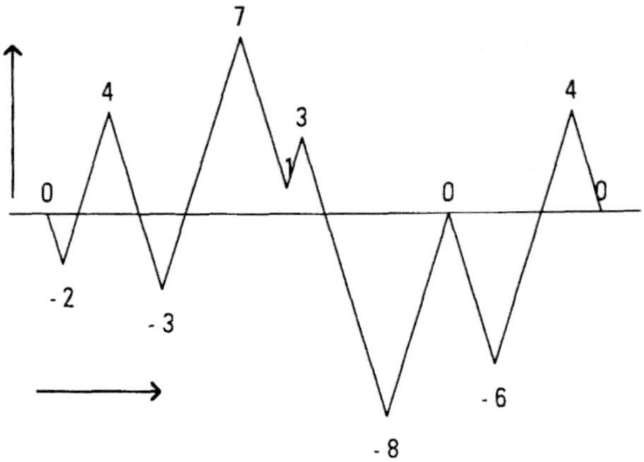

Figure 1.13: Example of a stress-time history

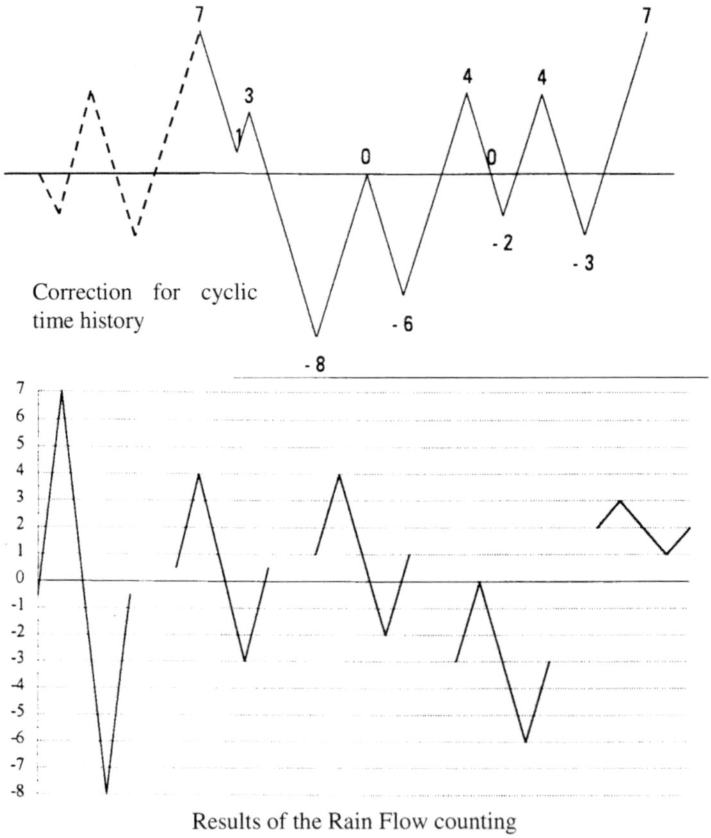

Figure 1.14: Rainflow counting

2. FATIGUE STRENGTH OF WELDED JOINTS

Welded joints are typical structural components which are critical in fatigue. The reasons are both the complex geometrical shape (that determines a stress concentration effect) and possible presence of defects (such as inclusions, lack of penetration etc.).

Fatigue cracks can initiate at the weld toe, at the weld root or from defects, as shown in fig. 2.1. The weld root criticity is particularly dangerous, since the crack cannot be seen until having completely crossed the weld bead, in such a way that the failure of the structure could precede the detection of the crack. This kind of failure is usually avoided by choosing a weld bead not too small.

The criticity due to defects depends on their dimensions and on the stress level in which they are embedded. For fusion welded joints this is not a typical cause of fatigue failure, since the welding technology gives a very sharp weld toe radius. Due to the acuity of its profile, the weld toe can be interpreted as a critical defect located in the most stressed part of the joint, as will be discussed later. Then other possible defects are not relevant in determining the fatigue strength of the joint.

For these reasons the fatigue strength of welded joints is usually analysed under the hypothesis of a crack starting from the weld toe, as it is the case for this note.

After all that, it is clear that the fatigue strength of a welded joint depends on the local stress field at the critical point, which is often located at the weld toe of the main plate, and not on the nominal stress level.

The evaluation of the local stress field intensity can be accomplished by experimental and by computational methods. In practical applications strain gauges are the most used tecnique among the experimental methods. Since the local stress field is characterised by a very steep gradient, its extension is limited to a region close to the weld toe. Then to be able to measure the local stress field it is necessary to use very small strain gauges located very near to the weld toe. They should never be located on the weld toe, as shown in fig. 2.2, because that zone is not smooth enough and the stress field under the strain gauge is not homogeneous. However this is a very special technique that is adopted only by well trained technicians. Instead it is much easier to be applied the experimental technique oriented to the measurement of the real nominal stress acting at the weld bead in a complex structure (which is often called "structural stress"). In such a case the evaluation can be performed by applying two strain gauges at different distances from the weld toe (usually something around t and 3t, being t the main plate thickness).

The most important advantage of the experimental measure of the stress state is that the real geometry (including misalignments) and the real applied loads are taken into account in such a way. The disadvantage is that the evaluation of the local stress is a difficult task and the results are questionable if the welded plates are not very thick.

Advantages and problems are on the opposite side when evaluating the stresses by computational techniques: the evaluation of the local stress field can be performed with the required resolution, but the finite element model has to be defined with reference to assumed geometry and loading condition. In particular geometrical imperfections introduced during manufacturing process are not included in the numerical model and the weld bead is simplified as shown in fig. 2.3. We prefer to adopt here for the numerical analysis a theoretical model, although also an idealised model has been often adopted.

As far as we are aware, DIN Standards were the first ones to propose a series of parallel fatigue design curves, which could be obtained one by the other simply by vertical translation in a double logarithmic diagram, as shown in fig. 2.4. The reason for such an assumption, which were not explained, were probably due to the practical usefulness of the concept and to its sufficient experimental validation. Only later some very extensive re-analyses of published fatigue tests results have been performed verifying the soundness of the hypothesis for welded joints in steel and aluminium alloys.

According to this hypothesis, the fatigue strength reduction of a welded joint in comparison with the base material can be interpreted as the superposition of two different reasons. The first one, represented by a fatigue strength reduction coefficient K'_f =($\sigma_A$base material/σ_A machined butt joint), is due to welding technology only and depends on the material and on the fatigue life. The second one, represented by a fatigue strength reduction coefficient K''_f =(σ_A machined butt joint/ $\sigma_A$welded joint), is due to the geometry of the joint only, as shown for some cases in fig. 2.4, and does not depend on the material and on the fatigue life. The total reduction factor of the fatigue strength will be obtained by multiplying the two coefficients.

For practical application the coefficient K'_f has little importance, since it is today recognized that, for welded joints, high cycle fatigue behaviour is not influenced by the particular type of steel or aluminium alloy.

The reason for the different fatigue strength of welded joints of different geometries in terms of nominal stress is due to the different relation between nominal and local stresses (which the fatigue strength depends on) as the geometry is varied. This is even more evident for the case of resistance spot welded joints, as shown in fig. 2.5, where the fatigue strength, in terms of applied load divided by the cross sectional area of the plate, is reported as a function of the ratio i/d, being i the distance between the spot weld axes and d the diameter of the spot welds. The fatigue strength curves for different geometries are similar, both for the dispersion of the results and for the slope k of the curves, but with very different absolute values (up to two degrees of magnitude). By taking into account the real cross sectional area of the plate interested by the stress flow near to the spot welds and the different amount of secondary bending between lap joints and contraposition joints, the different test series will show a very similar fatigue behaviour, collapsing into the same scatter band.

For a better understanding of the differences between nominal stress, structural stress (often called also geometrical stress, with a different meaning from the one used in this note) and local stress, two different interpretations of the local stress state in a cruciform welded joint are shown in fig. 2.6. The local stress state is there decomposed in two different contributes, the first one due to the overall shape of the joint and the second one due to the local shape in the most stressed area. Since the second contribution is characterised by a very steep stress concentration effect, the stresses due to the first contribution can be considered nearly constant within the small area where the local stresses exist. Due to this the first contribution can be considered as the correct nominal stress to which the sharp weld toe profile is subjected. Depending on the available fatigue strength data, the local stress field can be compared with the fatigue strength of a butt joint or with the fatigue strength of an open V notch, as shown in the figure. As a consequence the correct nominal stress, which will be called here "geometrical stress" since it takes into account the geometry of the joint, will have a different meaning and a different value in each case. Both the proposed models allow to calculate the real stress state with sufficient accuracy, as shown in fig. 2.7, but the values of the

geometrical and local stresses depends on the assumed model (and it should be remarked here that also the fatigue design strength, obviously, should be coherent with the assumed model).
The problem of the definition of an univocal nomenclature for the local stress around the weld toe is more complicate when a complex welded structure is analysed. In this case the terms "structural stress" and "geometric stress" are often used in the literature with reference to the same stress, due to the overall geometry of the structure but not to the local effect of the weld bead. This stress, which appears to be coincident with the nominal stress of fig. 2.6, will be called here structural stress. The higher stress, which has been previously defined and takes into account the presence of the weld bead, but not the effect of the weld toe, will be called instead geometrical stress. Then the local stress field is defined by the superposition of the stress field due to a sharp V-notch (which is the same as far as the weld throat thickness is the same) and of the geometrical stress (which depends on the applied nominal stress and on the joint geometry). The assumption of a unified scatter band, valid for all the welded joints in steel (fig. 2.8) and in aluminium alloys (fig. 2.9) appears then to be justified and not only acceptable from an experimental point of view.
The definition of a unified scatter band is very useful in practical applications. It can be extended also to variable amplitude load histories, as shown in fig. 2.10 for gaussian loading cumulatives characterised by different values of the ratio $p = \sigma_{amin}/\sigma_{amax}$. The defined scatter bands have been extended also to the case of resistance welding, as shown in fig. 2.11, and laser welding, as shown in fig. 2.12, although very well executed laser butt welds have exhibited lately a fatigue behaviour more similar to the base material than to the fusion welds. The application of the unified scatter bands is also particularly useful to reduce the number of fatigue tests in evaluating the influence of welding or geometrical parameters on the fatigue strength of a particular joint, as shown in fig. 2.13 for a spot-welded lap joint.
Although so far we did not consider the influence of residual stresses, it is well known that the usual welding technologies cause high residual stresses, in particular at the weld toe. In real structures the residual stresses are even higher and often not less than the yielding stress of the material, due to the decisive contribution of the welds in constraining the structure against the elastic relaxation. The Design Standards usually take into account the presence of residual stresses defining the fatigue strength as independent from the nominal stress ratio $R=\sigma_{min}/\sigma_{max}$, according to the curve C of fig. 2.14.

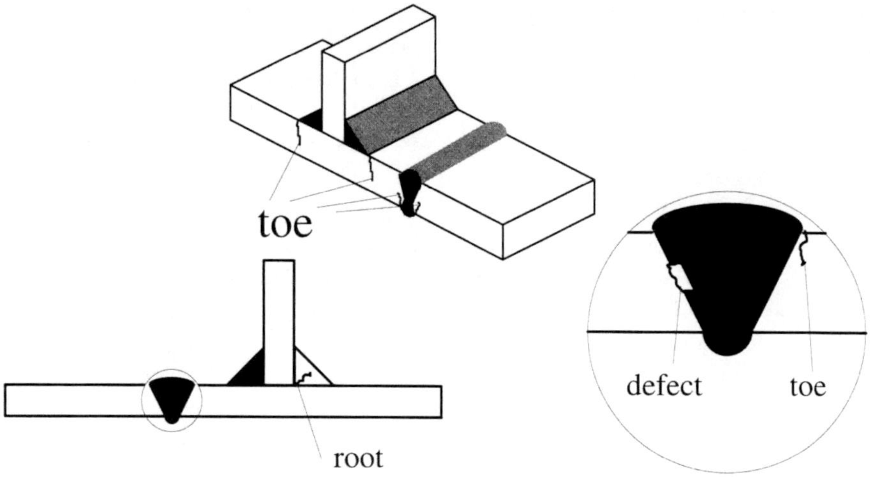

Figure 2.1: Critical points in fatigue of welded joints

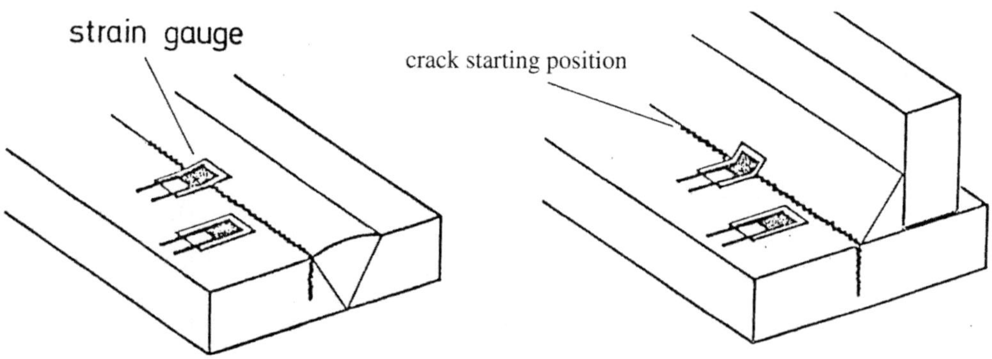

Figure 2.2: Correct and non correct positioning of strain gauges

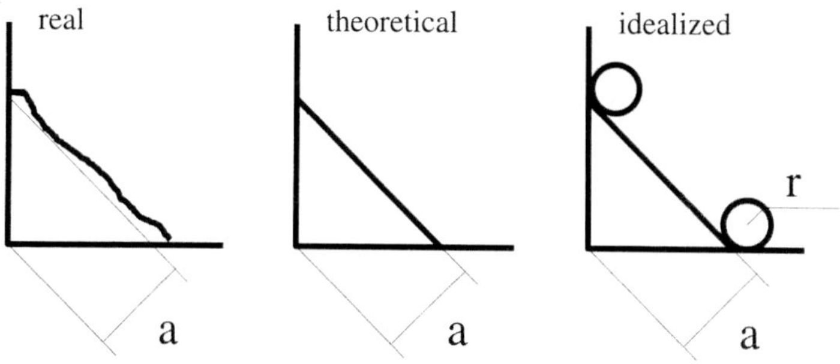

Figure 2.3. Theoretical and idealized weld profile

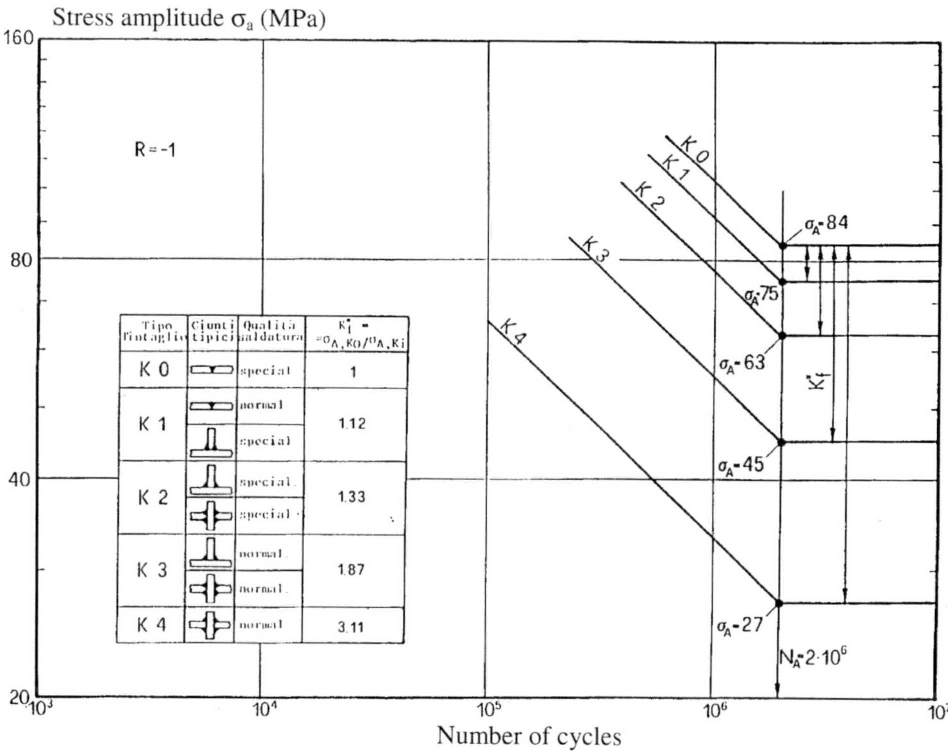

Figure 2.4: Fatigue design Woehler curves in terms of stress concentration coefficients

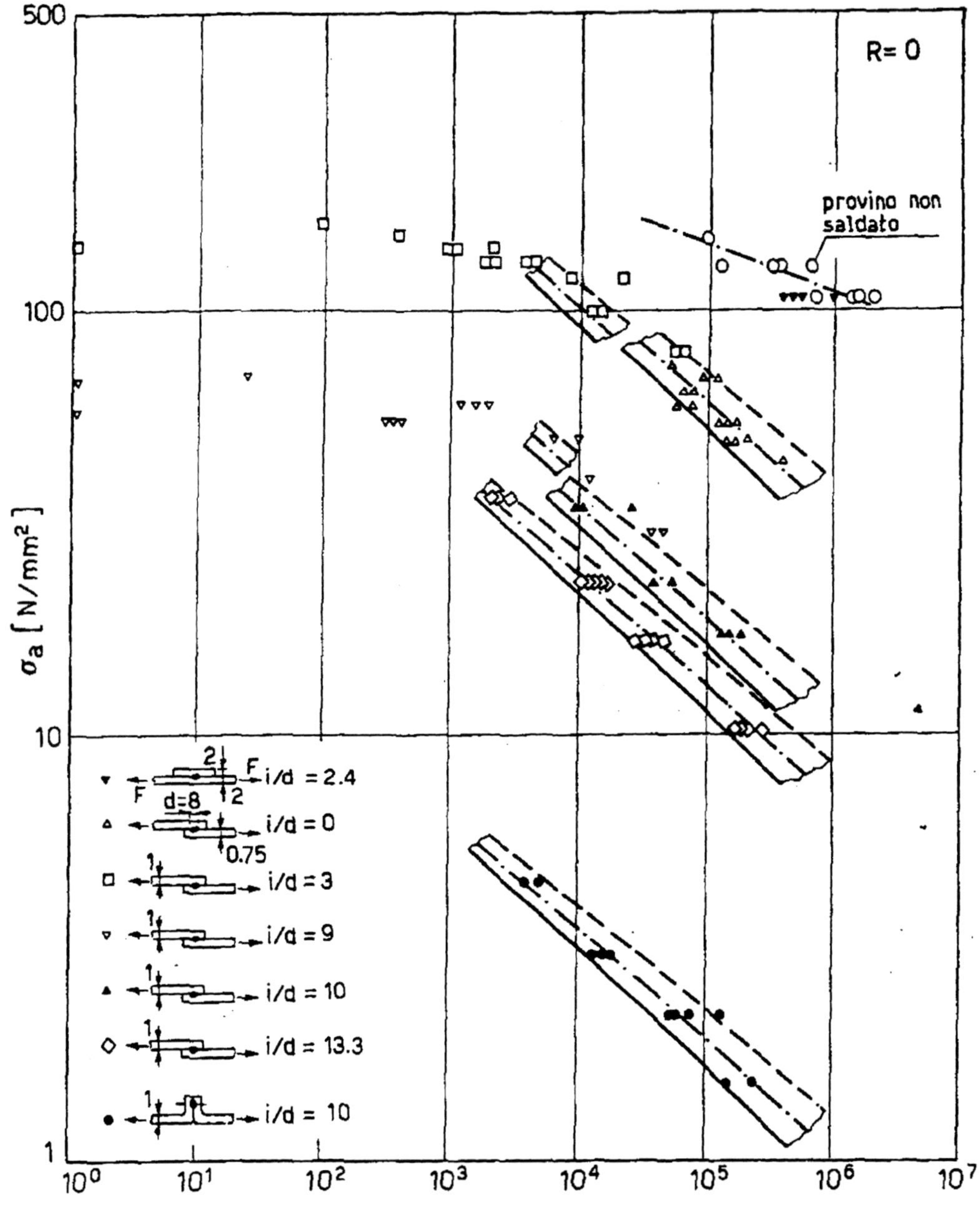

Figure 2.5: Spot welded joints: effect of an ill-defined nominal stress

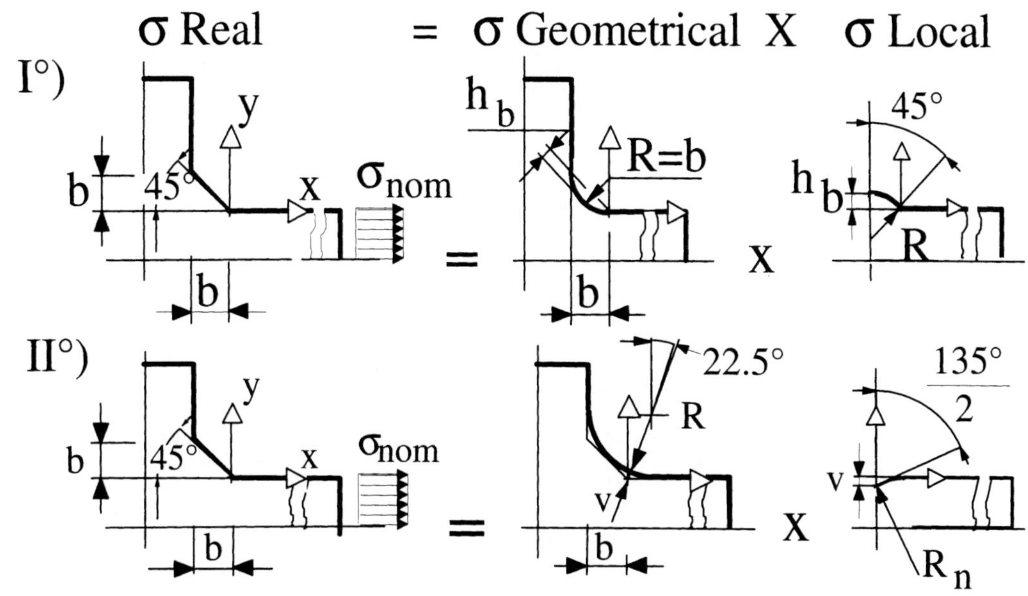

Figure 2.6: Interpretation of the notch effect at the weld toe

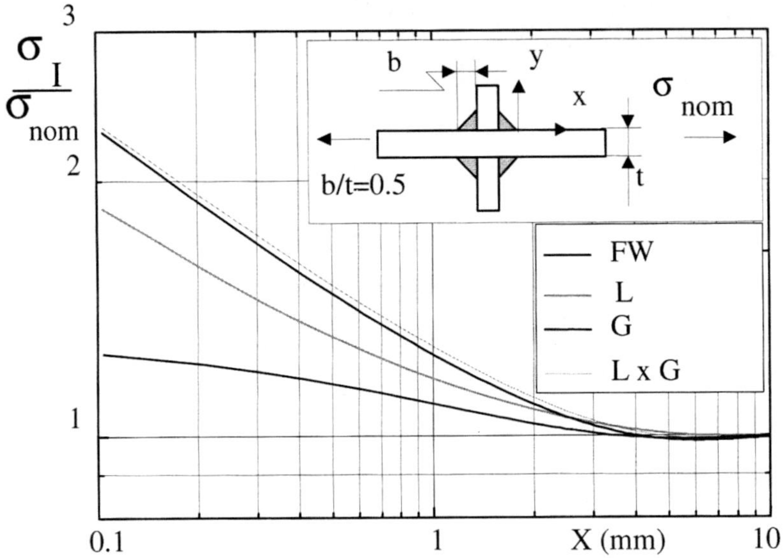

First interpretation of the notch effect

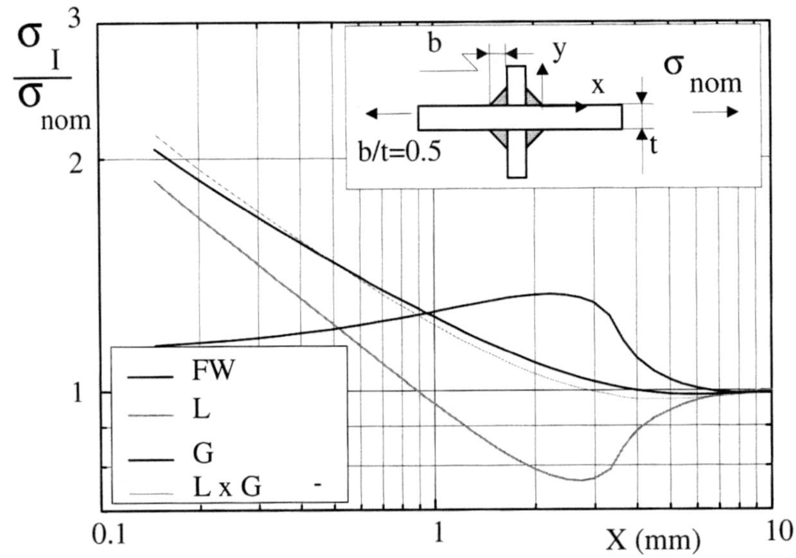

Second interpretation of the notch effect

Figure 2.7: Superposition of geometric and local stresses for transverse fillet welded joint

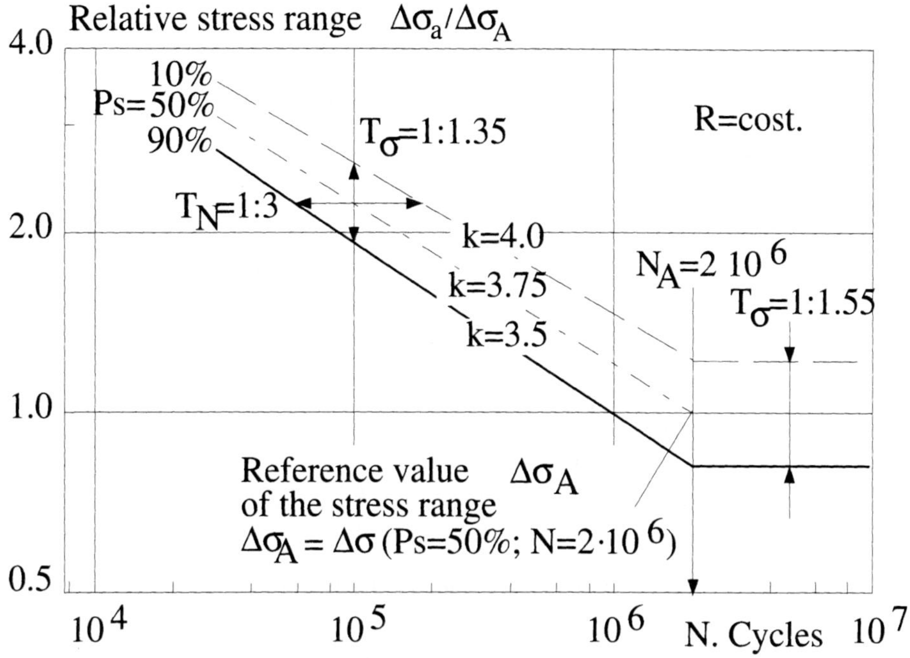

Figure 2.8: Unified scatter band for structural steels

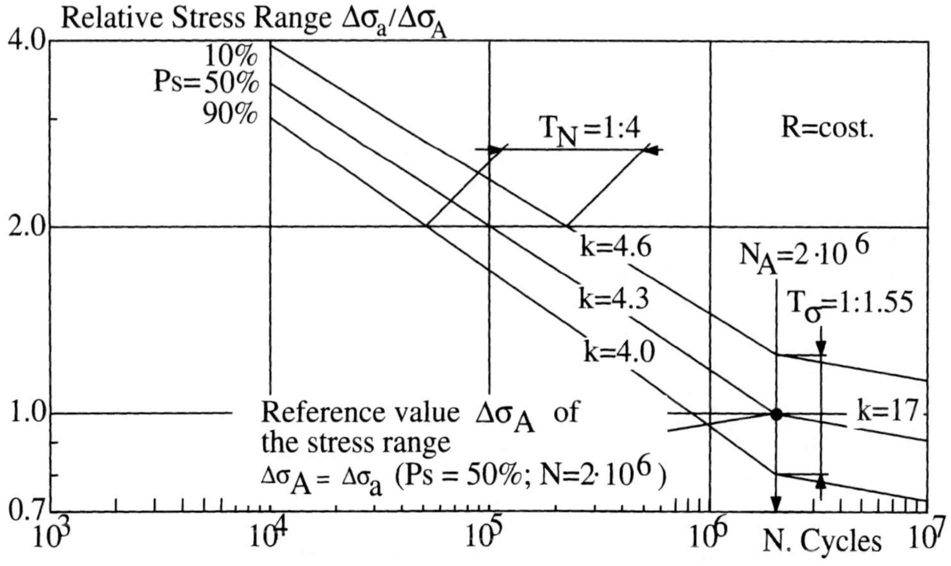

Figure 2.9: Unified scatter band for aluminium alloys

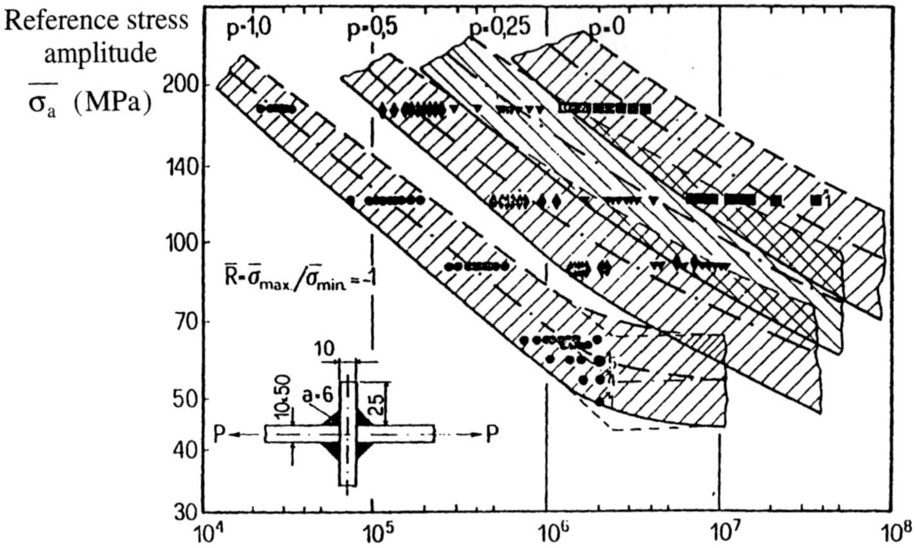

Figure 2.10: Comparison of theoretical and experimental Gassner curves

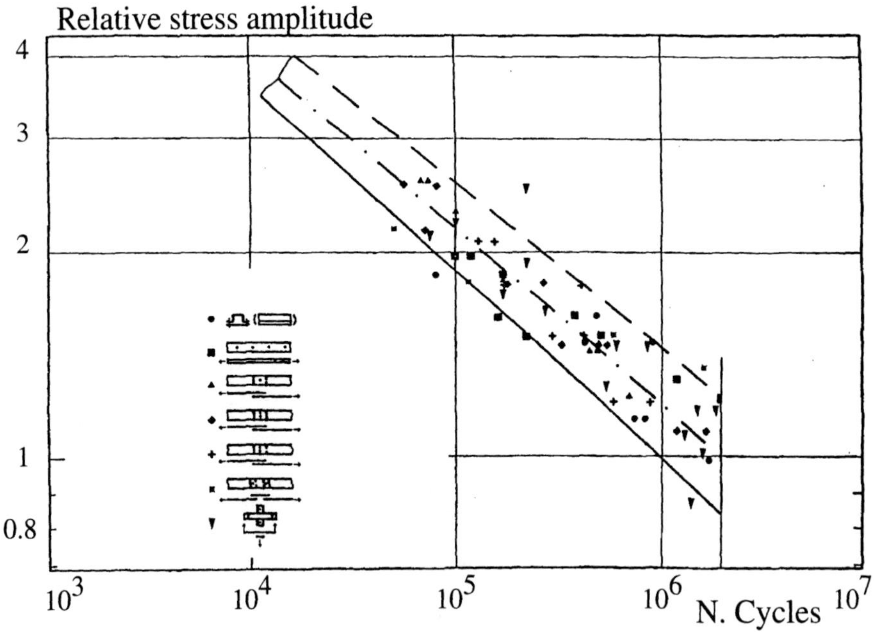

Figure 2.11: Application of the unified scatter band approach to test results for spot welded joints

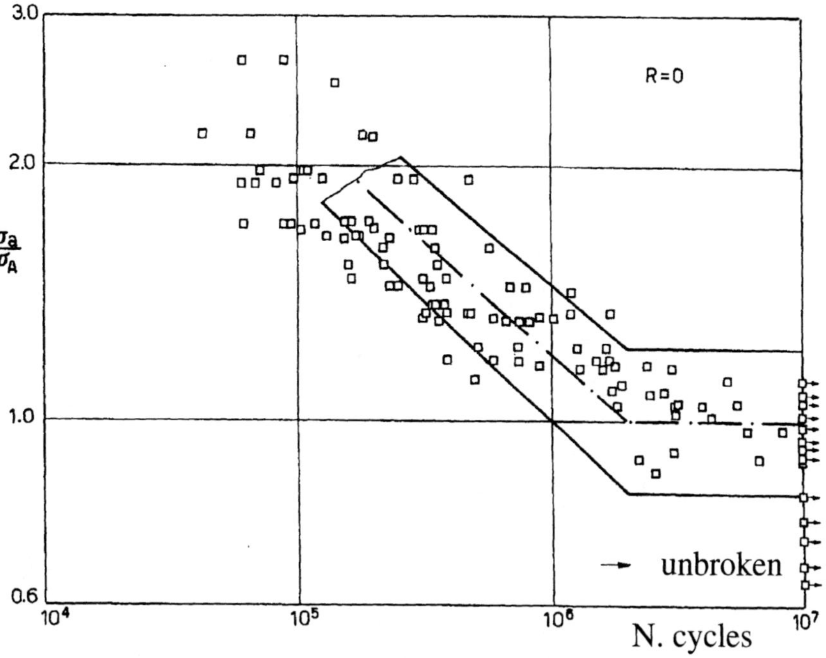

Figure 2.12: Application of the unified scatter band approach to laser welded joints

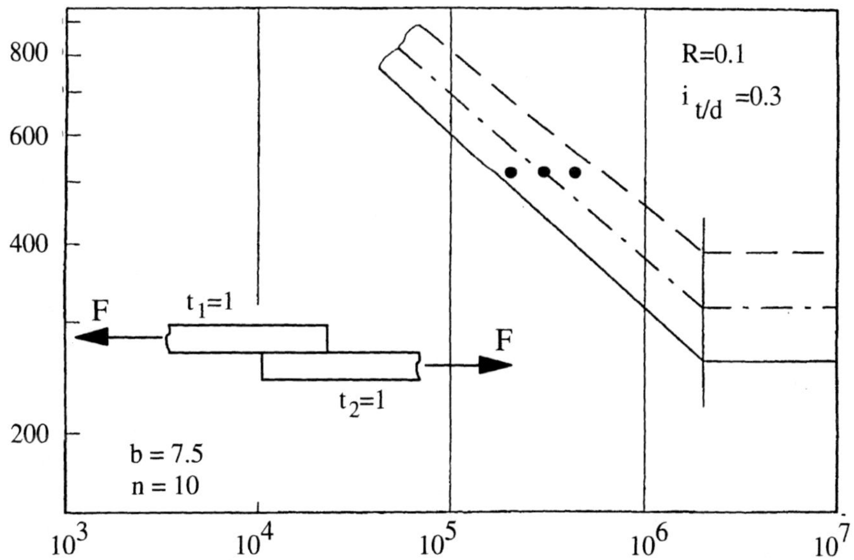

Figure 2.13: Application of the unified scatter band approach to a reduced number of test results

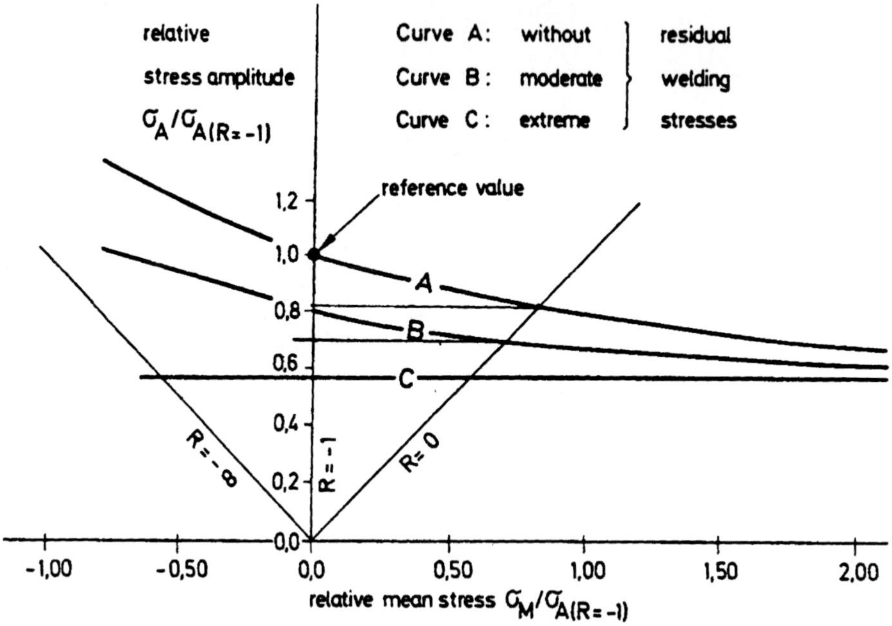

Figure 2.14: Combined effect of applied mean stress and residual stresses

3. FATIGUE DESIGN BY EUROCODE 9

Recently it has been developed an European Standard for the design of aluminium alloy structures, the Eurocode 9, which contains one part completely devoted to fatigue loaded structures (fig 3.1.). Since this Code is more new than existing national standards and it has been developed with the contribution of several european countries, only this one will be cited in this note, and its more innovative and interesting points will be sintetically exposed.

As far as the innovative part is concerned, it should be highlighted that among the design methods, beside the consolidated "safe life" approach, also the approaches "Damage tolerant" and "Design by testing" have been introduced, as shown in fig. 3.2. Noticeable is also the inclusion of a "Guidance on assessment by fracture mechanics" (fig. 3.3) and of a description of "Improvement techniques" (fig. 3.10).

Among the most interesting points it should be cited obviously the part where the design fatigue strength curves are given. The code offers two different approaches: the first based on the nominal stress evaluated on the main plate of the joint for different categories of joints and a second one based on the hot spot stress.

In the first case a fatigue strength curve is defined in a general way (fig. 3.4). The parameters that define each joint category (and thus each design curve) are given by means of suitable tables as the ones shown in figures 3.5, 3.6, 3.7, 3.8. For each design category the reference fatigue strength is given by means of the fatigue strength range $\Delta\sigma_c=(\sigma_{max}-\sigma_{min})_c$ for a life $N_c=2\cdot10^6$ cycles and the slope of the fatigue curve for a number of cycles respectively lower and higher than $N_D=5\cdot10^6$. The fatigue strength in correspondence of N_D defines the constant amplitude fatigue limit. For variable amplitude load histories, when the maximum applied stress range exceeds the value of $\Delta\sigma_D$, then the fatigue limit $\Delta\sigma_D$ disappears and a new variable amplitude fatigue limit $\Delta\sigma_L$ is defined at a life $N_L=10^8$ cycles.

In the second case (fatigue design approach based on hot spot stress) a table of values of the hot spot fatigue design strength is given, where the fatigue strength is a function only of the main plate thickness, as shown in fig. 3.9. In our opinion the application of this approach to general structures today is still a special approach and to specialists only it should be left.

DRAFT
prENV 1999 Part 2
CEN/TC250/SC9/N89
[PT3/N43]

EUROPEAN PRESTANDARD
PRÉNORME EUROPÉENNE
EUROPÄISCE VORNORM

June 1996

European Standard

English version

Eurocode 9 : Design of aluminium alloy structures

Part 2 : Structures susceptible to fatigue

CEN/TC 250/SC 9

Project team 3

Draft for Eurocode 9 : Part 2 - June 1996

CEN

European Committee for Standardisation
Comité Européen de Normalisation
Europäisches Komitee für Normung

Central Secretariat: Rue de Stassart 36, B-1050 Brussels

Figure 3.1.

2.1.6 Methods of fatigue design

(1) **Safe life design:** This method is based on the calculation of damage during the structure's design life using standard lower bound endurance data and an upper bound estimate of fatigue loading. This will provide a conservative estimate of fatigue life and in-service inspection shall not normally be considered essential for safety. The method is given in 2.2.

(2) **Damage tolerant design:** This method is based on limiting the growth of fatigue cracking by means of a mandatory inspection programme. Once a fatigue crack has reached a pre-determined size the part shall be repaired or replaced. The method is given in 2.3.

The method may be suitable to apply in certain applications where a safe life assessment shows that fatigue has a significant effect on design economy and a higher risk of fatigue cracking during the design life may be justified than is permitted using safe life design principles. Whilst the method is intended to result in a risk of ultimate structural failure comparable to that assumed for safe life design, it may result in a higher risk of temporary loss of serviceability.

(3) **Design by testing:** This method shall be resorted to where the necessary loading data, response data, fatigue strength data or crack growth data are not available from standards or other sources for a particular application. Test data shall only be used in lieu of standard data on condition that they are obtained and applied under controlled conditions. Guidance is given in 2.4 and Annex C.

Figure 3.2.

Annex B [Informative]:

GUIDANCE ON ASSESSMENT BY FRACTURE MECHANICS

B.1 Scope

(1) The objective of the annex is to provide information on the use of fracture mechanics for assessing the growth of fatigue cracks from sharp planar discontinuities. Main uses are in the assessment of:

- known flaws (including fatigue cracks found in service).
- assumed flaws (including consideration of the original joint or NDT detection limits).
- tolerance to flaws (including fitness for purpose assessment of fabrication flaws for particular service requirements).

(1) The method covers fatigue crack growth normal to the direction of principal tensile stress (Mode 1).

Figure 3.3.

DRAFT Page 14
prENV 1999 Part 2 : 1996
CEN/TC250/SC9/N89

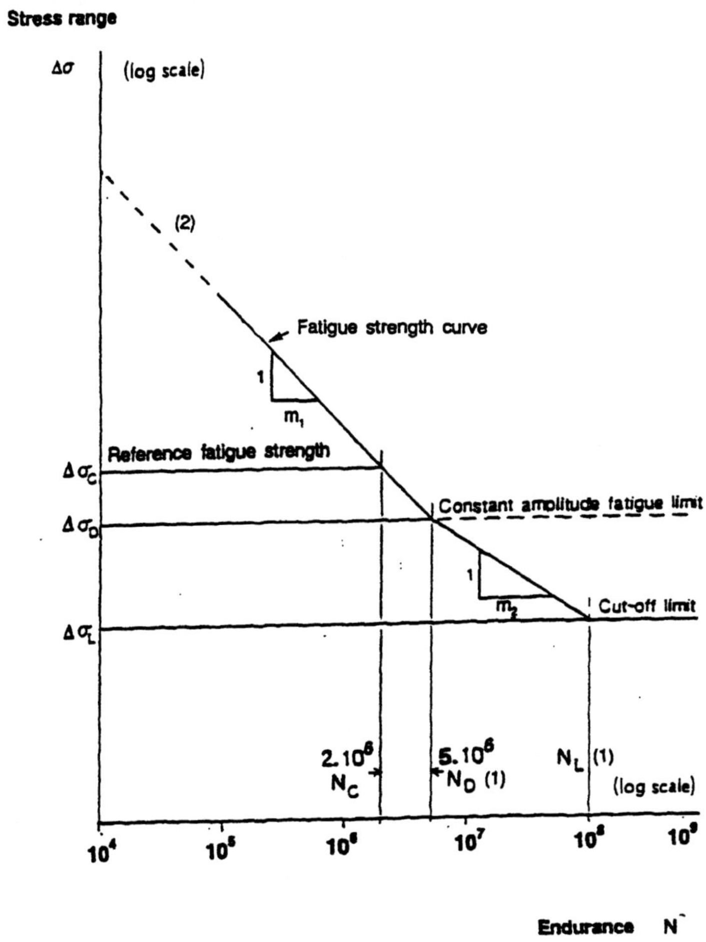

(1) for certain environmental conditions see 5.4.
(2) see 5.3.4 for $N < 10^5$

Figure 1.5.2. Fatigue strength curve

Figure 3.4.

DRAFT Page 38
prENV 1999 Part 2 : 1996
CEN/TC250/SC9/N89

Table 5.1.2(a) Detail Categories for Members with Welded Attachments - Transverse Weld Toe

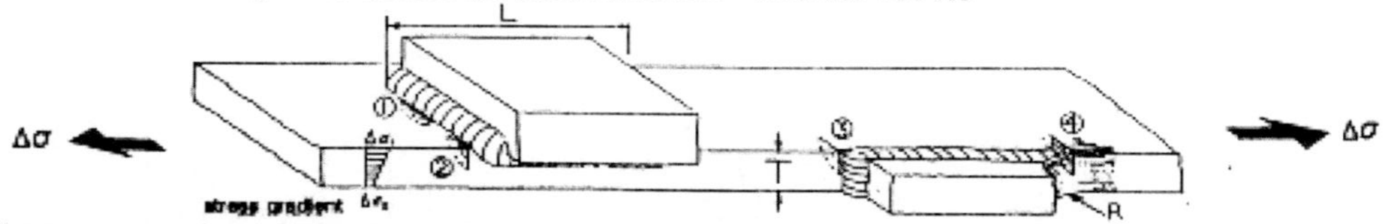

				2.1	2.2	2.3	2.4	2.5	2.6	2.7	2.8	2.9	2.10	2.11	2.12	2.13
Product Forms				Rolled, extruded and forged products												
Initiation site	Reference No.			1								2	3	4		
	Location			At transverse weld toe on stressed member						→	→	→	→	At longitudinal weld end		
				on surface away from edge				→	→	→	→	At corner	On edge	In ground weld toe on edge		
Stress orientation (see 4.3.4)				Normal to transverse weld toe				→	→	→	→	→	→	Parallel to weld axis		
Alloys				As table 1.1.1.		→	→	→	→	→	→	→	→	→	→	→
Particular Require-ments	Dimensional	Joint Geometry		Attachment on member surface				→	→	→	→	→	Attachment on member edge			→
				Weld on surface away from corner					→	→	→	Welded round corner	Weld on edge		→	→
		Length L(mm)	> ≤	0 10	10 20	20 30	30 50	50 80	80 120	120 200	200 ∞	L and T as for Types 2.1 to 2.8	No Radius	Corner radius R (mm)		
		Thickness T (mm)		See table below			→	→	→	→	→			10<R≤20	20≤R≤40	R>40
	Fabrication											Grind undercut smooth		Grind radius in direction of Δσ*		
	Inspection/testing			As Annex D table D1			→	→	→	→	→	→	→	→	→	→
	Quality standard			As Annex D table D2			→	→	→	→	→	→	→	→	→	→
Stress Analysis	Stress Parameter			Nominal stress at initiation site				→	→	→	→	→	→	→	→	→
	Stress concentrations already allowed for			Weld profile permitted by Annex D Table D2						→	→	→	→	→	→	→
				Stiffening effect of attachment				→	→	→	→	→	→	→	→	→
Type Number				2.1	2.2	2.3	2.4	2.5	2.6	2.7	2.8	2.9	2.10	2.11	2.12	2.13
Detail Category $\Delta\sigma_c$			T≤4	31	28	25	25	25	25	25	25	As Types 2.1 to 2.8, but reduced by one Detail Category	18	25	28	31
m_1 = 3,2 for all types			4<T≤10	31	28	25	22	22	22	22	22					
			10<T≤15	31	28	25	22	20	20	20	20					
			15<T≤25	31	28	25	22	20	18	18	18					
			25<T≤40	31	28	25	22	20	18	16	16					
			T > 40	31	28	25	22	20	18	16	14					
Adjustment for stress gradient				where $\Delta\sigma_1$ and $\Delta\sigma_2$ are of opposite sign increase by 2 Categories where T≤15mm or 1 Category where 15<T≤40mm												

Key: * Weld toe shall be fully ground out → requirement continuous from left to right

Figure 3.5.

DRAFT Page 39
prENV 1999 Part 2 : 1996
CEN/TC250/SC9/N89

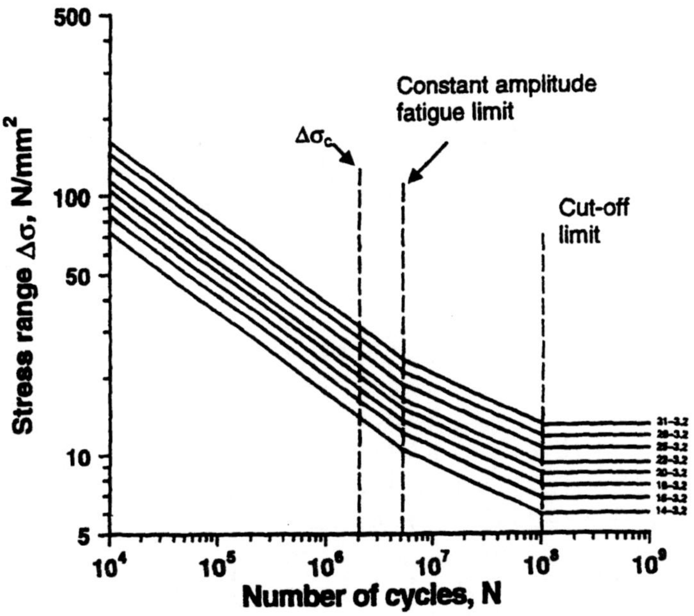

Fig.5.2.2(a) Δσ–N curves for members with welded attachments - transverse weld toe

Table 5.2.2(a) Numerical values of $\Delta\sigma$ (N/mm^2) for members with welded attachments - transverse weld toe				
Detail Category (N = 2 x 10^6)		N = 10^5	N_D = 5x10^6	N_L = 10^8
$\Delta\sigma_c$	m_1	$\Delta\sigma$	$\Delta\sigma_D$	$\Delta\sigma_L$
31	3,2	79,1	23,2	13,1
28	3,2	71,4	21,0	11,8
25	3,2	63,8	18,8	10,6
22	3,2	56,1	16,5	9,3
20	3,2	51,0	15,0	8,4
18	3,2	45,9	13,5	7,6
16	3,2	40,8	12,0	6,8
14	3,2	35,7	10,5	5,9

Figure 3.6.

DRAFT Page 42 prENV 1999 Part 2 : 1996 CEN/TC250/SC9/N69

Table 5.1.3. Detail Categories for Welded Joints Between Members

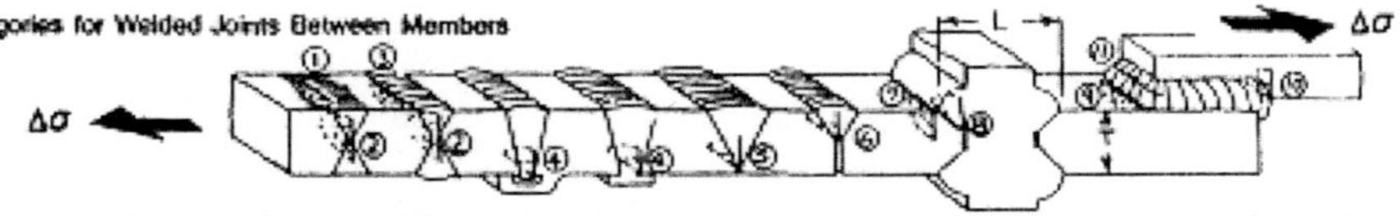

Product Forms			Rolled, extruded and forged products		→	→	→	→		→	→	→	Castings
Initiation sites	Reference No.		1, 2	3, 2	4	5	6	7		8	9, 10	11	as Types 3.1 to 3.10
	Locations		surface or embedded discontinuity	weld toe or embedded discontinuity	weld root	root discontinuity	unfused root	weld toe	weld toe	weld root	weld toe	weld root	
Stress orientation (see 4...)			Normal to weld axis		→	→	→	→		→	→	→	
Alloys			As tab 1.1.1	→	→	→	→	→		→	→	→	tab 1.1.2
Particular Require-ments	Dimensional	Joint type	In-line butt	→	→	→	→	Cruciform or Tee		→	Lap	→	as Types 3.1 to 3.10
		Weld type	Butt	→	→	→	→	Butt			Fillet	→	
		Preparation	Double sided	→	Single sided	→	→						
		Penetration	Full	→	→	→	partial	full	partial	→			
		Transition	Taper slope ≤1 in 4 at width or thickness change	→			→						
	Manufacturing	Root	Ground	→	Backed	Unbacked		Ground					
		Cap	Ground flush										
		Ends	Extension plates used on ends, cut off and ground flush in direction of Δσ					→	→	→			
	Inspection/Testing		As Annex D Table D.1		→	→	→	→		→	→	→	
	Quality Standard		As Annex D Table D.2		→	→	→	→		→	→	→	
Stress Analysis	Stress parameter		Net throat + specified misalignment stress (no overfill)				→	net section		net throat	net section	net throat	
	Stress concentration effects included in Detail Category		Profile, unspecified misalignment and discontinuities permitted by Annex D Table D.2						→	→	→	→	
								Stiffening effect of transverse element			stress peaks at weld ends		
Type Number			3.1	3.2	3.3	3.4	3.5	3.6	3.7	3.8	3.9	3.10	3.11→3.20
Detail Category $\Delta\sigma_c - m_1$		Flats, solid	(55-6) **44**-5	(39-4) **35**-4	} (35-4)	} (29-3,2)	}	as Table 5.1.2(a) Types 2.1 to 2.9		}	as Types 3.6 and 3.7	}	as Types 3.1 to 3.10 less 1 Detail Category
		Open shapes	(44-5) **28**-4	(35-4) **28**-4	} **25**-3,2	} **18**-3,2	} **14**-3,2			} **18**-3,2		} **14**-3,2	
		Hollow	NA	NA	}	}	}			}		}	

Key: NA = not applicable, → requirement continuous from left to right ( ) Before using values in brackets see 5.2.1(10)

Figure 3.7.

DRAFT Page 43
prENV 1999 Part 2 : 1996
CEN/TC250/SC9/N89

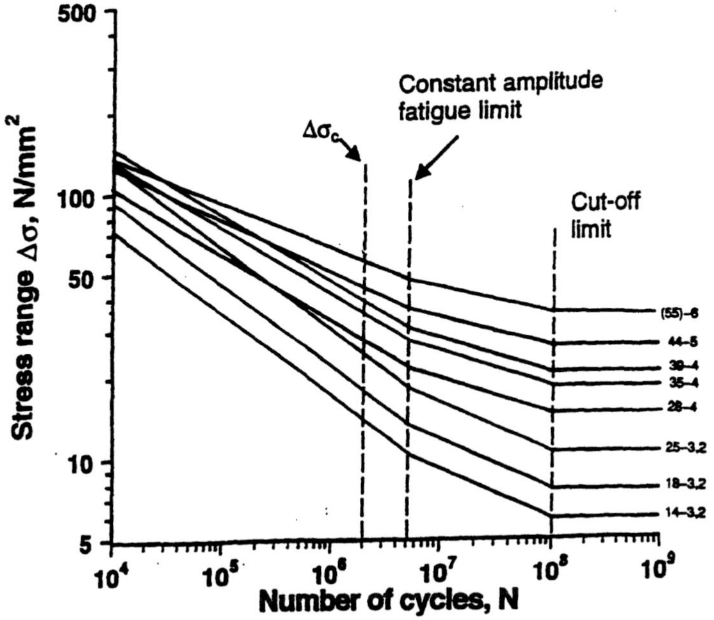

Fig.5.2.3 $\Delta\sigma$-N curves for welded joints between members

Table 5.2.3 Numerical values of $\Delta\sigma$ (N/mm²) for welded joints between members				
Detail Category ($N_c = 2 \times 10^6$)		$N = 10^5$	$N_D = 5\text{x}10^6$	$N_L = 10^8$
$\Delta\sigma_c$	m_1	$\Delta\sigma$	$\Delta\sigma_D$	$\Delta\sigma_L$
(55)	6	90,6	47,2	32,5
44	5	80,1	36,6	23,9
39	4	82,5	31,0	18,8
35	4	74,0	27,8	16,8
28	4	59,2	22,3	13,5
25	3,2	63,8	18,8	10,6
18	3,2	45,9	13,5	7,6
14	3,2	35,7	10,5	5,9

Figure 3.8.

5.2.4 Hot spot stress

Values of $\Delta\sigma_c$ for hot spot stress assessment of weld toes are given in table 5.2.7

Table 5.2.7 $\Delta\sigma_c$ for hot spot stress assessment	
Thickness of stressed member T (mm)	$\Delta\sigma_c$ N/mm²
$0 < T \leq 4$	44
$4 < T \leq 10$	39
$10 < T \leq 15$	35
$15 < T \leq 25$	31
$25 < T \leq 40$	28
$T > 40$	25

Figure 3.9.

5.5 Improvement Techniques

(1) The fatigue strength of certain detail types shown in tables 5.1.1 to 5.1.5 may be improved by the application of special manufacturing techniques. These are generally expensive to apply and present quality control difficulties. They should not be relied upon for general design purposes, unless fatigue is particularly critical to the overall economy of the structure, in which case specialist advice should be sought. They are more commonly used to overcome existing design deficiencies.

(2) The following techniques have been used on aluminium alloys and are most effective for high cycle applications.

(a) Introduction of compressive residual stresses at the location of crack initiation. This may be carried out at transverse weld toes by peening. At bolt holes the cold expansion method may be used.

(b) Reduction of stress concentration effect at the location of crack initiation. This may be carried out by grinding transverse weld toes to a smooth profile.

(3) Further information on improvement techniques is given in Annex E.

Figure 3.10.

4. FATIGUE DESIGN BY LOCAL AND STRUCTURAL STRESSES

As aforesaid dealing with the fatigue design of welded joints, fatigue is a local phenomenon which cannot be understood taking into account only the nominal stress, usually sufficient for the static design. For the evaluation of the stresses, several methods can be used, as shown in fig. 4.1. In the practical applications the most used techniques are at present the strain gauges, among the experimental techniques, and the finite elements (displacement method) among the computational methods. The local elastic peak stress of a structure can also be evaluated directly by multiplying the nominal stress by a stress concentration factor K_t, which is usually available in the literature for the most typical stress risers found in mechanical components.

The nomenclature used by different authors to define the various stresses which can be defined when analysing a stress riser is not univocal. For this reason attention should be paid in understanding and clarifying the meaning of each stress definition. Even the nominal stress is not always defined in an explicit way, because it can be seen from fig. 4.2 that also the very simple case of an hole in a strip can be confusing, since two different definitions of the nominal stress can be given: the gross stress σ_g and the net stress σ_n. Depending on which stress is adopted, the procedures to estimate the fatigue strength of the notched component from the fatigue strength of the base material are very different.

In real structures the situation can be much more complicate, due for example to the presence of a secondary bending, as shown in fig. 4.3 for some welded joints. As it can be seen, secondary bending can be directly caused by the shape and loading conditions of the component, or it can be the consequence of manufacturing imperfections. In both cases the reference nominal stress does not take into account the secondary bending. That is why the fatigue strength of some welded joints can appear to be very low, as already shown for spot welds in fig. 2.5.

The nomenclature problems are much more relevant when the local stresses are concerned. The stress gradient is localised and is governed by an exponential law, as shown in fig. 4.4. for a welded joint. Structural stresses are usual defined as the stresses due to the overall joint geometry and to the applied loads without taking into account local stress concentrations. In general the variation of such stresses is linear and in such a way it has been represented in fig. 4.4. Structural stresses are used to define the hot spot stress, as the stress evaluated by extrapolation of the structural stress field up to the weld toe (the origin of the x-axis in fig. 4.4). When the two stress fields (local and structural) are well defined and separated as in the shown example, the evaluation of the hot spot stress is easy and clear since the hot spot stress is actually the nominal stress evaluated at the weld toe. Since in practical situations the stress fields can be much more complicate, it is not always possible to clearly separate the structural stress field and so define the hot spot stress. For this reason it can be useful the evaluation of a "geometric stress" as defined in chapter 5.2. As far as the fatigue strength of the joint is involved, since the hot spot stress is simply a nominal stress evaluated at the weld toe, it can be argued that for a given material the fatigue strength is influenced both by the joint geometry and absolute dimensions, while only the fatigue strength in terms of local stress is always the same for the same material.

The elastic peak stress can reach a peak value (notch stress) or not, depending on the tip radius being different from zero or not. In the latter case the stress field is singular and it can be represented in a log σ-log x diagram with a straight line. Since the radius at the weld toe is very small for the usual welding techniques, in this note the problem has been simplified by assuming a radius $\rho=0$. The accurate evaluation of the local stress field for welded joints requires a very fine mesh, as shown for example in figures 4.5.e 4.6. These refined model are usually not intended for practical application but rather for the development of design methods.
The assumption of a tip radius equal to zero makes the stress field around the weld toe in some way similar to the stress field around a crack and then fatigue design criteria based on local stresses, as is usual in fracture mechanics, can be developed.
As it is well known in fracture mechanics, a crack in a strip subjected to tension produces a singular and exponential stress field around the tip (for distances less than one tenth of a millimetre) as shown in fig. 4.7.When the tip radius is small but different from zero the stress field is not singular anymore but again exponential, as shown in the same figure. The governing equations for the case of $\rho=0$ are shown in fig. 4.8. Then the critical condition is not found by imposing a limiting value of the stress, but by imposing a limiting value for the whole stress field, as shown again in fig. 4.8. The limiting value of the stress field should be found by appropriate experimental testing and is constant for a given material. When instead of a crack the notch is a slot with parallel edges but with a tip radius not so small, the stress field can be represented in a schematic way as shown in fig. 4.9 (in reality the transition from one zone to another is gradual). In this case the fatigue assessment can be accomplished by using one of the three methods shown in the figure. The choice of the method depends on the available experimental data on the fatigue strength of the material. The evaluation of the stress which is necessary to apply one of the three methods, requires different analyses: the nominal stress method requires just a simple stress analysis, the peak stress method requires a more detailed stress analysis. In each case the diagram given in the figure clearly indicates the distances from the notch tip where the stresses are to be evaluated. Similar considerations can be applied also to notches with an opening angle greater than zero. In fact the stress field is still exponential but with an exponent lower than 0.5, which is valid for a crack (for example for a notch opening angle equal to 135°, which is typical for welded joints, the stress field exponent is equal to 0.326). As a consequence also the values of x_p and x_g will be different from the ones shown in fig. 4.9.

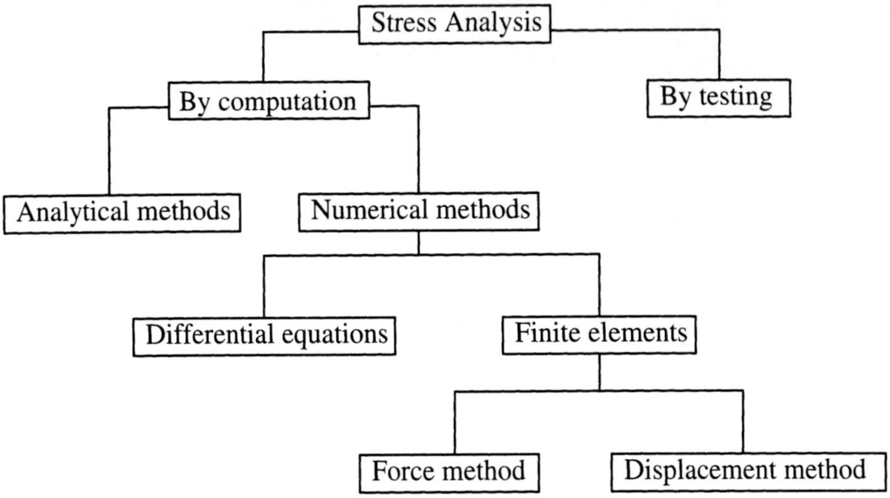

Figure 4.1: Different methods for stress analysis

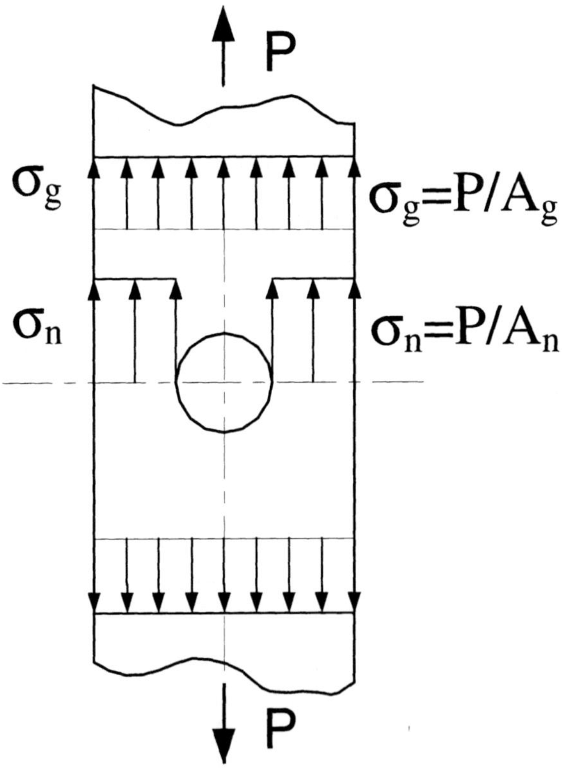

Figure 4.2: Definition of nominal stresses for a simple case

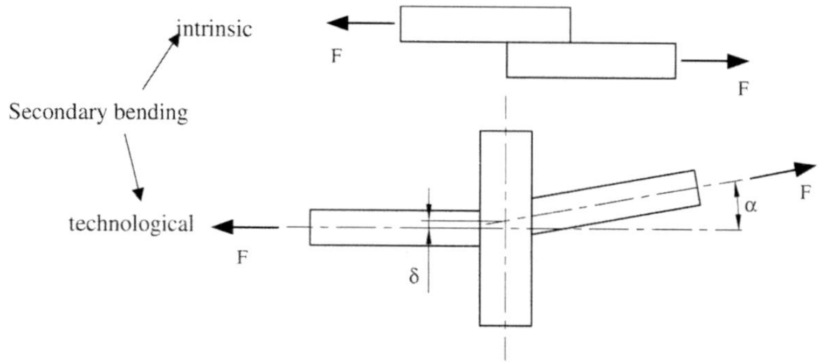

Figure 4.3: Causes of secondary bending effects

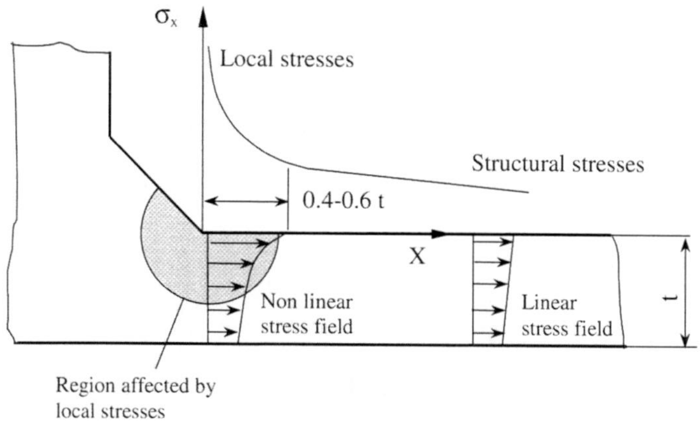

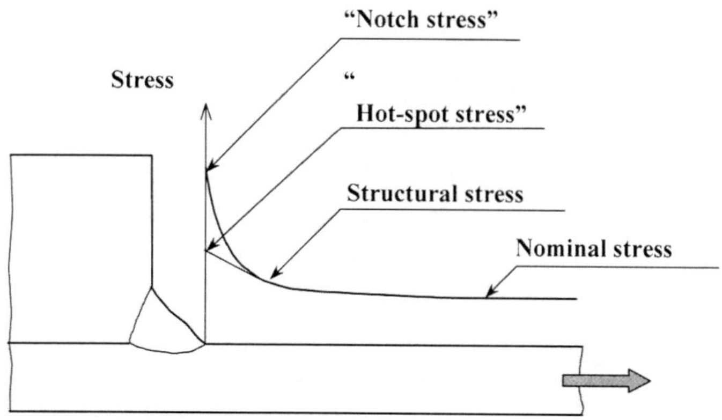

Figure 4.4: Local and Structural Stresses

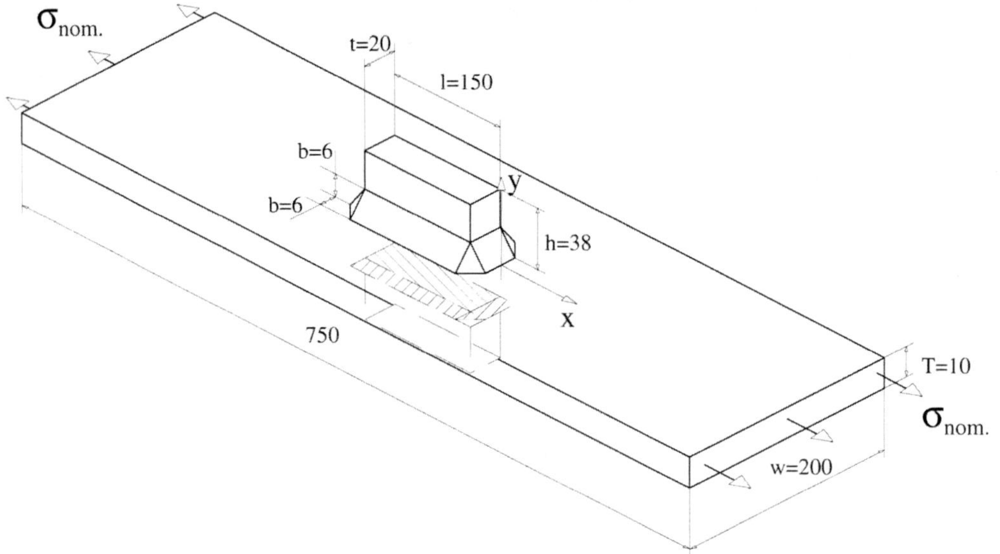

Figure 4.5: Example of a welded structural detail to be analysed

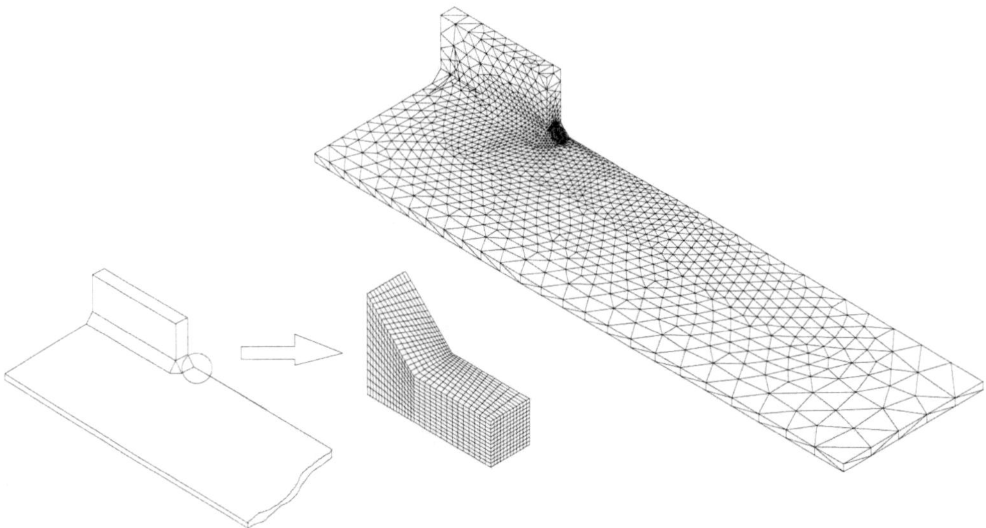

Figure 4.6: Example of the FEM mesh to obtain the local stress field

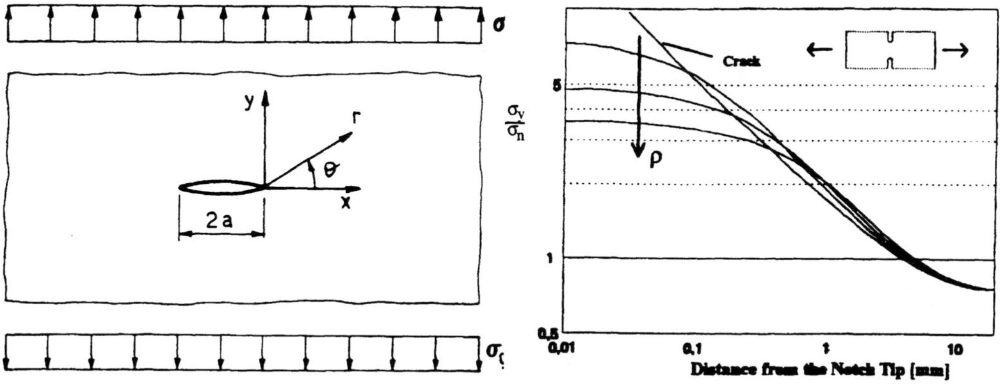

Figure 4.7: Crack in an infinite plate subject to opening loads and local stress field

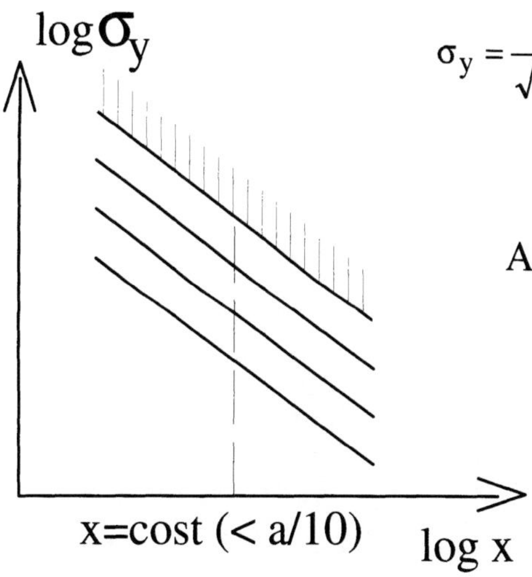

For a crack in an infinite plate:

$$\sigma_y = \frac{K_I}{\sqrt{2 \cdot \pi \cdot r}} \cdot f_y(\theta) \qquad \text{being} \qquad K_I = \sigma_g \cdot \sqrt{\pi \cdot a}$$

$$\Rightarrow \sigma_y = \sigma_g \cdot \sqrt{\frac{a}{2 \cdot r}} \cdot f_y(\theta)$$

Along the x axis $f_y(\theta) = 1$ then:

$$\sigma_y = \sigma_g \cdot \sqrt{\frac{a}{2 \cdot x}}$$

For a general geometry:

$$\sigma_y = \alpha \cdot \sigma_g \cdot \sqrt{\frac{a}{2 \cdot x}}$$

Figure 4.8: Fracture Mechanics: fundamental concepts

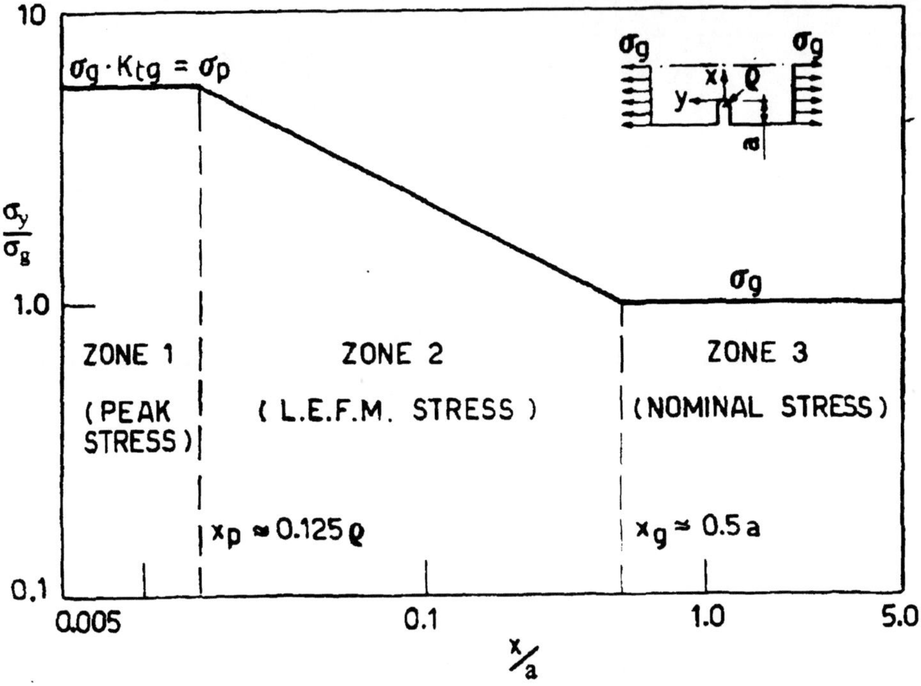

Figure 4.9: Local stresses necessary to the application of different design approaches

5. FATIGUE DESIGN OF WELDED JOINTS BY LOCAL APPROACHES: COMPARISON BETWEEN COMPUTATIONAL AND EXPERIMENTAL APPROACHES

The approaches based on local or structural stresses for fatigue design of welded joints have been reviewed in the previous chapter. Since the fatigue strength depends on the local stress state at the critical point, then it was demonstrated that only the local approach is able to interpret the fatigue test results of welded joints of different geometries and absolute dimensions. Now, since the local stress field can be experimentally or numerically evaluated, the differences between these two techniques, the advantages and disadvantages and the correlation between theoretical calculation and experimental measurements will be discussed. Since a re-analysis of experimental results on joints in aluminium oriented to this issue is not available, we will refer to a similar analysis relative to welded joints in structural steels [Atzori 01].

As it was explained in the previous chapter, it is possible to define a notch-stress intensity factor (N-SIF) for welded joints, similarly to the stress intensity factor adopted in the field of fracture mechanics. By using the N-SIF approach it is possible to define a unique scatter band able to fit the experimental data of cruciform welded joints characterised by different geometries and absolute dimensions, as illustrated in fig. 5.1 and 5.2. The data shown in the figures refer to structural steel joints and the complete re-analysis is reported in [Lazzarin 98].

Even if fig. 5.1 can be used in fatigue assessment of welded joints in structural steel, in view of a comparison between the local stress field estimation by means of numerical and experimental analyses, the same data are reported in fig. 5.3 in terms of local strain, which were numerically evaluated on the basis of the applied nominal strain. The numerical analyses were performed by assuming a "nominal" joint geometry without any geometrical imperfection. Fig. 5.3 contains also other data not included in fig. 5.1. The amount of available data allowed to define a new unified scatter band reported in fig. 5.4, which is more accurate than that proposed in the past without any theoretical support [Haibach 68].

The agreement between the unified scatter band and the experimental results is shown in fig. 5.5, which demonstrates the agreement with the experimental results.

Since the obtained scatter band is valid for steel welded joints with transverse continuous welds subjected to both tension and bending loads, it was applied also to experimental results obtained by means of strain-gauged joints, where the local strains are experimentally measured. Both V-type joints subjected to bending load [Atzori 79], reported in fig. 5.6, and cruciform joints subjected to tension load [Dattoma 91, Atzori 92], reported in fig. 5.7 were considered. In the former case no secondary bending effects are present (due to the applied bending load), while in the latter case, due to the tension load, the secondary bending stresses can be very high, as the strain-gauge measurements reported in the upper part of fig. 5.7 demonstrate. The experimental results of the latter series, analysed by means of the "nominal geometry" (without any geometrical imperfection) are in agreement with the experimental fatigue strength reported in fig. 5.5, which refer also to axially loaded joints. The agreement is also highlighted in fig. 5.7. Conversely the fatigue strength of V-type joints subjected to bending load is quite higher, as can be seen again from fig. 5.7. Now if the local strain fields numerically evaluated are re-calibrated on the basis of the experimentally measured strain, which necessarily accounts for the secondary bending effects, close to the weld toe (where fatigue cracks initiate) then the local strain at 0.01 mm far from the weld toe in the case of axial load is in excellent agreement with the same value in the case of bending load, as shown in fig. 5.7. This demonstrates that the different fatigue strengths in tension and bending are due to the presence of a secondary bending effect in the former case, which the numerical analyses do not take into account, but the experimental strain measurements do. This is an advantage of the experimental determination of the "local" strain field. On the other hand, the exact intensity of the local strain field cannot be derived from the experimental measurements, both due to the excessive strain gauge dimensions related to the local strain field extension (see fig. 5.7) and to difficulty to measure its distance from the weld toe because the weld bead is irregular and a weld toe radius different from zero is always present.

Since the fatigue strength depends on the local stress field intensity, the strain experimentally measured at a fixed distance from the weld toe will vary for a given fatigue life as the joint geometry and absolute dimensions vary, as illustrated in fig. 5.8. In fact, being the local strain field the same (and so the fatigue life), the strain value at the distance of 2.5 mm from the weld toe, where strain gauges are usually located, is different for different geometries and absolute dimensions, being higher for joints of smaller dimensions. For the same reason a reference strain value at 2.5 mm from the weld toe will produce a more critical local strain field (and so a lower fatigue life) for a joint of bigger dimensions. Due to that, a fatigue assessment procedure based on reference strain values experimentally determined at a given distance from the weld toe will on the be safe side if applied in fatigue design of welded joints characterised by absolute dimensions equal to or lower than those adopted to experimentally derive the reference strain value. Otherwise the fatigue assessment will not be safe.

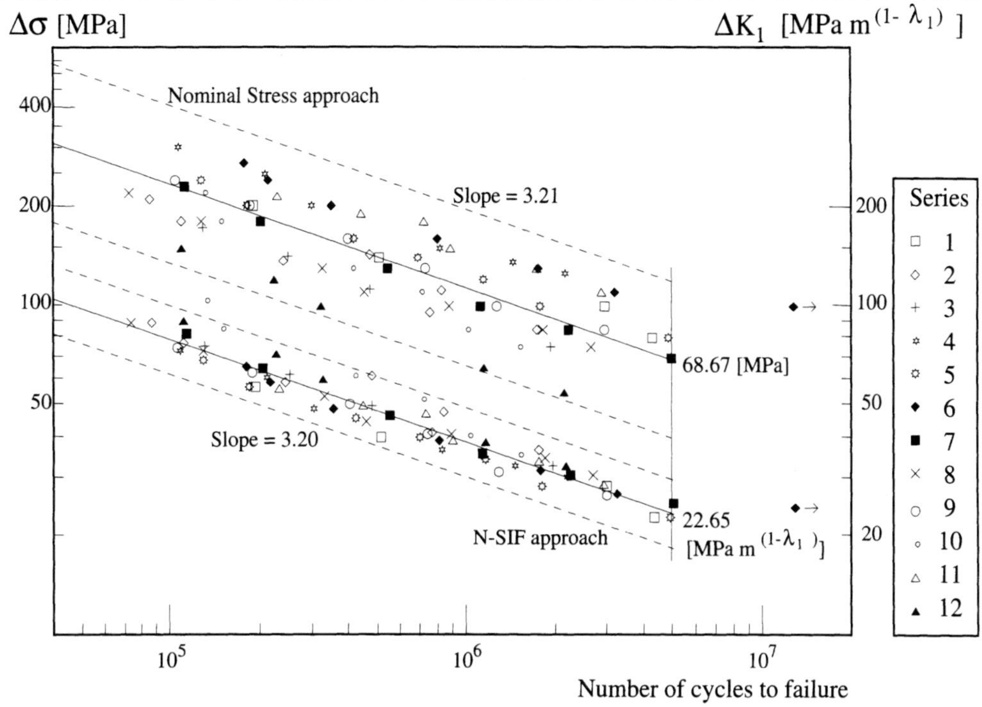

Figure 5.1: Results of fatigue tests on cruciform welded joints in terms of nominal stress or local stress field

Series	t (mm)	$2h/t$	L/t
1	13	1.231	0.769
2	50	0.640	1.0
3	100	0.320	0.5
4	13	0.769	0.231
5	13	1.538	0.615
6	25	0.4	0.120
7	25	0.720	1.28
8	25	1.2	8.80
9	38	0.421	0.342
10	38	0.789	5.789
11	100	0.1	0.030
12	100	0.3	2.200

Figure 5.2: Geometrical data of the test series reported in fig. 5.1

t=main plate thickness; h=weld bead dimension; L=transverse attachment thickness

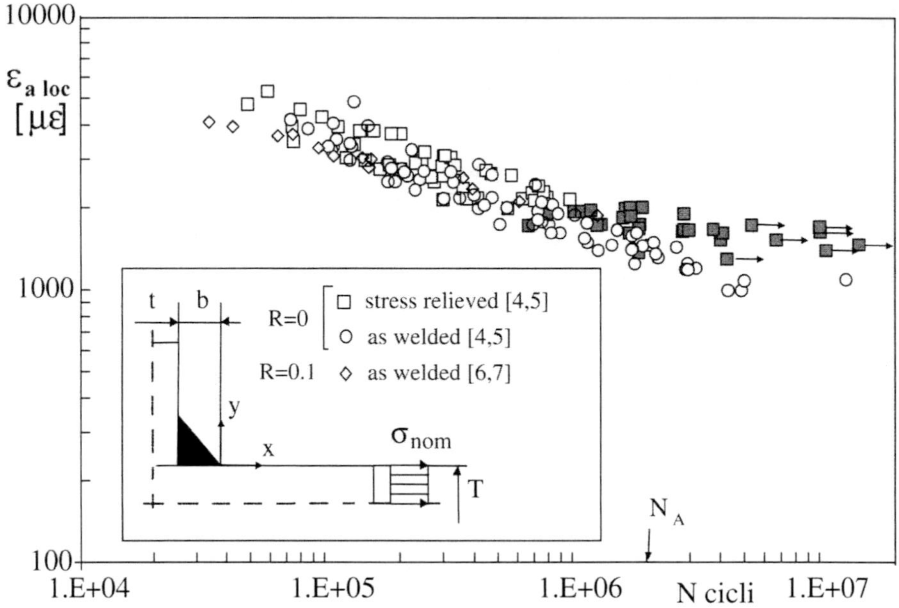

Figure 5.3: Experimental results taken from literature in terms of local strain amplitude calculated at 0.01 mm from the weld toe

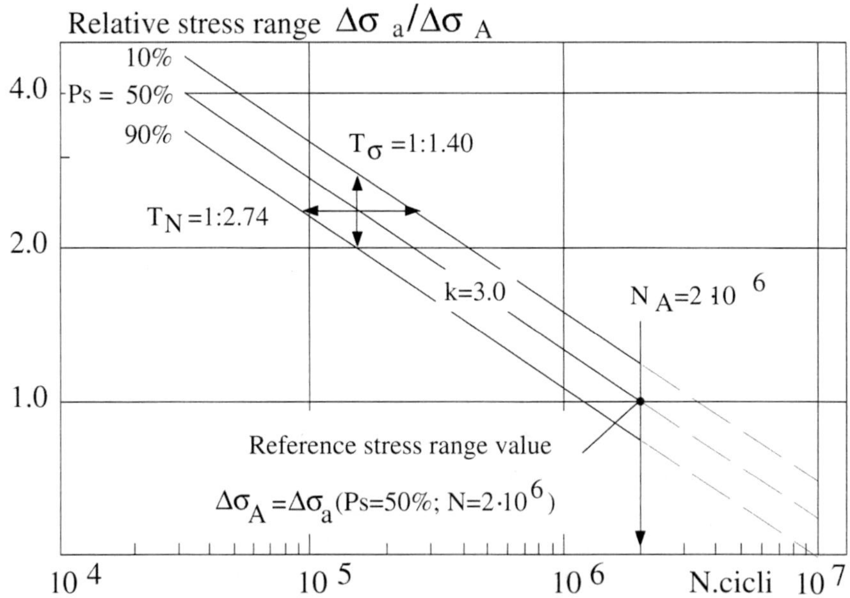

Figure 5.4: Uniform scatter band for welded joints in steel based on local stress analysis

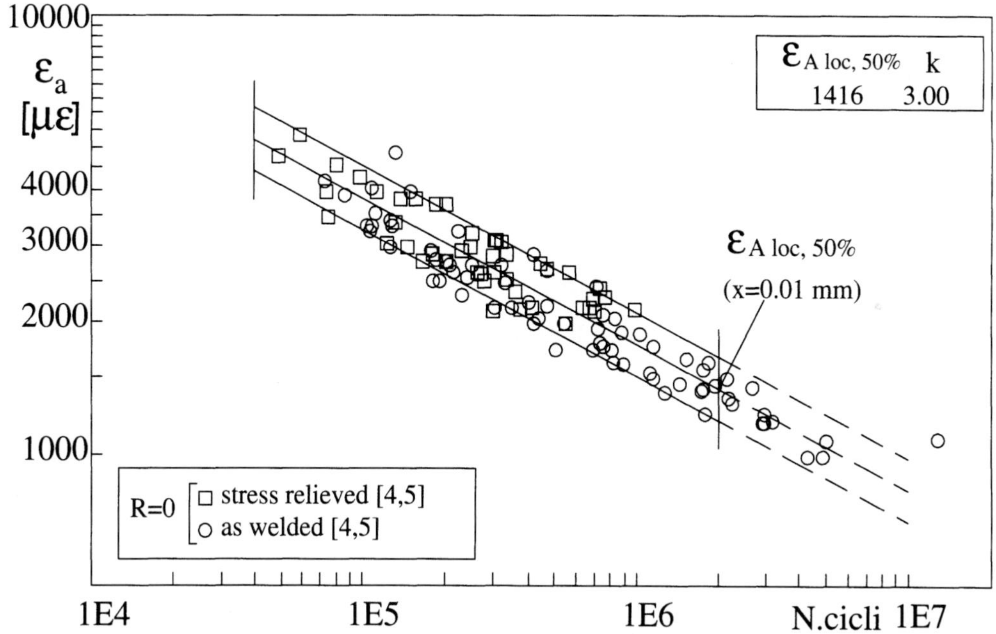

Figure 5.5: Uniform fatigue strength for welded joints in steel when evaluated on the basis of local strain

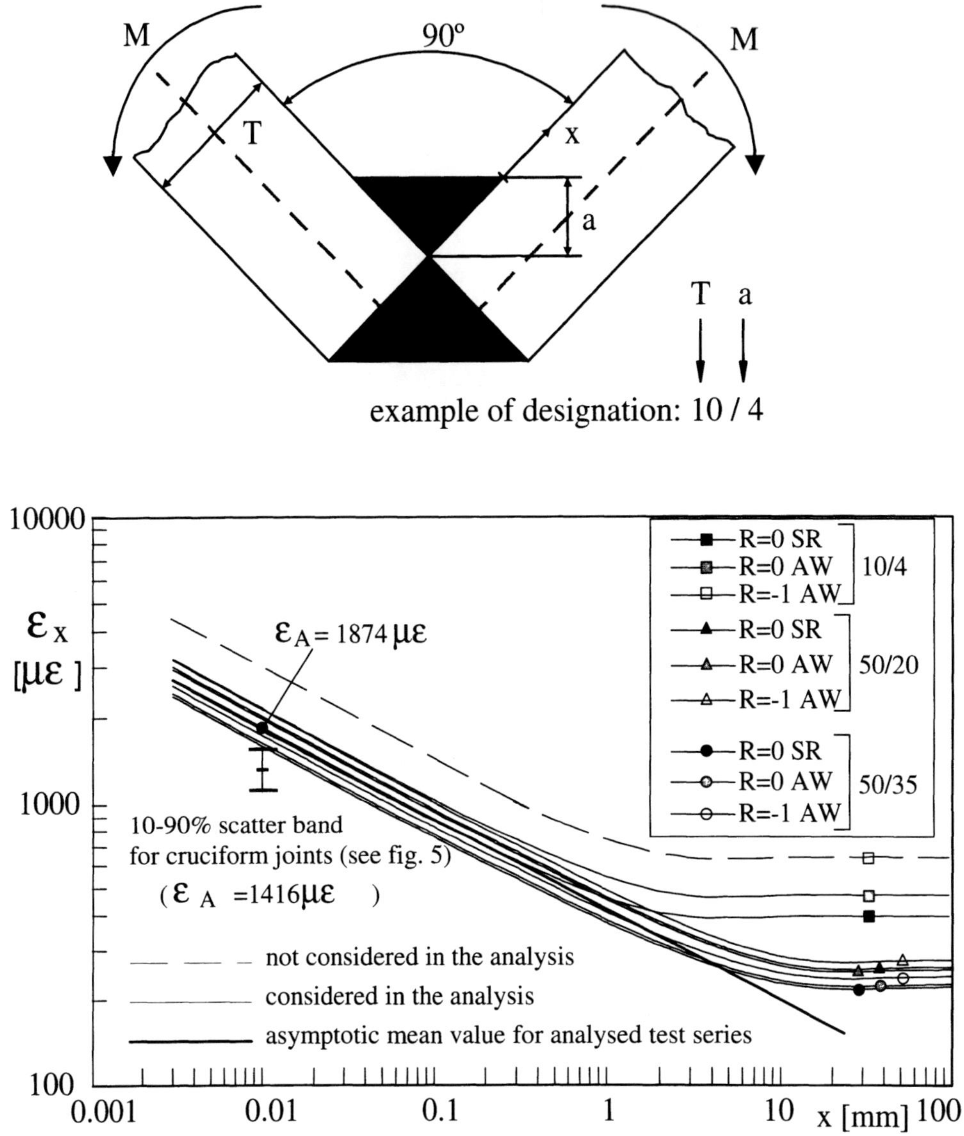

Figure 5.6: Local strain fields for V welded joints at 2 million cycles life

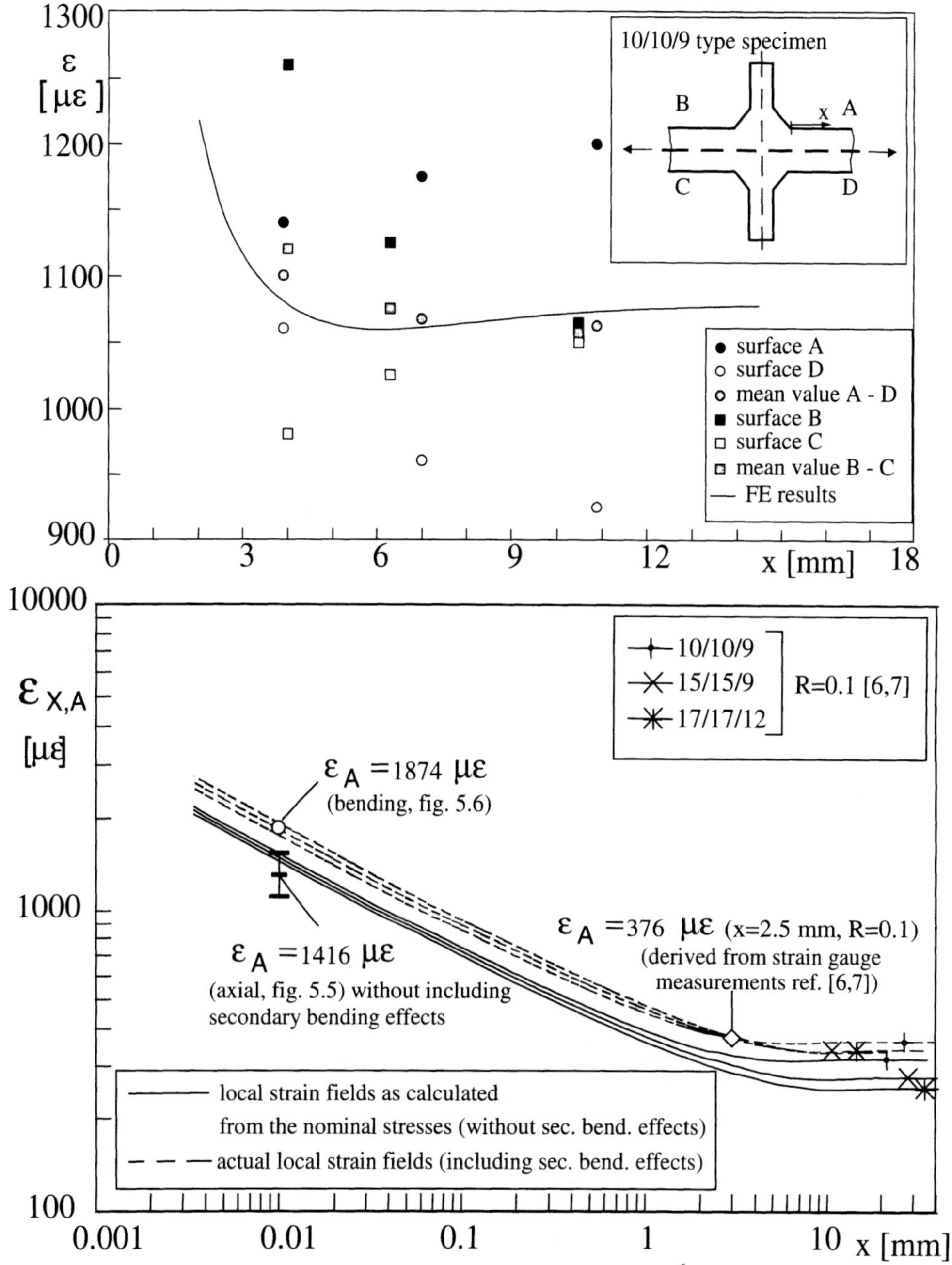

Figure 5.7. Secondary bending effects on local strain fields at $2 \cdot 10^6$ cycles

$\varepsilon_{x,A}$: numerical strain fields evaluated for a fatigue life of $2 \cdot 10^6$ cycles

ε_A: local strain amplitude for $N=2 \cdot 10^6$ cycles and 50% probability of survival

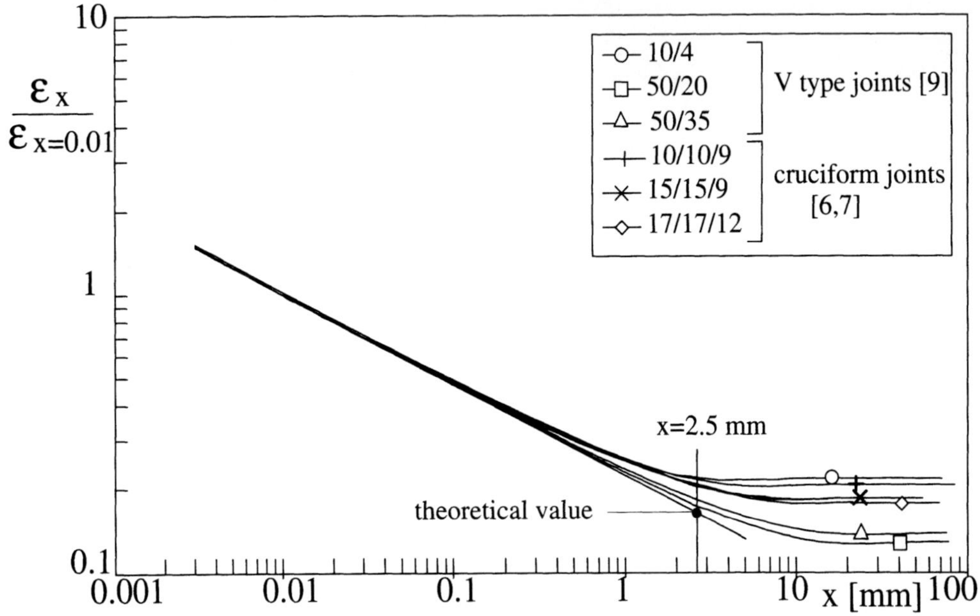

Figure 5.8: Local strain fields normalised with respect to the strain value at x=0.01 mm

6. NOTCH SENSITIVITY AND DEFECT SENSITIVITY FOR CAST ALLOY COMPONENTS

It is well-known that the mechanical properties of cast aluminium alloys are strongly influenced by the metallurgical and microstructural features. The main parameters, that have to be considered to correctly understand their mechanical behaviour, are: the shape and the size of grains, the dendritic arm spacing (DAS), the microstructural characteristics of the phases and the presence, the amount, the size and the shape of porosity and defects [Bonollo 99, Gall 00]. Almost all the above mentioned parameters depend mainly on the casting manufacturing process: as a consequence if the process is varied both the static and the fatigue material properties can substantially change.
The static properties of AA356-T6 are mainly influenced by the grain size, by DAS and by the size and number of precipitates (Mg_2Si), whereas the influence on the static strength of porosity and the inclusions seems to be negligible [Stanzl 95].
Figure 6.1 presents a re-analysis of the data collected from literature [Stanzl 95, Rosso 97, Torti 96, Baumel 91, Beretta 97, Mahoney 89, Odegard 94, Stephens 89, Shiozawa 97]. First of all, several authors highlight that the material static strength can be improved by obtaining grains of size lower than 800μm. Moreover, figure 6.1 shows that the tensile strength σ_T decreases from 300 MPa to 220 MPa and the yield strength σ_y from 230 MPa to 160 MPa as the DAS increases from 20μm up to 90μm. The tensile strength values are characterised by a lower dispersion with respect to that showed by the yield strength data. Finally, the DAS influence on the fracture toughness K_{IC} seems not to be crucial: a value for K_{IC} of 17.4 MPa·m½ can be suggested as an average reference value [Gall 00, Koh 89] for AA 356-T6, whatever the casting process.
Fatigue crack initiation in AA356-T6 generally occurs in AA356-T6 near to the biggest pores, near to the inclusions greater than 100μm in size and near to the defects. In general terms, the fatigue behaviour is affected by DAS and by amount and size of pores [Bonollo 99, Gall 00] located close to the component surface [Seniw 00]. In particular, the fatigue strength increases as the DAS decreases [Couper 90-a, Honma 84] and a DAS value lower than 40μm is always advised. Moreover, in order to improve the medium-cycle fatigue strength, some authors (for example [Wickberg 84]) highlight the importance to reduce the Fe amount down to 0.2% or less.
The first diagram reported in figure 6.2 shows the fatigue strength $\Delta\sigma_0/2$ as a function of the DAS (where $\Delta\sigma_0/2$ is the stress amplitude at 2·106 cycles to failure and for a 50% probability of survival): a reduction of DAS slightly increases the fatigue strength, but the data dispersion does not allow to establish a sound $\Delta\sigma_0/2$ vs. DAS relationship. Similarly, figure 6.2 highlights that a direct relation between DAS and the inverse slope of the Wöhler curve k does not exist, since high cycle fatigue behaviour is mostly influenced by the size of the pores.
Unfortunately, the available data did not allow us to perform a systematic study of the porosity influence on the fatigue strength. Then the experimental data have been subdivided in two different groups according to the casting manufacturing process, which the obtained porosity depends on. Table 6.1 shows that a fatigue strength $\Delta\sigma_0/2$ of 115 MPa and an inverse slope of 7 can be suggested when the casting process allows to obtain a pore size lower than 180μm and a porosity amount less than 0.4%. Conversely, when the amount of porosity is greater than 0.4% and pore size ranges between 300μm and 800μm, the fatigue strength $\Delta\sigma_0/2$ drops to an average value of 76

MPa, while concerning the inverse slope k the data dispersion does not allow to give a significant reference value.
Figure 6.2 and table 6.1 demonstrate that the microstructure strongly influences the fatigue properties of AA356-T6 alloy and a critical analysis of the metallurgical aspects is the unique way to correctly understand the correlation between foundry process and fatigue behaviour. Due to the lack of information about the microstructural parameters and in order to suggest some practical formulae to estimate $\Delta\sigma_0$ and k values, the fatigue ratio $\Delta\sigma_0/(2\sigma_T)$ has been introduced. The diagrams plotted in figure 6.3 show that the fatigue ratio allows all the experimental data to collapse in a restricted scatter band. In particular this holds true for $\Delta\sigma_0/2$ values.
The parameters which influence the threshold value of the stress intensity factor ΔK_{th} for long cracks were also analysed. Figure 6.4 shows that ΔK_{th} increases as the ratio $\Delta K_{th}/\sigma_T$ increases, being the load ratio R ($R=K_{min}/K_{max}$) unchanged. Figure 6.4 highlights both that ΔK_{th} decreases as R increases, when R is greater than 0, and that it does not vary when $-1 \le R \le 0$, at least according to the few available data. This figure also shows that two different ΔK_{th} vs. R trends exist, due to the different material tensile strengths. The available data for R=0.1 allowed to verify that if $250<\sigma_T<276$ MPa then $\Delta K_{th}=6\div7 \text{MPa}\cdot\text{m}^{1/2}$, whereas if $290<\sigma_T<310$ MPa then $\Delta K_{th}=4\div5 \text{MPa}\cdot\text{m}^{1/2}$. However, the DAS influence on the threshold stress intensity factor seems to be negligible.
The Kitagawa diagram highlights the lower threshold nominal stress range exhibited by short cracks with respect to that expected if they behaved as long cracks.
Recently, Atzori and Lazzarin [Atzori 00] modified the Kitagawa diagram in order to account for both the sensitivity to defects (or short cracks) and the notch sensitivity. The link between the two fatigue behaviours has been established by considering the variation in the fatigue limit of a notch for a scale variation in a wide range of its absolute dimensions, i.e. being the elastic stress concentration factor K_{tg} unchanged (figure 6.5). The fatigue limit will be always in between the material fatigue limit $\Delta\sigma_0$ and the blunt notch fatigue limit $\Delta\sigma_0/K_{tg}$ characterised by full notch sensitivity. The sensitivity to small defects and the notch sensitivity in fatigue can then be represented in the same diagram. Subsequently, Atzori and Lazzarin [Atzori 01] proposed to define an equivalent notch depth ($\alpha^2\cdot a$) to fit with the same sloping curve different fatigue test results obtained with specimens of different geometry. Moreover, in order to correctly describe the real fatigue behaviour near to the a_0 value, they modified the classical Topper's relationship [El Haddad 79] as follows [Atzori 01]:

$$\Delta K_{th}=\Delta\sigma_0[\pi(\alpha^2\cdot a+a_0)]^{1/2} \qquad (1)$$

Doing so the a_0 definition does not depend on the shape coefficient α and coincides with that derived from Topper's equation by imposing $\alpha=1$.
In order to verify the Atzori-Lazzarin diagram applicability to the AA356-T6 aluminium cast alloy, a set of 50 rotating bending tests has been performed on notched cylindrical specimens obtained from the spokes of a number of wheels produced by a permanent mould casting process. The average values of the experimentally determined mechanical properties are: $\sigma_T=260$ MPa, $\sigma_y=192$ MPa, E=74800 MPa, $\Delta K_{th}=3.95$ MPa·m½ (R=-1), amount of porosity 0.3% and DAS=41μm. The notch geometry is reported in figure 6.6 and the main dimensions of each tested series and the shape coefficients α are listed in table 6.2. The shape coefficients α, calculated by

means of accurate finite element analyses, refer to a specimen containing a crack of the same depth as the notch. In table 6.2 the fatigue results are summarised in terms of fatigue strength range $\Delta\sigma_0$ (referred to the gross area) corresponding to $2\cdot10^6$ cycles to failure (for a 50% probability of survival), inverse slope k and ratio $T_\sigma=\Delta\sigma_{0,PS=90\%}/\Delta\sigma_{0,PS=10\%}$. Finally, figure 6.7 shows the sound agreement between the Atzori-Lazzarin diagram and the experimental fatigue test results.

The Atzori-Lazzarin diagram shown in figure 6.5 is applicable to different materials and technological processes, as shown by the diagram in figure 6.8, where experimental data for two carbon steels are plotted together. The diagram clearly demonstrates that the sensitivity to defects and the sensitivity to notches are different concepts, linked to different physical phenomena and to different absolute dimensions of the geometrical discontinuities of the material. The diagram shows also that both of them depend on the characteristic dimension a_0, which is often called "intrinsic crack length".

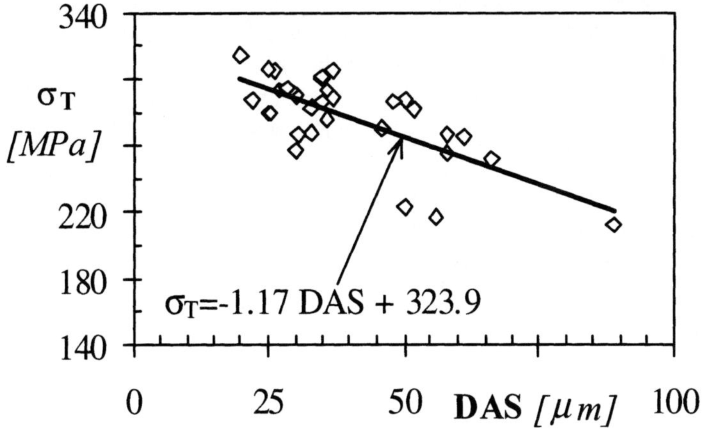

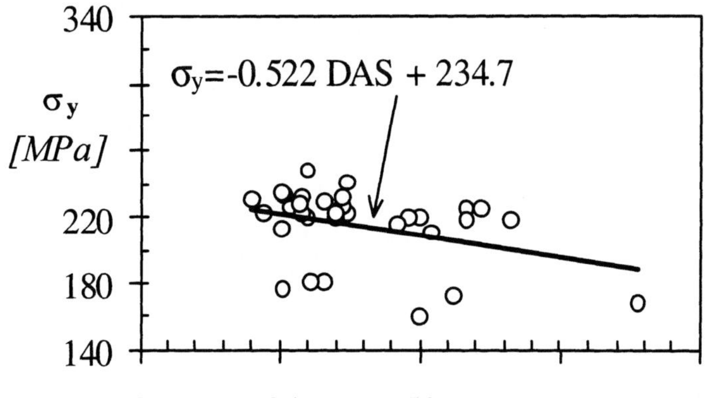

Figure 6.1: DAS influence on tensile strength σ_T and on yield strength σ_y [3-11]

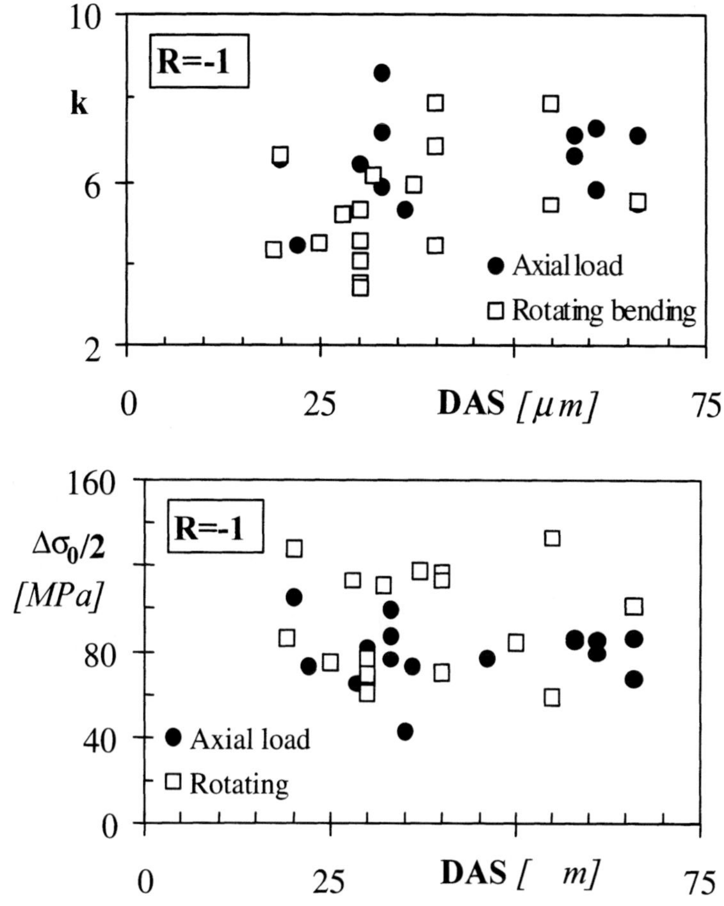

Figure 6.2: DAS influence on fatigue strength at $2 \cdot 10^6$ cycles to failure and on inverse slope k of the Woehler curve [3, 7, 9, 10, 11, 16, 17, 18]

Casting process	Porosity		$\Delta\sigma_0/2$ *[MPa]*	k	References
	Size	**Amount**			
Permanent- mould casting Sand-casting Die-casting	>300 μm and <800 μm	>0.4%	76±10	4.5÷7	3, 7, 9, 10, 15, 17, 19, 20
Direct-Chill casting Squeeze casting	<180 μm	<<0.4%	115±10	7	3, 9, 10, 11, 16, 18, 19, 20, 21, 22

Table 1: Average fatigue strength at $2 \cdot 10^6$ cycles to failure for R=-1 and inverse slope for different casting manufacturing processes

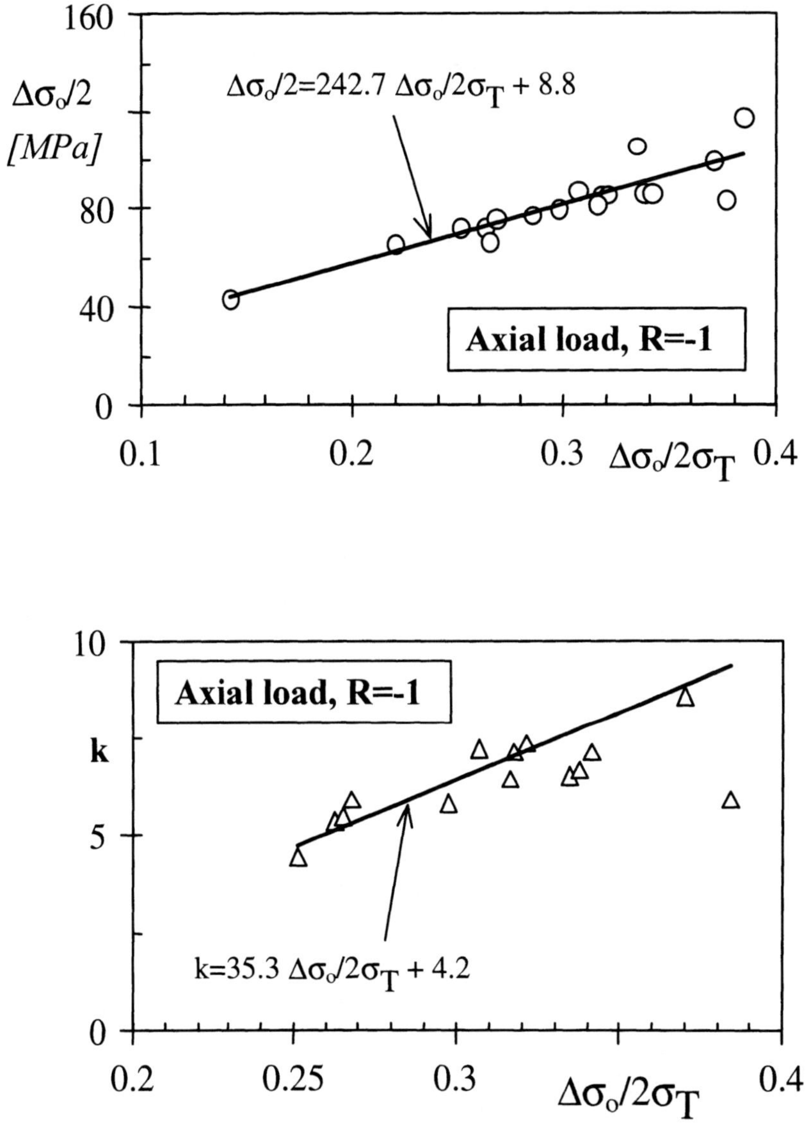

Figure 6.3. Fatigue ratio $\Delta\sigma_0/2\sigma_T$ influence on fatigue strength at $2 \cdot 10^6$ cycles to failure and on inverse slope k [3, 7, 9, 10, 11, 16, 18]

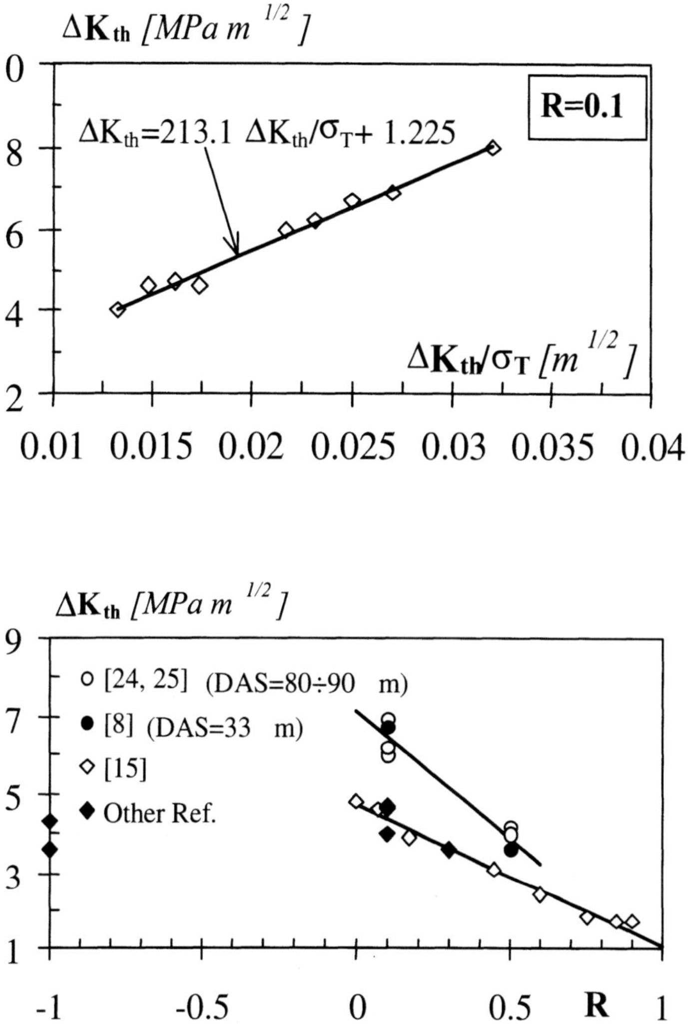

Figure 6.4: Tensile strength σ_T and load ratio R influence on the long crack threshold stress intensity factor ΔK_{th} [3, 8, 15, 19, 23, 24, 25, 26, 27]

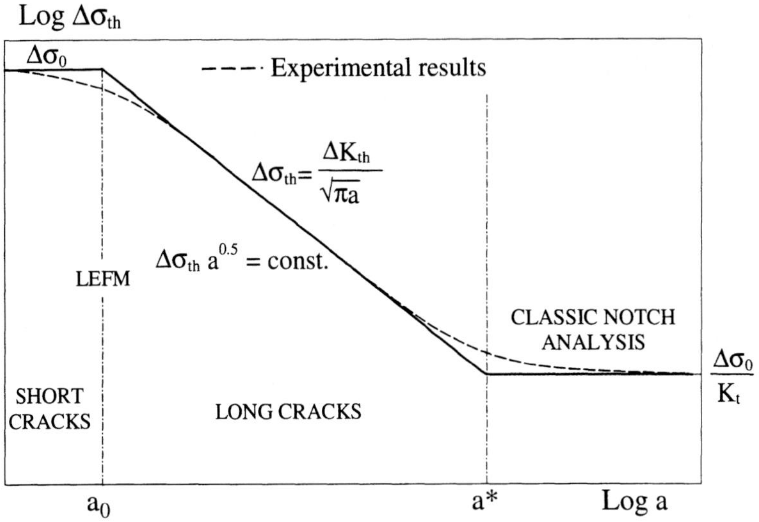

Figure 6.5: The Atzori-Lazzarin diagram

Code	d_g *[mm]*	ρ *[mm]*	a *[mm]*	β *[°]*	α	K_{tg}	$\Delta\sigma_0$ *[MPa]*	k	T_σ
SS	7.51	∞	0	180	1	1	231.0	6.88	1.306
NS02	7.56	0.10	0.24	79.9	1.141	4.05	114.8	4.51	1.188
NS04	9.04	0.18	0.63	59.2	1.201	5.17	109.8	5.86	1.276
NS1	9.03	0.09	1.30	60.3	1.557	14.25	44.8	3.57	1.596
NS3	11.98	0.08	2.90	68.3	2.780	39.50	17.7	4.49	1.337

Table 6.2: Specimen dimensions and fatigue test results

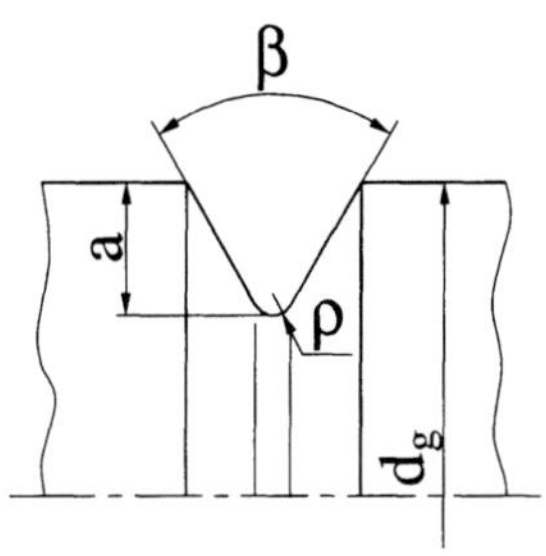

Figure 6.6: Notch geometry

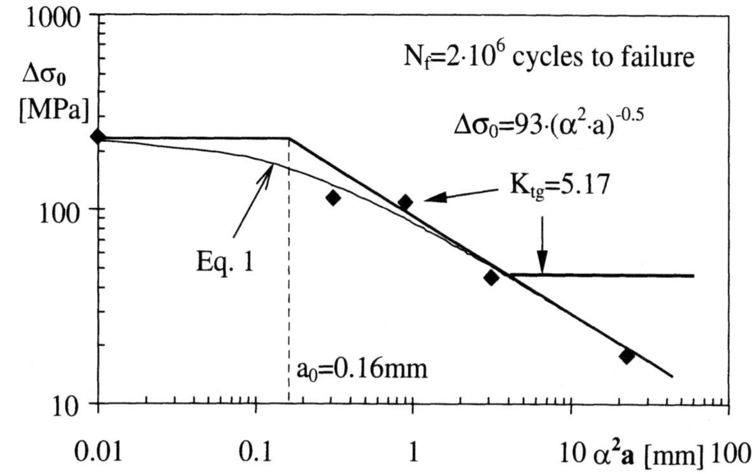

Figure 6.7: Fatigue strength of AA356-T6 weakened by notches

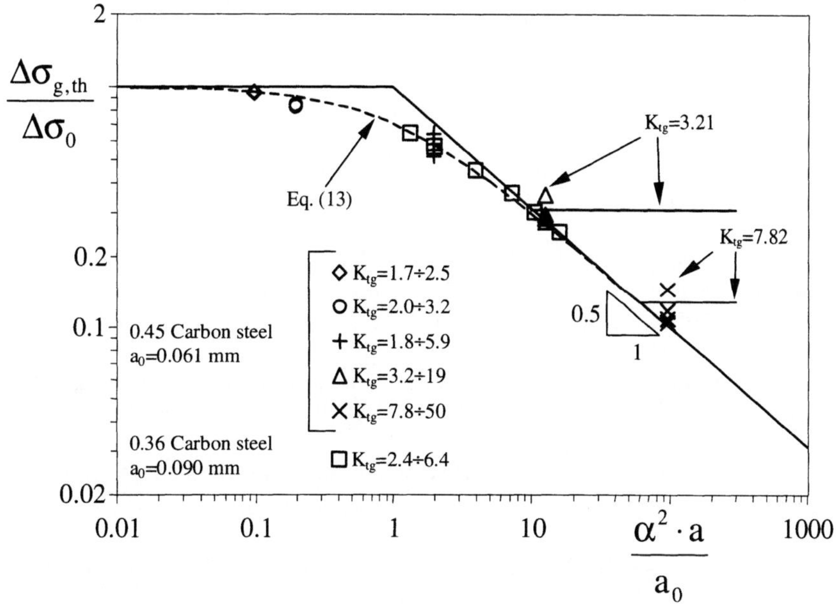

Figure 6.8: Universal diagram and comparison with experimental data from Nisitani and Endo [5] for carbon steels

References

ASM Handbook (1996). *Fatigue and Fracture*. ASM International, vol. 19.
ASTM E 647 – 88a, *Standard test method for measurement of fatigue crack growth rates*.
Atzori B., Haibach E. (1979). Local strains at the toe of fillet welds and relationship with the fatigue life. *Proceedings of the VII AIAS National Congress*, Cagliari (in italian).
Atzori, B. (1979). *Metodi e Procedimenti di Calcolo nella Progettazione Meccanica*. Laterza, Bari.
Atzori, B. (1985). Meccanica della frattura o effetto di intaglio nella progettazione a fatica. Atti del XIII Convegno Nazionale AIAS, Bergamo.
Atzori B., Dattoma V., Demelio G. (1992). Relationship between the fatigue strength and the local strain at the toe of fillet welded joints. *Proceedings of the XXI AIAS National Congress*, Genova (in italian).
Atzori B., Meneghetti G. (1998). The fatigue strength of fillet welded joints in structural steel. *Proceedings of the XIV IGF National Congress*, Trento (in italian).
Atzori B., Lazzarin P., Tovo R. (1999). Stress field parameters to predict fatigue strength of notched components. *J Strain Anal Eng*. 34(6): 1-17.
Atzori B., Lazzarin P. (2000). Notch sensitivity and defect sensitivity under fatigue loading: two sides of the same medal. *Int. J. Fracture,* 107: L3-L8.
Atzori, B. (2000). *Appunti di Costruzione di Macchine*. Libreria Cortina, Padova.
Atzori B., Lazzarin P. (2001). Meccanica della frattura e sensibilità all'intaglio. *XV AIMETA Conference*, Taormina (Italy).
Atzori B., Meneghetti G. (2001). Fatigue strength of fillet welded structural steels: finite elements, strain gauges and reality. *Int. J. Fatigue,* 23(8): 713-721.
Baumel A., Seeger T. (1991). *Materials data for cyclic loading*. Elsevier Amsterdam.
Beretta S., Clerici P. (1997). Stima della resistenza a fatica in getti di AlSi7Mg. *AIM National Conference*, Padova.
Bonollo F., Muffato G., Tovo R. (1997). Effetto della microstruttura sulle caratteristiche a fatica di getti in lega leggera d'alluminio. *AIM National Conference*, Padova.
Bonollo F., Tovo R. (1999). *Fatigue in Al casting alloys: metallurgical aspects*. TALAT Lecture 1254, 3-16.
Broek, K.D. The Practical use of fracture mechanics. Kluwer Academic, Dordrecht, Germany.
BS7608 (1993): *Fatigue design and assessment of steel structures*. London: British Standards Institution.
Couper M. J., Griffiths J. R. (1990). Effects of crack closure and mean stress on the threshold stress intensity factor for fatigue of an aluminium casting alloy. *Fatigue Fract. Engng. Mater. Struct.*, 13: 615-624.
Couper M. J., Neeson A. E., Griffiths J. (1990). Casting defects and the fatigue behaviour of an aluminium casting alloy. *Fatigue Fract. Engng. Mater. Struct.* 13: 213-227.
Dario Y. (1994). *Statistical analysis of fatigue strength of specimens for aluminium alloy wheels*. Degree Thesis, University of Padova.
Dattoma V., Demelio G., Pappalettere C. (1991). Local strain for fatigue design of cruciform welded joints. *Proceedings of the Spring Conference on Experimental Mechanics*, Milwaukee.
El. Haddad M. H., Topper M. H., Smith K. N. (1979). Prediction of Non-Propagating Cracks. *Eng. Fract. Mech.* 11: 573-584.

Eurocode 3 (1992). *Design of steel structures. Part 1-1: general rules and rules for buildings.* ENV 1993-1-1. European Committee for Stardardisation.

Frost, N.E., Marsh K.J., Pook, L.P. (1974). *Metal Fatigue.* Oxford University Press.

Fuchs, H. O., Stephens R. I. (1980). *Metal Fatigue in Engineering.* John Wiley & Sons.

Gall K., Yang N., Horstemeyer M., McDowell D. L., Fan J. (2000). The influence of Modified intermetallics and Si particles on fatigue crack paths in a cast A356 alloy. *Fatigue Fract. Engng. Mater. Struct.* 23: 159-172.

Guerini M. (1996). Previsione di resistenza a fatica di componenti automobilistici in lega leggera da fusione. Degree Thesis, University of Milano.

Gurney T. R. (1991). *The fatigue strength of transverse fillet welded joints - A study of the influence of joint geometry.* Abington Publishing, Cambridge.

Gurney, T.R. (1984). *Fatigue of welded structures.* Cambridge University Press.

Haibach E. (1968). *Die Schwingfestigkeit von Schweissverdindungen aus der Sicht einer örtlichen Beanspruchungsmessung.* Laboratorium für Betriebsfestigkeit, Darmstadt, Bericht N° FB-77.

Honma U., Yokohama I., Kitaoka S. (1984). Fatigue Strength and Mechanical Properties of Aluminium alloy castings of different structural fineness. *ALUMINIUM*, 60, (12): 917-920.

Koh S. K., Stephens R. I. (1989). Fracture toughness of A356-T6 cast aluminium alloy. SAE SP-760: 61-69.

Lazzarin P., Tovo R. (1998). A notch intensity factor approach to the stress analysis of welds. *Fatigue Fract Eng Mater Struct*, 21(9): 1089-1104.

Lazzarin P., Tovo R., Meneghetti G. (1997). Fatigue crack initiation and propagation phases near notches in metals with low notch sensitivity. *Int. J. Fatigue,* 19: 647-657.

Lazzarin, P., Tovo, R. (1998). A Notch Intensity Factor Approach to the Stress Analysis of Welds. *Fatigue and Fracture of Engineering Materials and Structures*, 21: 1089-1103.

Maddox S. J. (1987). *The effect of plate thickness on the fatigue strength of fillet welded joints.* The Welding Institute, Abington, Cambridge.

Mahoney B. J., Stephens R. I. (1989). *Fatigue Crack Growth of A35 -T6 cast aluminium alloy.* SAE SP-760: 49-59.

Niemi, E. (1993). *Stress Determination for Fatigue Analysis of Welded Components.* IIW1221-93. Abington Publishing, Cambridge.

Nisitani H., Endo M. (1988). Unified treatment of deep and shallow notches in rotating bending fatigue. *ASTM 924 Basic Questions in Fatigue,* 1: 136-153.

Ødegård J.A., Pedersen K. (1994). Fatigue properties of an A356 aluminium alloy for automotive applications. *Metal Matrix Composites*, SP-1010: 25-32.

Radaj, D. (1996). Review of Fatigue Strength Assessment of non Welded and Welded Strucutres based on local parameters. *International Journal of Fatigue*, 18: 153-170.

Radaj, D., Sonsino, C. (1998). *Fatigue assessment of welded joints by local approaches.* Abington Publishing, Cambridge.

Rosso M., Scavino G., Albertinazzi M., Zanda A. (1997). Caratterizzazione Microstrutturale meccanica e a fatica della lega d'alluminio ipoeutettica A356 in un getto a scalini. *AIM National Conference*, Padova.

Seniw M. E., Conley J. G., Fine M. E. (2000). The effects of microscopic inclusion locations and silicon segregation on fatigue lifetimes of aluminium alloy A356 castings. *Mat. Sc. . 2* , 1-2: 43-48.

Shiozawa K., Tohda Y., Sun S. M. (1997). Crack initiation and small fatigue crack growth behaviour of squeeze-cast Al-Si aluminium alloy. *Fatigue Fract. Engng. Mater. Struct.* 20: 237-247.

Skallerud B., Iveland T., Härkegård G. (1993). Fatigue life assessment of aluminium alloys with casting defects. *Eng. Fracture Mech.* 44: 857-874.

Stanzl-Tscegg S. E., Mayer H. R., Beste A., Kroll S. (1995). Fatigue and fatigue crack propagation in AlSi7Mg Cast alloys under in-service loading conditions. *Int. J. Fatigue,* 17: 149-155.

Stephens R. I., Wigant C.C. (1989). *Low cycle fatigue of A356-T6 cast aluminium alloy- A Round-Robin test program.* SAE SP-760: 2-28.

Suresh, S. (1991). *Fatigue of Materials.* Cambridge University Press.

Suzuki H., Weimin G., Dewa H., Tomota Y., Ito Y. (1990). An influence of cooling rate on the fatigue strength of aluminium alloy castings for the wheel of automobile. *Proceedings of Fatigue 90,* 1, 273-278.

Torti C. (1996). *Previsione di resistenza a fatica di componenti in lega leggera di fusione.* Degree Thesis, Politecnico di Milano.

Verreman Y., Nie B. (1996). Early development of fatigue cracking at manual fillet welds. *Fatigue Fract Eng Mater Struct,* 19(6): 669-681.

Wickberg A., Gustafsson G., Larsson L. E. (1984). Microstructural Effects on the Fatigue Properties of a Cast AlSi7Mg alloy. *SAE Technical Paper N. 840121,* 93: 1728-1735.

Wigant C. C., Stephens R. I. (1987). Fatigue crack growth behaviour of A356-T6 aluminium alloys. *Proc. Int. Conf. Fatigue 87,* 165-173.

Wigant C. C., Stephens R. I. (1989). Fatigue Crack growth of A356-T6 cast aluminium alloy. *SAE SP-760*: 49-59.

Wiliams G., Fisher K. M. (1983). *Squeeze forming of aluminium-alloy components.* In Production to Near Net Shape: Source Book, ASM, Metals park OH.

Chapter 6

CRASHWORTHINESS OF ALUMINIUM STRUCTURES

Professor Magnus Langseth and Arve Grønsund Hanssen, Dr. ing.

Structural Impact Laboratory (SIMLab)
Department of Structural Engineering
Norwegian University of Science and Technology

1. INTRODUCTION

Lightweight materials such as aluminium offer the automotive industry an opportunity to design and manufacture high-performance vehicles that are safe, energy-efficient and environmentally friendly, and much lighter than traditional designs. The introduction of these materials will challenge the automotive design engineers to explore and develop new solutions in design and production technology in order to fully realize the potential that can be gained in the interaction between these materials, product/structural design and the manufacturing process. Even though aluminium is an "old" material, it is relatively new as a load-carrying material in the automotive industry. This implies that material producers and parts suppliers have to develop new knowledge about these materials to gain an increased market share.

In a design process, computational mechanics is an indispensable tool. It enables new design concepts to be evaluated without extensive testing and provides a framework for combining new knowledge in the modelling of materials and structures with experimental data on components and systems behaviour. However, in an efficient design process the designer must still possess knowledge about the physical mechanisms controlling the response, as well as have tools available for initial design studies. This will give guidelines to design solutions as well as what kind of simplifications that can be made in the numerical modelling of material and structural parts and still retain sufficient reliability and accuracy.

In this part of the course we will try to provide basic knowledge related to energy absorption and crashworthiness of aluminium structures that is relevant to engineers in the automotive industry. Focus is placed on material behaviour under static and dynamic loading conditions with special emphasis on plastic flow and tensile instabilities of thin-walled components. With regard to structural component behaviour, emphasis is placed on the behaviour of laterally and axially loaded members as well as the behaviour of aluminium foam and foam-filled components. Here, simple design methods are given relevant for thin-walled aluminium extrusions subjected to impact loading conditions.

2. MATERIAL BEHAVIOUR AND MODELLING

2.1. Uniaxial behaviour

2.1.1. True stress and true strain

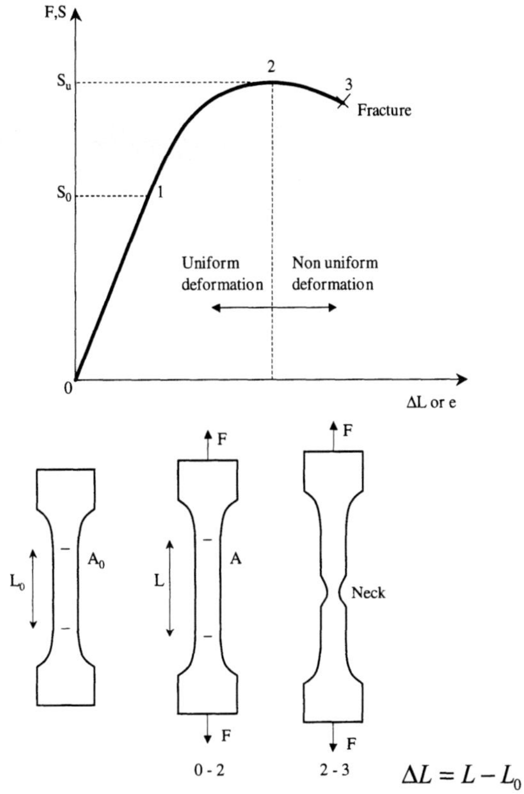

$$\Delta L = L - L_0$$

0-1	Elastic behaviour, $E_t = d\sigma/d\varepsilon$=constant
1-2	Elastic-plastic behaviour, E_t varies
2-3	Localized deformation, necking
S_0	Yield stress
S_u	Ultimate stress

Figure 1: Stress-strain relationship

A uniaxial test is mainly used to establish the relationship between load and extension (stress and strain). The basic information from such a test is the force, *F*, and the corresponding extension or displacement *ΔL*. Values of stress and strain are computed from these data since they provide more general information. Figure 1 is a schematic presentation of a typical load-extension curve for aluminium where the elastic and elastic-plastic domains are defined. The data from a load-extension curve can be converted into nominal or engineering stress and strain as follows

$$S = \frac{F}{A_0} \tag{2.1}$$

$$e = \frac{\Delta L}{L_0} = \frac{L - L_0}{L_0} \tag{2.2}$$

Furthermore, the tensile or ultimate strength is defined as

$$S_u = \frac{F_u}{A_0} \tag{2.3}$$

where F_u is the maximum or ultimate load carried by the specimen.

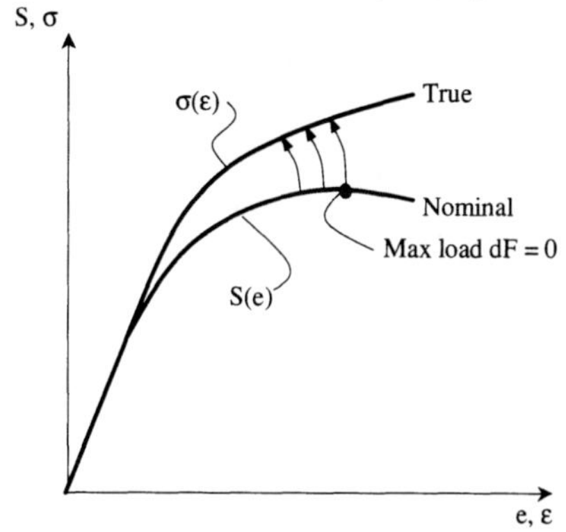

Figure 2: Comparison of nominal and true stress-strain curves

Figure 2 shows a nominal stress-strain curve and its counterpart based upon calculations of true stress and strain as follows:

Incremental true strain,

$$d\varepsilon = \frac{dL}{L}.$$

Integration gives the total true strain

$$\varepsilon = \int_{L_0}^{L} \frac{dL}{L} = \ln\frac{L}{L_0} = \ln(1+e) \tag{2.4}$$

True stress,

$$\sigma = \frac{F}{A} = \frac{F}{A_0}\frac{L}{L_0} = S\left(\frac{L_0 + \Delta L}{L_0}\right)$$

or

$$\sigma = S(1+e) \tag{2.5}$$

In the calculation of the true stress or Cauchy stress it is assumed that the elastic deformations are small and that plastic deformations occur without volume change, i.e. $AL = A_0 L_0$. The above expression for the true stress and true strain are only useful up to the ultimate load (*dF=0*). After the onset of diffuse necking, the length changes are localized in the neck. Consequently, the nominal strain, *e*, which involves a measurement using the entire gauge section, cannot be used to calculate the true strain ε. An alternative expression for the true strain is to use the initial area A_0 and the current area A,

$$\varepsilon = \ln\frac{L}{L_0} = \ln\frac{A_0}{A} \tag{2.6}$$

After necking, the expression $\sigma = F/A$ but not $\sigma = S(1+e)$, gives the average true stress in the neck for the direction of loading. This is no longer the effective stress, σ_e, since the stress in the neck is triaxial.

2.1.2. Modelling of the stress-strain curve

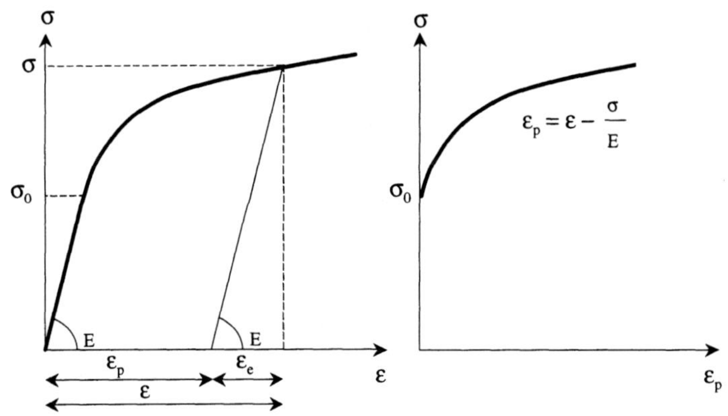

Figure 3. Elastic ε_e and plastic ε_p strains

From a uniaxial test the elastic ε_e and plastic ε_p strain can be assumed to be additive, see Figure 3

$$\varepsilon = \varepsilon_e + \varepsilon_p = \sigma / E + \varepsilon_p \tag{2.7}$$

When using a computer code the input stress-strain curve is normally the σ-ε_p curve. Having performed a tensile test the above expression can be used to generate the true stress vs. plastic-strain curve denoted by $\sigma = Y(\varepsilon_p)$. An approximation to the σ-ε_p curve can be expressed by the following power-law relation

$$\sigma = K(\varepsilon_p)^m \quad , \quad m = \frac{1}{n} \tag{2.8}$$

or

$$\varepsilon_p = \left(\frac{\sigma}{K}\right)^{\frac{1}{m}} \tag{2.9}$$

This gives the well-known Ramberg-Osgood equation

$$\varepsilon = \frac{\sigma}{E} + \left(\frac{\sigma}{K}\right)^n \tag{2.10}$$

where

$$K = \frac{\sigma_{0.2}}{(0.002)^{1/n}} \tag{2.11}$$

The stress $\sigma_{0.2}$ corresponds to a permanent deformation after unloading of 0.2 %. Eq. (2.10) is not able to give an accurate description of the uniaxial stress-strain curve in both the small and large strain domains. Hence, the model must always be calibrated in view of the strain range that is expected to be important in a particular analysis. A multicomponent strain hardening model has been proposed in the form

$$\sigma = \sigma_0 + \sum_{i=1}^{n} Q_i (1 - \exp(-C_i \cdot \varepsilon_p)) \tag{2.12}$$

It has been observed that this expression can give an almost exact representation of the whole stress-strain curve for aluminium alloys. In Eq. (2.12), σ_0 defines the proportionality limit corresponding to point 1 in Figure 1, while Q_i and C_i are the constants that are to be determined.

2.1.3. Tensile instability

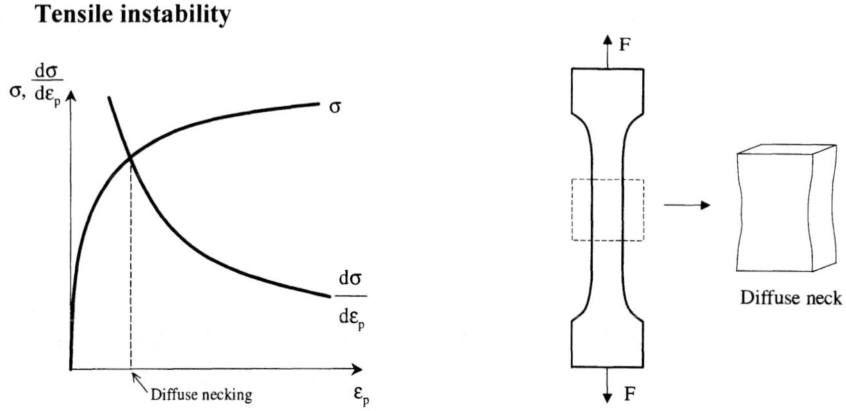

Figure 4: Diffuse necking

For ductile metals subjected to plastic deformation in tension, the load-carrying capacity of the test-piece reaches a maximum, and then decreases. The onset of tensile instability, named diffuse necking, coincides with this maximum load or applied force. In terms of true stress and true strain

$$F = \sigma A \tag{2.13}$$

When the maximum load is reached, *dF=0*. Thus

$$dF = d\sigma A + \sigma dA = 0 \tag{2.14}$$

or

$$\frac{d\sigma}{\sigma} + \frac{dA}{A} = 0 \tag{2.15}$$

No volume change during plastic deformation gives

$$d(AL) = 0 \tag{2.16}$$

or

$$\frac{dA}{A} = -\frac{dL}{L} = -d\varepsilon_p \tag{2.17}$$

Thus

$$\frac{d\sigma}{\sigma} = \frac{dL}{L} = d\varepsilon_p \tag{2.18}$$

or

$$\frac{d\sigma}{d\varepsilon_p} = \sigma \tag{2.19}$$

A schematic representation of the condition given in Eq. (2.19) is shown in Figure 4. Hence, during a tensile test , load F may be increased to a value at which the material hardening balances the decrease in the cross-section area. Using Eq. (2.8) as a representation of the stress-plastic strain curve, Eq.(2.19) becomes

$$\frac{d\sigma}{d\varepsilon_p} = K \cdot m \cdot \varepsilon_p^m \cdot \frac{1}{\varepsilon_p} = \frac{\sigma \cdot m}{\varepsilon_p} = \sigma \tag{2.20}$$

or

$$\varepsilon_{pu} = m = \frac{1}{n} \tag{2.21}$$

Figure 5 shows typical engineering stress-strain curves for the aluminium alloy AA6060 tempers T4 and T6. The n value in Eq. (2.8) for temper T4 is $n \approx 10$ which gives $\varepsilon_{pu} = 0.1$ when using Eq.(2.21). The corresponding values for temper T6 are $n \approx 25$ and $\varepsilon_{pu} = 0.04$ respectively. As a rule of thumb, aluminium alloys with significant strain hardening are preferable in order to avoid tensile instabilities. After diffuse necking, localized necking will take place. This will be discussed in more detail in Section 2.6.

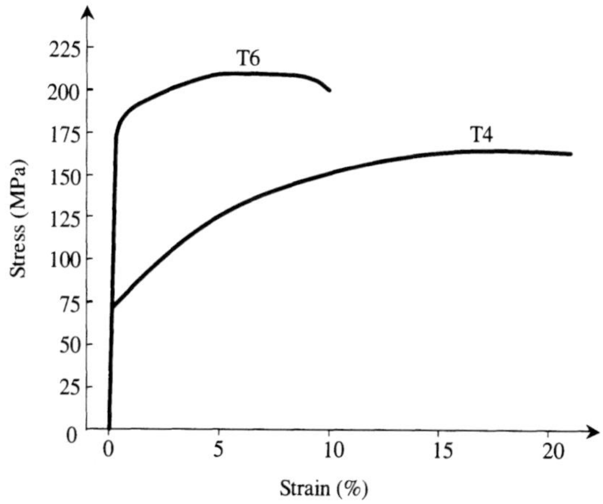

Figure 5: Typical stress-strain curves for tempers T4 and T6

2.1.4. Behaviour after diffuse necking (cylindrical test specimens)

As previously discussed, once the necking begins, the true stress $\sigma = F / A$ is no longer equal to the effective stress, σ_e, or the uniaxial flow stress Y. Thus, the measured curve will not correctly describe the strain hardening. The reason for this is the triaxial stress state that develops in the neck. The smallest section in the neck is assumed to be elongated under uniform displacement (strain). Thus, the state of stress at each point in the minimum section is just an axial tension with a varying superposed hydrostatic stress (hydrostatic stresses give no contribution to yielding). Figure 6 illustrates the stress situation in the neck.

It can be shown by assuming a uniform radial strain that $\sigma_r = \sigma_\theta$ at $z=0$. Using cylindrical coordinates, the von Mises yield criterion is

$$\sigma_e^2 = \sigma_z^2 + \sigma_r^2 + \sigma_\theta^2 - \sigma_z \sigma_r - \sigma_z \sigma_\theta - \sigma_r \sigma_\theta$$

and with $\sigma_r = \sigma_\theta$

$$\sigma_e = \sigma_z - \sigma_r \tag{2.22}$$

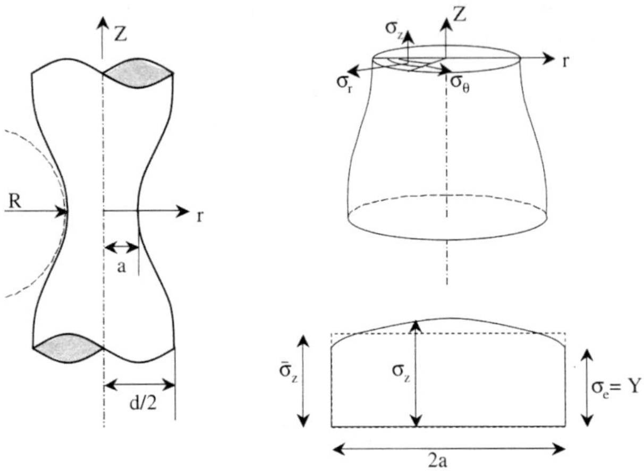

Figure 6: Stress situation in the neck

According to Bridgman [Hill (1989)] the axial stress in the neck can be expressed as

$$\frac{\sigma_z}{\sigma_e} = 1 + \ln\left(\frac{a^2 + 2 \cdot a \cdot R - r^2}{2 \cdot a \cdot R}\right) \tag{2.23}$$

Be aware that at the surface in the neck σ_z is the only stress so that $\sigma_r=\sigma_\theta=0$, thus the stress state here is purely uniaxial. Having determined the average stress $\overline{\sigma}_z = F/(\pi \cdot a^2)$ the equivalent uniaxial tension stress can be calculated as

$$\sigma_e = Y = \frac{\overline{\sigma}_z}{\left(1 + \frac{2 \cdot R}{a}\right) \cdot \ln\left(1 + \frac{a}{2 \cdot R}\right)} \tag{2.24}$$

The strain in the neck reads

$$\varepsilon_p = \ln\frac{A_0}{A} \tag{2.25}$$

where

$$A_0 = \pi \cdot d^2 / 4 \text{ and } A = \pi \cdot a^2$$

Figure 7 shows a schematic representation of the difference between the nominal stress-strain curve (S-e_p), the stress-strain curve calculated based on an average stress $\overline{\sigma}_z$, ($\overline{\sigma}_z - \varepsilon_p$), and the stress-strain curve corrected by Bridgeman ($\sigma_e - \varepsilon_p$). Note that the Bridgeman correction takes place after the maximum load is reached.

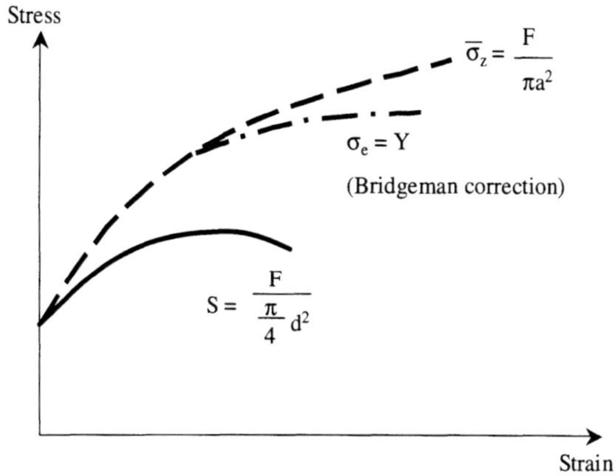

Figure 7: Stress-strain curves

2.1.5. Effect of triaxiality on failure strain (cylindrical specimens)

When carrying out tensile tests, the failure strain is a decreasing function of the triaxiality, σ_H, in the neck

$$\sigma_H = \frac{1}{3}(\sigma_z + \sigma_\theta + \sigma_r). \tag{2.26}$$

Introducing Eq. (2.22) into Eq. (2.26) and utilizing the fact that $\sigma_r = \sigma_\theta$ yields

$$\sigma_H = \frac{1}{3}(\sigma_z + 2\sigma_r) = \sigma_z - \frac{2}{3}\sigma_e \tag{2.27}$$

The triaxiality ratio $\sigma*$ is now defined as

$$\sigma^* = \frac{\sigma_H}{\sigma_e} = \frac{\sigma_z}{\sigma_e} - \frac{2}{3} \tag{2.28}$$

where the maximum value of σ*, can be found for $r=0$, see Eq. (2.23). Thus

$$\sigma^*_{\max} = \frac{1}{3} + \ln\left(1 + \frac{a}{2 \cdot R}\right) \tag{2.29}$$

This means that in a tensile test where the material follows the von Misies yield criterion, fracture will start at the centre of the necked region. Using smooth and notched specimens it is possible to determine the fracture strain ε_f as a function of σ^*, see Børvik (2001). This is shown schematically in Figure 8.

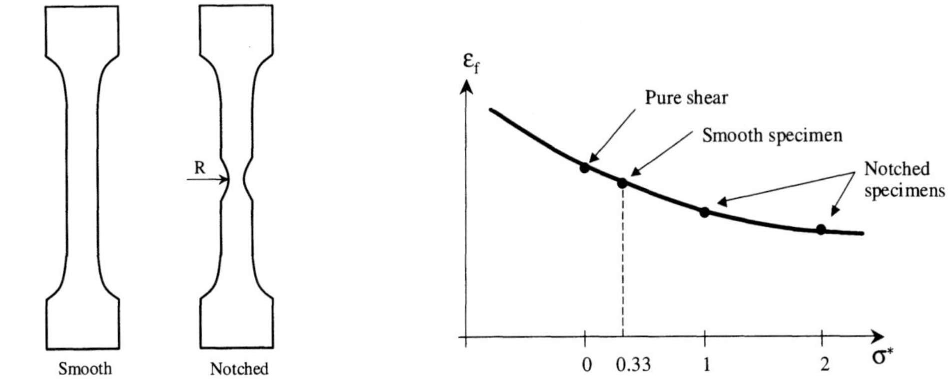

Figure 8: Fracture strain vs. triaxiality ratio

2.2. Plastic anisotropy

Aluminium is a polycrystalline aggregate in which the individual grains have a crystallographic orientation that is different from that of its neighbours. Usually, the orientations of the grains have a non-random distribution; i.e. they have certain preferred orientations, which are called texture. Texture in aluminium profiles is developed in all its manufacturing processes and is dependent on the alloying ingredients, mechanical and thermal treatments etc. The two main reasons for texture development in extruded profiles are the extrusion process and subsequent annealing.

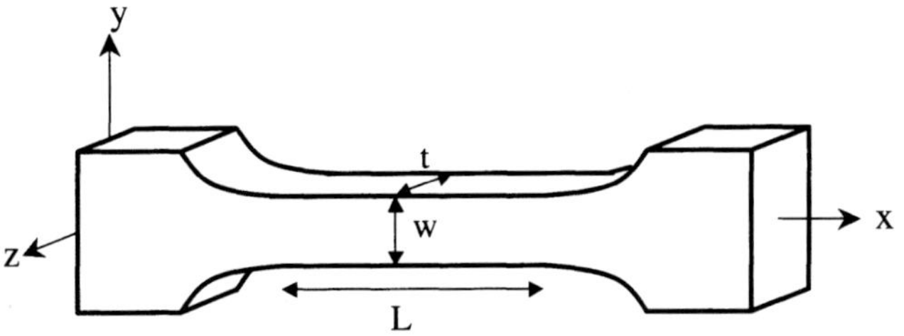

$$d\varepsilon_L^p = \frac{dL}{L}$$

$$d\varepsilon_w^p = \frac{dw}{w}$$

$$d\varepsilon_t^p = \frac{dt}{t}$$

Figure 9: Definitions of strain indices; ε_t, ε_w and ε_L

The width and longitudinal strains can be determined when using two extensiometers in a tensile test (see Figure 9). The plastic strain increment $d\varepsilon_t^p$ in the thickness direction can be calculated applying the assumption of plastic incompressibility

$$d\varepsilon_L{}^p + d\varepsilon_t{}^p + d\varepsilon_w{}^p = 0$$

or

$$d\varepsilon_t{}^p = -\left(d\varepsilon_L{}^p + d\varepsilon_w{}^p\right) \tag{2.30}$$

Definition of the *R*-value

$$= \frac{d\varepsilon_W{}^p}{d\varepsilon_t{}^p} = \frac{d\varepsilon_y{}^p}{d\varepsilon_z{}^p} = \frac{d\varepsilon_W{}^p}{-\left(d\varepsilon_L{}^p + d\varepsilon_W{}^p\right)} \tag{2.31}$$

Thus

$$d\varepsilon_W{}^p = -\frac{R}{R+1} d\varepsilon_L{}^p \tag{2.32}$$

and

$$d\varepsilon_t{}^p = -\frac{1}{R+1} d\varepsilon_L{}^p \tag{2.33}$$

If tensile tests are performed in different directions with respect to the extrusion direction the R ratio can be defined as

$$R_\alpha = \left.\frac{d\varepsilon_W{}^p}{d\varepsilon_t{}^p}\right|_\alpha \tag{2.34}$$

If $R_\alpha = 1$, the material is said to have isotropic flow properties.

2.3. Yield criteria

2.3.1. Von Mises

A general stress state can be divided into two parts, one called the spherical or the hydrostatic stress tensor and the other called the deviatoric stress tensor. The hydrostatic stress tensor is the tensor whose elements are $p\delta_{ij}$, where p is the mean stress

$$p = \frac{1}{3}\sigma_{kk} = \frac{1}{3}\left(\sigma_x + \sigma_y + \sigma_z\right) \tag{2.35}$$

The stress deviator tensor σ_{ij}' is defined as

$$\sigma_{ij}' = \sigma_{ij} - p \cdot \delta_{ij} \tag{2.36}$$

For pure uniaxial stress, $p = \sigma_x / 3$ and $\sigma_x' = 2 \cdot \sigma_x / 3$. The von Mises yield criterion is based on the strain energy of distortion, i.e. the deviator stress tensor. Thus, the yielding is assumed to be independent of pressure. Equating the distortion energy for a multiaxial stress state to that found for uniaxial behaviour gives

$$\left[U_{dist}\right]_{multiaxial} = \left[U_{dist}\right]_{uniaxial} \tag{2.37}$$

In a more general form, the criterion can be written as

$$(\sigma_x-\sigma_y)^2+(\sigma_y-\sigma_z)^2+(\sigma_z-\sigma_x)^2+6\cdot(\tau_{xy}{}^2+\tau_{yz}{}^2+\tau_{zx}{}^2)=2Y^2$$

or

$$(\sigma_1-\sigma_2)^2+(\sigma_2-\sigma_3)^2+(\sigma_3-\sigma_1)^2=2Y^2 \tag{2.38}$$

For thin-walled aluminium members, plane stress conditions will be the case, i.e. $\sigma_z=\sigma_3=0$. Thus, the yield criterion can then be written as

$$\sigma_y{}^2+\sigma_x{}^2+(\sigma_x-\sigma_y)^2+6\tau_{xy}=2Y^2$$

or

$$\sigma_1^2+\sigma_2^2-\sigma_1\sigma_2=Y^2 \tag{2.39}$$

Von Misses is an isotropic criterion and can be calibrated by a single uniaxial tension test. Y is the current yield stress of the material.

2.3.2. Tresca

The Tresca criterion postulates that yielding will occur when the largest shear stress reaches a critical value

$$\left[\frac{1}{2}|\sigma_1-\sigma_2|,\frac{1}{2}|\sigma_2-\sigma_3|,\frac{1}{2}|\sigma_3-\sigma_1|\right]=\frac{Y}{2} \tag{2.40}$$

For plane stress conditions the criterion reads:

$$Max\left[\frac{1}{2}|\sigma_1-\sigma_2|,\frac{1}{2}|\sigma_2|,\frac{1}{2}|\sigma_1|\right]=\frac{Y}{2} \tag{2.41}$$

Figure 10 shows a comparison of the Tresca and von Mises criteria for $\sigma_3=0$.

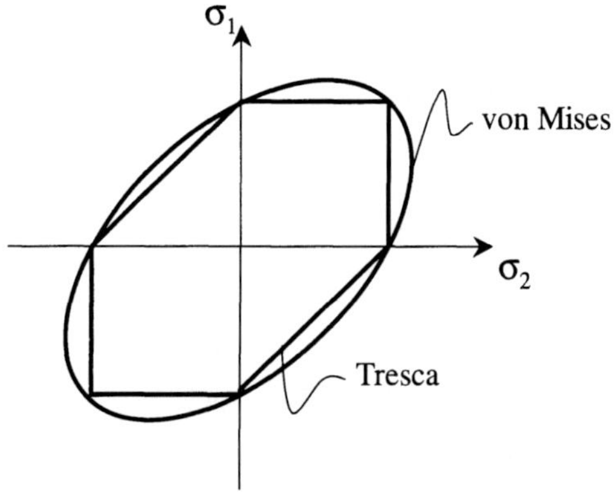

Figure 10: Tresca and von Mises yield loci for plane stress conditions

2.3.3. Hill's anisotropic plasticity criterion

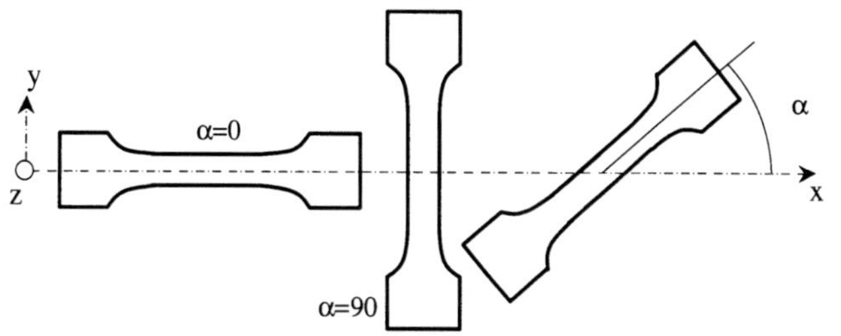

Figure 11: Tensile test specimens in different directions with respect to the principal axis

Hill (1989) has modified the von Mises yield criterion to take anisotropy into account. For plane stress conditions this criterion reads

$$F \cdot \sigma_y^{\ 2} + G \cdot \sigma_x^{\ 2} + H \cdot \left(\sigma_x - \sigma_y\right)^2 + 2N \cdot \tau_{xy}^{\ 2} = 2Y^2 \tag{2.42}$$

When $F=G=H=1$ and $N=3$ we get the von Mises criterion. If we assume associated plastic flow and that the principal stress directions are coincident with the coordinate system, see Figure 11, the criterion can be rewritten in such a way as to exactly satisfy the initial R-values in the direction of the principal stresses (x and y axes).

$$R_0\sigma_y^{\ 2} + R_{90}\sigma_x^{\ 2} + R_0R_{90}\left(\sigma_x - \sigma_y\right)^2 = R_{90}\left(1+R_0\right)Y^2 \tag{2.43}$$

Here, R_0 and R_{90} are obtained from tensile tests in the x and y directions, respectively. For definition of the R ratios, see Section 2.2. The yield loci for materials characterized by $R=R_0=R_{90}$ are sketched in Figure 12.

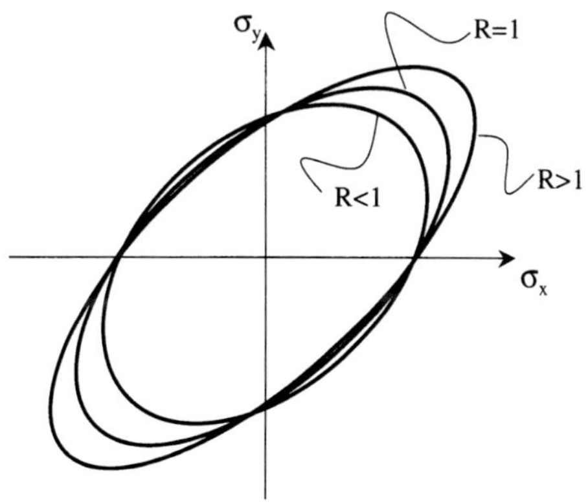

Figure 12: Schematic yield loci for different R ratios

The process of strain localization is dependent on the chosen constitutive model. It is therefore important to remember that several other and better yield criteria have been developed for aluminium alloys. A summary of yield criteria for aluminium alloys to be used in crash and forming analysis is given in Lademo (1999).

2.4. Flow rules and plastic stress-strain relations

For elastic deformations, Hooke's law relates to the stress and strain, for example

$$e_1 = \frac{1}{E}\left[\sigma_1 - \nu(\sigma_2 + \sigma_3)\right] \tag{2.44}$$

$$e_2 = \frac{1}{E}\left[\sigma_2 - \nu(\sigma_1 + \sigma_3)\right] \tag{2.45}$$

$$e_3 = \frac{1}{E}\left[\sigma_3 - \nu(\sigma_1 + \sigma_2)\right] \tag{2.46}$$

Similar relations can be obtained for plastic deformations by using the flow rule. This rule states that the direction of the plastic strain vector is defined as the gradient of a flow potential. When the flow potential equals the yield surface, the plastic flow is said to be associated. In general, the associated flow rule can be stated as

$$d\varepsilon_{ij}^{\ p} = d\lambda \frac{\partial f}{\partial \sigma_{ij}} \tag{2.47}$$

where f is the function of σ_{ij} that describes yielding (e.g. the yield criterion). If the von Mises criterion is used, f becomes

$$f = J_2 - \left(Y / \sqrt{3}\right)^2 = 0 \tag{2.48}$$

where

$$J_2 = \frac{1}{6}\left[(\sigma_1 - \sigma_2)^2 + (\sigma_2 - \sigma_3)^2 + (\sigma_3 - \sigma_1)^2\right] = \frac{1}{2}\sigma'_{ij}\sigma'_{ij} \tag{2.49}$$

while σ'_{ij} is the deviatoric stress tensor. Performing the appropriate differentiation gives

$$d\varepsilon_{ij}^{\ p} = d\lambda \sigma'_{ij} \tag{2.50}$$

On a component level:

$$d\varepsilon_1^{\ p} = d\lambda \frac{2}{3}\left[\sigma_1 - \frac{1}{2}(\sigma_2 + \sigma_3)\right] \tag{2.51}$$

$$d\varepsilon_2^{\ p} = d\lambda \frac{2}{3}\left[\sigma_2 - \frac{1}{2}(\sigma_1 + \sigma_3)\right] \tag{2.52}$$

$$d\varepsilon_3^{\,p} = d\lambda \frac{2}{3}\left[\sigma_3 - \frac{1}{2}(\sigma_1 + \sigma_2)\right] \tag{2.53}$$

The parameter $d\lambda$ can now be determined by squaring Eq. (2.50)

$$\left(d\varepsilon_{ij}^{\,p} d\varepsilon_{ij}^{\,p}\right) = (d\lambda)^2 \left(\sigma_{ij}' \sigma_{ij}'\right) \tag{2.54}$$

Taking the square root on both sides and introducing the von Mises effective stress

$$\sigma_e = \sqrt{3J_2} = \sqrt{\frac{3}{2}\sigma'_{ij}\sigma'_{ij}} \tag{2.55}$$

and the effective strain increment

$$d\varepsilon_e^p = \sqrt{\frac{2}{3} d\varepsilon_{ij}^p \varepsilon_{ij}^p} \tag{2.56}$$

yields

$$d\lambda = \frac{3}{2}\frac{d\varepsilon_e^p}{\sigma_e} \tag{2.57}$$

Then the plastic strain increments can be expressed as

$$d\varepsilon_1^{\,p} = \frac{d\varepsilon_e^{\,p}}{\sigma_e}\left[\sigma_1 - \frac{1}{2}(\sigma_2 + \sigma_3)\right] \tag{2.58}$$

$$d\varepsilon_2^{\,p} = \frac{d\varepsilon_e^{\,p}}{\sigma_e}\left[\sigma_2 - \frac{1}{2}(\sigma_1 + \sigma_3)\right] \tag{2.59}$$

$$d\varepsilon_3^{\,p} = \frac{d\varepsilon_e^{\,p}}{\sigma_e}\left[\sigma_3 - \frac{1}{2}(\sigma_1 + \sigma_2)\right] \tag{2.60}$$

Assuming proportional loading, the incremental theory reduces to the deformation theory. In proportional loading the stress increases in the ratio

$$\sigma_{ij} = k \cdot \sigma_{ij}^{\,0} \tag{2.61}$$

where *k(t)* is a monotonically increasing proportionality factor and $\sigma_{ij}^{\,0}$ is an arbitrary reference state. From this we can obtain

$$\sigma_{ij}' = k \cdot \sigma_{ij}'^{0} \tag{2.62}$$

and

$$\sigma_e = k \cdot \sigma_e^{\,0} \tag{2.63}$$

Substitution of Eq. (2.57), Eq. (2.62) and Eq. (2.63) into Eq. (2.50) gives

$$d\varepsilon_{ij}^{\;p} = d\lambda\sigma'_{ij} = \frac{3}{2}\frac{d\varepsilon_e^{\;p}}{\sigma_e}\sigma'_{ij}$$

or

$$d\varepsilon_{ij}^{\;p} = \frac{3}{2}\frac{d\varepsilon_e^{\;p}}{\cdot\sigma_e^{\;0}} \cdot \sigma'^{0}_{ij}$$

or

$$d\varepsilon_{ij}^{\;p} = \frac{3}{2}\frac{d\varepsilon_e^{\;p}}{\sigma_e^{\;0}}\sigma'^{0}_{ij} \tag{2.64}$$

Integration gives

$$\varepsilon_{ij}^{\;p} = \frac{3}{2}\frac{\varepsilon_e^{\;p}}{\sigma_e^{\;0}}\sigma'^{0}_{ij} = \frac{3}{2}\frac{\varepsilon_e^{\;p}}{\sigma_e}\sigma'_{ij} \tag{2.65}$$

and on a component level

$$\varepsilon_1^p = \frac{\varepsilon_e^p}{\sigma_e}\left[\sigma_1 - \frac{1}{2}(\sigma_2 + \sigma_3)\right] \tag{2.66}$$

where

$$\varepsilon_e^p = \sqrt{\frac{2}{3}\varepsilon^p{}_{ij}\varepsilon^p{}_{ij}} \tag{2.67}$$

The strains ε_2^p and ε_3^p are expressed by a proper change of subscripts.

The flow rule can be used to determine the yield criterion of plane strain conditions, i.e. when $d\varepsilon_3^p = 0$. Eq. (2.50) together with the von Mises yield criterion gives that

$$d\varepsilon_3^{\;p} = d\lambda\sigma_3' = 0 \;\Rightarrow\; \sigma_3' = 0 \tag{2.68}$$

The deviatoric stresses can now be expressed as

$$\sigma_1' = \sigma_1 - \frac{1}{3}(\sigma_1 + \sigma_2 + \sigma_3) \tag{2.69}$$

$$\sigma_2' = \sigma_2 - \frac{1}{3}(\sigma_1 + \sigma_2 + \sigma_3) \tag{2.70}$$

$$\sigma_3' = \sigma_3 - \frac{1}{3}(\sigma_1 + \sigma_2 + \sigma_3) = 0 \tag{2.71}$$

Combination of the above equations gives

$$\sigma_3 = \frac{\sigma_1 + \sigma_2}{2}, \sigma_1' = \frac{\sigma_1 - \sigma_2}{2}, \sigma_2' = \frac{\sigma_2 - \sigma_1}{2} \tag{2.72}$$

Introducing these expressions into the von Mises yield criterion, Eq. (2.48) becomes

$$\sigma_1 - \sigma_2 = \frac{2}{\sqrt{3}} Y \tag{2.73}$$

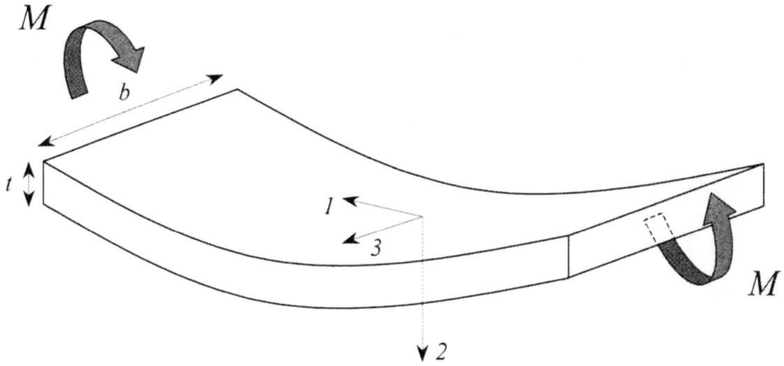

Figure 13: Plate subjected to bending

Figure 13 shows a plate bent by end moments. Plane strain conditions are assumed, i.e. the plastic strain increment in the width direction $d\varepsilon_3^p = 0$. Furthermore, the plate is thin-walled, i.e. $\sigma_2 = 0$. Assuming a rigid perfect plastic material, the plastic bending moment of the plate (per unit width) can then be expressed as

$$M_p = \sigma_1 \frac{t^2}{4} = \frac{2Y}{\sqrt{3}} \frac{t^2}{4} \tag{2.74}$$

2.5. Strain hardening

Strain hardening is often described through assumptions regarding changes in the size and/or position of the yield surface in the stress space as a function of some kind of equivalent uniaxial strain. Figure 14 shows a uniaxial case where loading and unloading of the specimen have been performed. From the figure it can be found that

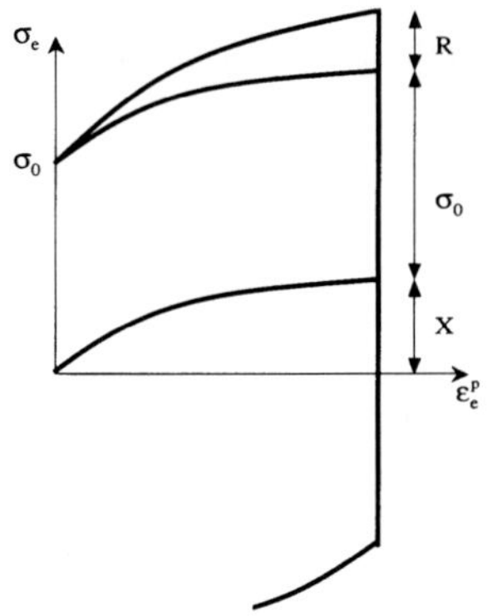

Figure 14. Uniaxial test, definition of isotropic and kinematic hardening

$$\sigma = \sigma_0 + X + R \tag{2.75}$$

or

$$f = (\sigma - X) - R - \sigma_0 \tag{2.76}$$

Here $X(\varepsilon_e^p)$ is kinematic hardening and $R(\varepsilon_e^p)$ is isotropic hardening. A non-linear expression for these terms may be written as

$$R = \sum_{i=1}^{n} \alpha Q_i \left[1 - \exp\left(-C_i \varepsilon_e^p\right)\right] \tag{2.77}$$

$$X = \sum_{i=1}^{n} (1-\alpha) Q_i \left[1 - \exp\left(-C_i \varepsilon_e^p\right)\right] \tag{2.78}$$

where α*=1.0* is pure isotropic hardening and α*=0* is pure kinematic hardening. Tests on aluminium alloy AA6060 have shown that α*=0.7*, see Figure 15. In the multiaxial case the loading function can be expressed as

$$f = \sqrt{\frac{3}{2}\left(\sigma'_{ij} - X'_{ij}\right)\left(\sigma'_{ij} - X'_{ij}\right)} - R(\varepsilon_e{}^p) - \sigma_0 \tag{2.79}$$

An illustration of pure isotropic and kinematic hardening is shown in Figure 16.

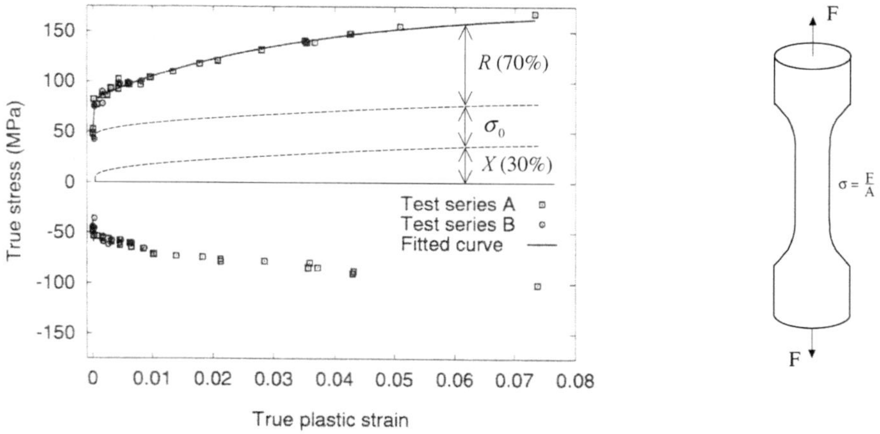

Figure 15: Stress-strain curve [Opheim (1996)]

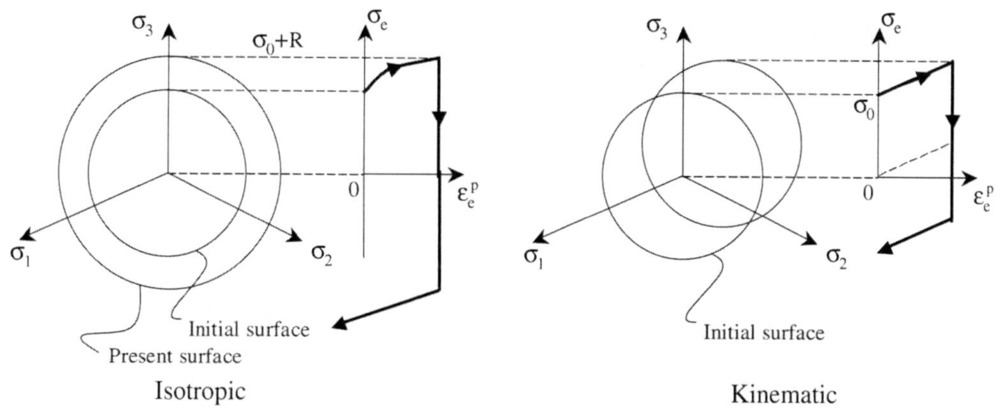

Figure 16: Isotropic and kinematic hardening

2.6. Failure modes

2.6.1. Introduction

Diffuse and localized necking as shown in Figure 17 are two important instability phenomena that are of tremendous importance in the study of failure of metals such as aluminium. A diffuse neck occurs at the instant of the formation of non-uniform strains. At this instant the capacity of the member is reached. The localized neck occurs through further straining which produces a reduction in the section thickness and subsequent failure. The formation of these instabilities is an important precursor to failure of components, and the computational prediction of the onset and growth of these instabilities are indispensable in calculations of the ultimate ductility and strength of the members and structures.

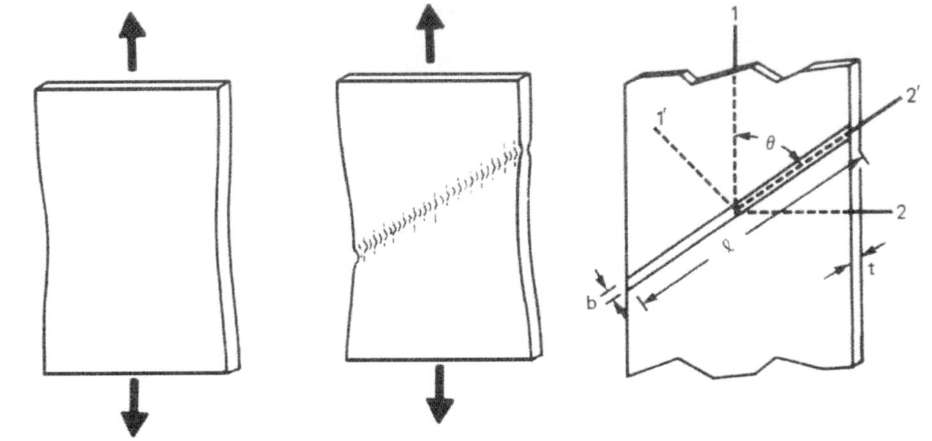

a) Diffuse neck b) Localized neck

Figure 17: Development of diffuse and localized neck [Hosford and Caddell (1993)]

2.6.2. Diffuse necking

Swift's (1952) criterion will now be used to illustrate diffuse necking of a plate element subjected to a biaxial stress state ($\sigma_3 = 0$), see Figure 18.

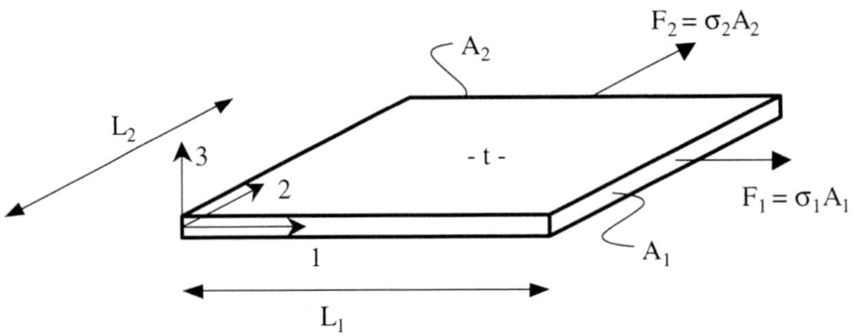

Figure 18. Diffuse necking of plate element subjected to biaxial loading

The total load acting on the plate element is

$$F = F(\sigma_1, A_1, \sigma_2, A_2) \tag{2.80}$$

The maximum load in an arbitrary direction is found by demanding that *dF=0*.

$$dF = \frac{\partial F}{\partial \sigma_1} d\sigma_1 + \frac{\partial F}{\partial A_1} dA_1 + \frac{\partial F}{\partial \sigma_2} d\sigma_2 + \frac{\partial F}{\partial A_2} dA_2 = 0 \tag{2.81}$$

where

$$F = \sqrt{F_1^2 + F_2^2} = \sqrt{(A_1\sigma_1)^2 + (A_2\sigma_2)^2} \tag{2.82}$$

Differentiation gives

$$dF = (\sigma_1 A_1)_{A_1} \cdot d\sigma_1 + (\sigma_1 A_1)_{\sigma_1} \cdot dA_1 + (\sigma_2 A_2)_{A_2} \cdot d\sigma_2 + (\sigma_2 A_2)_{\sigma_2} \cdot dA_2 = 0$$

or

$$F_1^2 \left[\frac{d\sigma_1}{\sigma_1} + \frac{dA_1}{A_1} \right] + F_2^2 \left[\frac{d\sigma_2}{\sigma_2} + \frac{dA_2}{A_2} \right] = 0 \tag{2.83}$$

Plastic incompressibility gives

$$\mathrm{d(A_1L_1)=d(A_2L_2)=0} \tag{2.84}$$

Thus

$$dA_1 L_1 + dL_1 A_1 = 0$$

or

$$\frac{dA_1}{A_1} = -\frac{dL_1}{L_1} = -d\varepsilon_1^p \tag{2.85}$$

and

$$dA_2 L_2 + dL_2 A_2 = 0$$

or

$$\frac{dA_2}{A_2} = -\frac{dL_2}{L_2} = -d\varepsilon_2^p \tag{2.86}$$

Introducing Eq. (2.85) and Eq. (2.86) into Eq. (2.83) gives

$$F_1^2 \left[\frac{d\sigma_1}{\sigma_1} - d\varepsilon_1^p \right] + F_2^2 \left[\frac{d\sigma_2}{\sigma_2} - d\varepsilon_2^p \right] = 0 \tag{2.87}$$

In the case that neither F_1 nor F_2 are zero, the criterion for plastic instability $dF = 0$ may thus be written

$$\frac{d\sigma_1}{d\varepsilon_1^p} = \sigma_1 \text{ and } \frac{d\sigma_2}{d\varepsilon_2^p} = \sigma_2 \tag{2.88}$$

Inspection of Eq. (2.88) reveals a pronounced dependency on the constitutive behaviour. Since the state of the stress is no longer uniaxial, it is necessary to apply the effective stress and strains. In addition to the hardening, an exact description of the yield criterion and the flow potential is essential for good predictions. For illustration purpose, the von Mises yield criterion is used together with an associate flow rule and a power law of hardening. More complicated yield surfaces and flow potentials can be applied, but numerical solution techniques may then become necessary. From Eqs. (2.51) and (2.52) we have ($\sigma_3 = 0$)

$$d\varepsilon_1^p = d\lambda \frac{2}{3}\left[\sigma_1 - \frac{1}{2}\sigma_2\right] \tag{2.89}$$

$$d\varepsilon_2^p = d\lambda \frac{2}{3}\left[\sigma_2 - \frac{1}{2}\sigma_1\right] \tag{2.90}$$

Assuming a constant stress ratio $\alpha = \sigma_2/\sigma_1$ (proportional loading) as well as a constant strain ratio $\rho = d\varepsilon_2^p / d\varepsilon_1^p$, the relationship between α and ρ can be expressed as

$$\alpha = \frac{1+2\rho}{2+\rho} \tag{2.91}$$

From Eq. (2.55) the effective stress can be written as

$$\sigma_e^2 = \sigma_1^2 + \sigma_2^2 - \sigma_1\sigma_2 \tag{2.92}$$

and the increment

$$d\sigma_e = \frac{(2\sigma_1 - \sigma_2)d\sigma_1 + (2\sigma_2 - \sigma_1)d\sigma_2}{2\sigma_e} \tag{2.93}$$

Introducing Eq. (2.88), the expressions for plastic instability, as well as α and ρ gives

$$d\sigma_e = \frac{2 - \alpha + (2\alpha - 1)\rho\alpha}{2\sigma_e} d\varepsilon_1^p \sigma_1^2 \tag{2.94}$$

The effective strain increment $d\varepsilon_e^p$ can be derived by requiring an equivalent increment of plastic work per unit volume

$$\sigma_e d\varepsilon_e^p = \sigma_1 d\varepsilon_1^p + \sigma_2 d\varepsilon_2^p \tag{2.95}$$

Rearranging and substitution of α and ρ into Eq. (2.95) gives

$$\sigma_e d\varepsilon_e^p = \sigma_1 d\varepsilon_1^p \left[1 + \frac{\sigma_2}{\sigma_1}\frac{d\varepsilon_2^p}{d\varepsilon_1^p}\right] = \sigma_1 d\varepsilon_1^p\,[1 + \alpha\rho] \tag{2.96}$$

Inserting for $d\varepsilon_1^p$ in Eq. (2.94) gives

$$\frac{d\sigma_e}{d\varepsilon_e^{\,p}} = \frac{2-\alpha+(2\alpha-1)\rho\alpha}{2\dfrac{\sigma_e}{\sigma_1}} \frac{\sigma_e}{1+\alpha\rho} \tag{2.97}$$

The stress ratio in the denominator σ_e / σ_1 can be expressed through the von Mises effective stress, Eq. (2.92), as follows

$$\sigma_e = \sigma_1 \left[1-\alpha+\alpha^2\right]^{1/2} \tag{2.98}$$

Combination of Eqs. (2.97) and (2.98) gives

$$\frac{d\sigma_e}{d\varepsilon_e^p} = \frac{2-\alpha+(2\alpha-1)\rho\alpha}{(1+\alpha\rho)\left(1-\alpha+\alpha^2\right)^{1/2}}\sigma_e \tag{2.99}$$

The strain that satisfies this equation can be found using the power law equation

$$\sigma_e = K\left(\varepsilon_e^{\,p}\right)^m \tag{2.100}$$

where the effective strain increment is defined as

$$d\varepsilon_e^{\,p} = \left[\frac{2}{3}\left(d\varepsilon_1^{p^2} + d\varepsilon_2^{p^2} + d\varepsilon_3^{p^2}\right)\right]^{1/2} \tag{2.101}$$

Plastic incompressibility gives

$$d\varepsilon_1^{\,p} + d\varepsilon_2^{\,p} + d\varepsilon_3^{\,p} = 0 \tag{2.102}$$

or

$$d\varepsilon_3^{\,p} = -\left(d\varepsilon_1^{\,p} + d\varepsilon_2^{\,p}\right)$$

Hence

$$d\varepsilon_e^{\,p} = d\varepsilon_1^{\,p}\frac{2}{\sqrt{3}}\sqrt{1+\rho+\rho^2} \tag{2.103}$$

Integration of Eq. (2.103) gives

$$\varepsilon_e^{\,p} = \varepsilon_1^{\,p}\frac{2}{\sqrt{3}}\sqrt{1+\rho+\rho^2} \tag{2.104}$$

Combination of Eqs. (2.99) and (2.100) yields

$$m\sigma_e \frac{1}{\varepsilon_e^{\,p}} = \frac{2-\alpha+(2\alpha-1)\rho\alpha}{(1+\alpha\rho)(1-\alpha+\alpha^2)^{1/2}}\sigma_e \tag{2.105}$$

After substitution of Eq. (2.104) and Eq. (2.91), Eq. (2.105) becomes

$$\varepsilon_1^p = \varepsilon_{1u}^p = m\frac{2(1+\rho+\rho^2)}{2+\rho+\rho^2+2\rho^3} = m\frac{2(1+\rho+\rho^2)}{(\rho+1)(2\rho^2-\rho+2)} \tag{2.106}$$

As seen, diffuse necking is dependent on the strain ratio ρ, i.e. the gradient to the yield surface. Thus, a correct description of the shape of the yield surface is crucial for the prediction of an instability phenomenon.

2.6.3. Localized necking

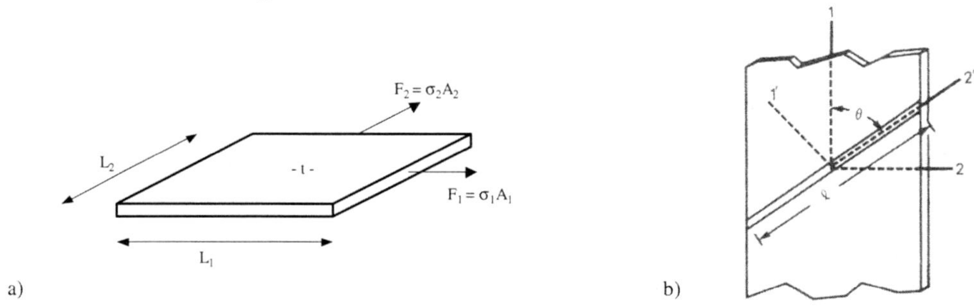

Figure 19: Localized necking of thin plate [Hosford and Caddell (1993)]

Hill (1952) was the first to study localised necking as a bifurcation from a stable homogeneous mode of deformation to a non-homogeneous one. The following presentation is based on Hosford & Caddel (1993) and Lademo (1999).

Figure 19a shows a thin isotropic plate subjected to proportional loading, i.e. a constant stress ratio $\alpha=\sigma_2/\sigma_1$ is assumed. A local neck can form when $dF_1=0$. Since $F_1=\sigma_1 A_1$,

$$dF_1 = 0 = d\sigma_1 A_1 + dA_1\sigma_1 \tag{2.107}$$

or

$$\frac{d\sigma_1}{\sigma_1} = -\frac{dA_1}{A_1} \tag{2.108}$$

As shown in Figure 19b, the cross-sectional area of the neck is $A' = \ell \cdot t$. Since ℓ is constant $dA' = \ell \cdot dt$ or $dA'/A' = dt/t = d\varepsilon_3^p$. The area perpendicular to the 1-axis is $A_1 = A'\cos\theta$, but θ is constant so

$$\frac{dA'}{A'} = \frac{dA_1}{A_1} = d\varepsilon_3^{\,p} \tag{2.109}$$

Plastic incompressibility gives

$$d\varepsilon_1^p + d\varepsilon_2^p + d\varepsilon_3^p = 0$$

or

$$d\varepsilon_3^p = -(1+\rho)d\varepsilon_1^p \tag{2.110}$$

where $\rho = d\varepsilon_2^p / d\varepsilon_1^p$, i.e. a constant strain ratio is assumed. Introducing Eqs. (2.109) and (2.110) into Eq. (2.108) gives

$$\frac{d\sigma_1}{\sigma_1} = -d\varepsilon_3{}^p = (1+\rho)d\varepsilon_1{}^p \tag{2.111}$$

A power law expression for the stress-strain curve is assumed in the form

$$\sigma_e = K\left(\varepsilon_e^p\right)^m \tag{2.112}$$

Introducing Eq. (2.98) and Eq. (2.104) into Eq. (2.112) gives

$$\sigma_1 = \bar{K}(\alpha,\rho)(\varepsilon_1^p)^m \tag{2.113}$$

where

$$\bar{K}(\alpha,\rho) = \frac{K}{\left(1-\alpha+\alpha^2\right)^{\frac{1}{2}}}\left[\frac{2}{\sqrt{3}}\sqrt{1+\rho+\rho^2}\right]^m \tag{2.114}$$

After substitution of Eq. (2.113), Eq. (2.111) becomes

$$\frac{1}{\sigma_1}\frac{d\sigma_1}{d\varepsilon_1^p} = \frac{\bar{K}}{\sigma_1} m(\varepsilon_1^p)^{m-1} = 1+\rho \tag{2.115}$$

or

$$\varepsilon_1^p = \varepsilon_{1u}^p = \frac{m}{1+\rho} \tag{2.116}$$

Like the criterion for diffuse necking, the above criterion depends on the gradient to the yield surface. Accordingly, the use of a different yield criterion alters the prediction of the localized neck.

The local neck forms under the constraint $d\varepsilon_{2'}^p = 0$, i.e. along 2' axis, see Figure 19b. It can be shown by transformation that

$$d\varepsilon_{2}^{p'} = d\varepsilon_{1}^{p}\cos^{2}\theta + d\varepsilon_{2}^{p}\sin^{2}\theta = 0 \tag{2.117}$$

Thus

$$\varepsilon_{1}^{p}\cos^{2}\theta + \rho\varepsilon_{1}^{p}\sin^{2}\theta = 0 \tag{2.118}$$

or

$$\tan\theta = \frac{1}{\sqrt{-\rho}} \tag{2.119}$$

The angle θ will have a real value only if ρ is negative (i.e. ε_2^p is negative). Physically this means that if ε_2^p is positive, there is no angle at which a local neck can form and still fulfil the requirement that $d\varepsilon_{2'}^p = 0$ (i.e. if ρ is constant and >0, local necks cannot form). The criteria for diffuse and localized necking are plotted in Figure 20.

Even if the above theory does not give any localization for $\rho>0$, experiments show the opposite. The theories of Marciniak et al (1973) for instance account for this. However, space does not permit any further discussion of this theory here.

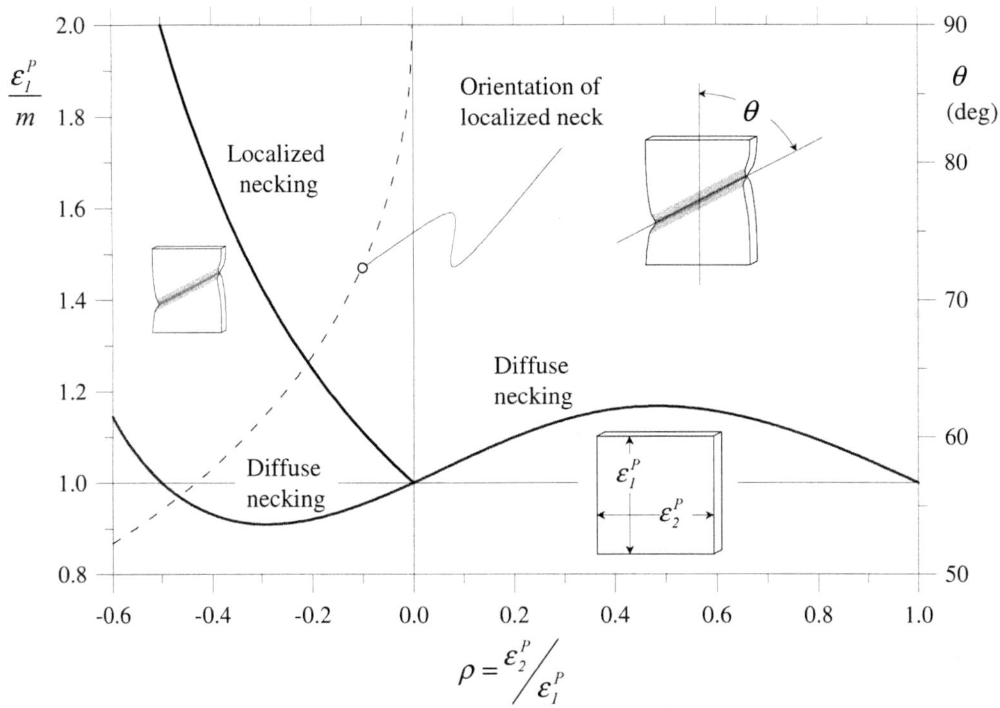

Figure 20: Critical strains for diffuse and localized necking

2.6.4. Effect of welding

Welding of heat-treated aluminium alloys introduces a heat-affected zone (HAZ) with a reduction of flow stress and an increase in ductility. If a welded component is subjected to tensile forces/stresses, the strains will tend to localize in the HAZ and failure may occur.

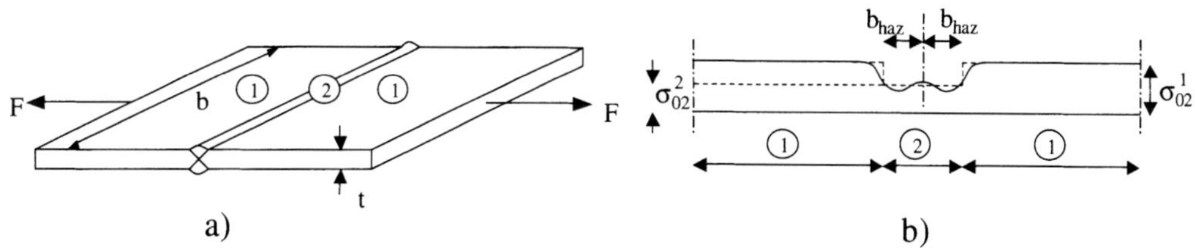

Figure 21: Butt-welded plate subjected to a tensile force.

Figure 21a shows two plates denoted 1 welded together by a butt weld. The plates have width b and thickness t and are subjected to a tensile force F. Due to welding, a heat affected zone, 2, is introduced with a reduction of the flow stress, Figure 21b. The width of HAZ is $2b_{haz}$. Further information about the mechanical properties of aluminium with respect to welding can be found in Matusiak (2000).

By using the Ramberg-Osgood equation an estimate of the localization of the strain in the HAZ will now be shown. The Ramberg-Osgood equation reads

$$\varepsilon = \frac{\sigma}{E} + 0.002\left(\frac{\sigma}{\sigma_{0.2}}\right)^n \tag{2.120}$$

Force equilibrium of the plate strip

$$F = \sigma_1 bt = \sigma_2 bt \tag{2.121}$$

or

$$\sigma = \sigma_1 = \sigma_2 = k \cdot \sigma^1_{0.2} \tag{2.122}$$

Here, k is a constant and $\sigma^1_{0.2}$ is the proof stress in the parent material. The strain in the HAZ becomes

$$\varepsilon_2 = \frac{k\sigma^1_{0.2}}{E} + .002\left(\frac{k \cdot \sigma^1_{0.2}}{\sigma^2_{0.2}}\right)^{n_2} \tag{2.123}$$

and in parent material

$$\varepsilon_1 = \frac{k\sigma^1_{0.2}}{E} + .002\left(\frac{k \cdot \sigma^1_{0.2}}{\sigma^1_{0.2}}\right)^{n_1} = \frac{k\sigma^1_{0.2}}{E} + 0.002(k)^{n_1} \tag{2.124}$$

A plot of the ratio $\varepsilon_2/\varepsilon_1$ as a function of k is shown in Figure 22 when using material data for alloy EN AW 6082-T6. The proof stress in the parent material and HAZ is taken as $\sigma^1_{0.2}$=250 N/mm^2 and $\sigma^2_{0.2}$=125 N/mm^2 respectively, while the corresponding n values are taken as n_1=25 and n_2=12.5. Figure 22 clearly demonstrates that strong localization of the strains will take place in HAZ even for low stress ratios k. The line for uniaxial diffuse necking in HAZ is shown and represents a strain ratio of $\varepsilon_2/\varepsilon_1$=33. Be aware that this ratio represents pure-elastic deformations in the parent material. This is a value of k that is equal to 0.67.

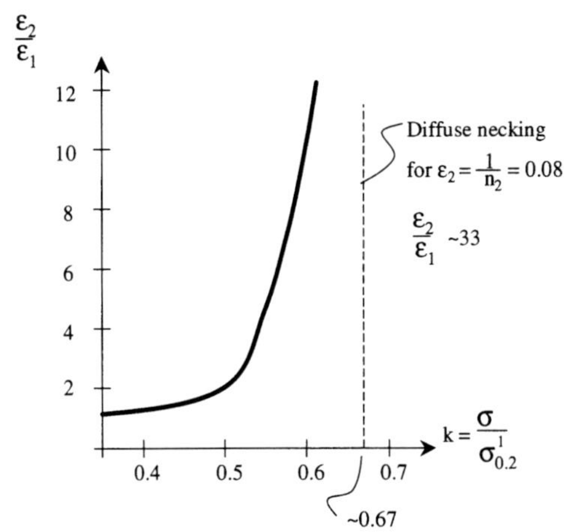

Figure 22: Localization of strains in HAZ

2.7. Strain-rate behaviour

2.7.1. Definition of strain rate

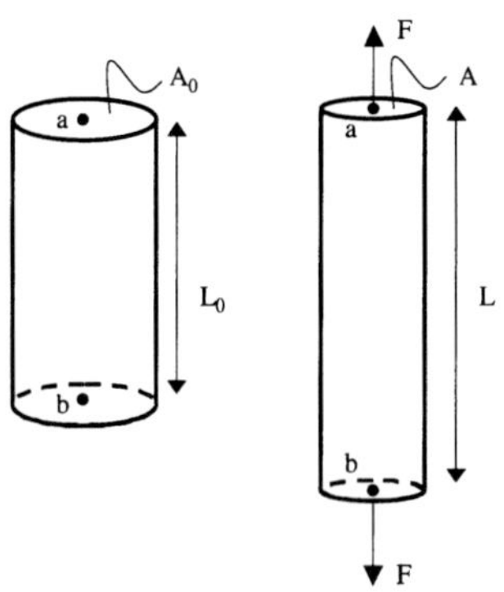

Figure 23: Tensile test specimen

A portion of a uniform homogeneous bar of metal, length L_0 and cross-sectional area A_0 in the unloaded state is subjected to a slowly applied axial force F which increases its length to L and reduces its cross-sectional area to the current value A, see Figure 23. The nominal and true strain increments are defined as $de=dL/dL_0$ and $d\varepsilon=dL/L$ respectively. The strain-rate values can be obtained from these values. The tensile engineering strain rate $\dot{e}$ is defined as

$$\dot{e} = \frac{de}{dt} = \frac{dL/dt}{L_0} = \frac{v}{L_0} \tag{2.125}$$

where v is the speed of separation at points a and b at the ends of the chosen portion of the bar of length L_0, see Figure 23. The tensile true strain rate is defined as

$$\dot{\varepsilon} = \frac{d\varepsilon}{dt} = \frac{dL/dt}{L} = \frac{v}{L} \tag{2.126}$$

If the gauge length is 44 mm and the velocity is 2 mm/min, $\dot{e} = 7.6 \cdot 10^{-4} s^{-1}$, i.e. a quasi-static tensile test.

2.7.2. Test methods

Collisions will normally cause strain rates in the range of $10^{-4} < \dot{\varepsilon} < 500 s^{-1}$, while impact and penetration problems may lead to strain rates well above 10^4-10^5 s^{-1}. This means that test methods must be available in order to characterize the material at elevated rates of strain. Hydraulic machines are often used for moderate to low strain rates ($\dot{\varepsilon} < 100 s^{-1}$). For strain rates in the range $10^2 < \dot{\varepsilon} < 10^4 s^{-1}$ the well-known Split Hopkinson Bar (SHB) is utilized, while at higher strain rates plate impact tests may be used. A material test at elevated rates of strain is not straightforward to carry out and often we see that it can be quite difficult to interpret the test data, as the material response interacts with the response of the test equipment.

2.7.3. Split Hopkinson Bar

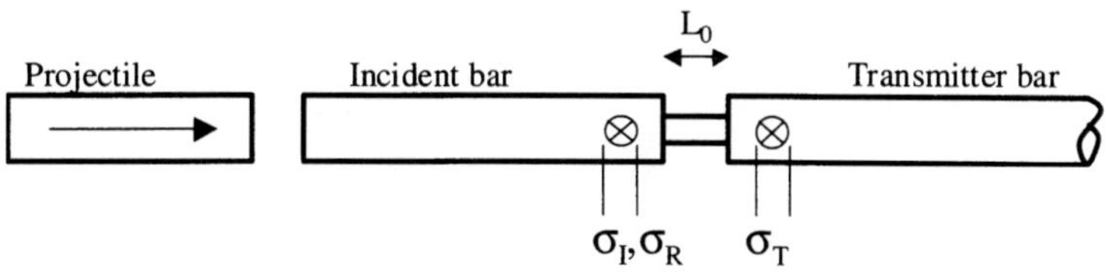

Figure 24: Split Hopkinson Pressure Bar

The Split Hopkinson Bar has found a wide acceptance as the instrument for intermediate strain-rate testing (10^2-10^4 s^{-1}). Figure 24 shows a compressional SHB where a projectile (striker bar) impacts the incident bar. A pulse is produced which has a considerable length compared to the length of the specimen L_0. This elastic wave travels through the incindent bar and reaches the specimen, which is sandwiched between the incident bar and transmitter bar. The strain gauges attached to the bars, see Figure 24, measure the incident pulse, the reflected pulse and the transmitted pulse. The basic equations that provide the stress, strain and strain rate in a test specimen when we know the strain in the bars can now be obtained.

Figure 25 schematically shows the displacement and velocity situation at the interface between the specimen and the bars. Since the displacement $u_A > u_B$, the specimen length decreases with time and the specimen will undergo plastic deformation.

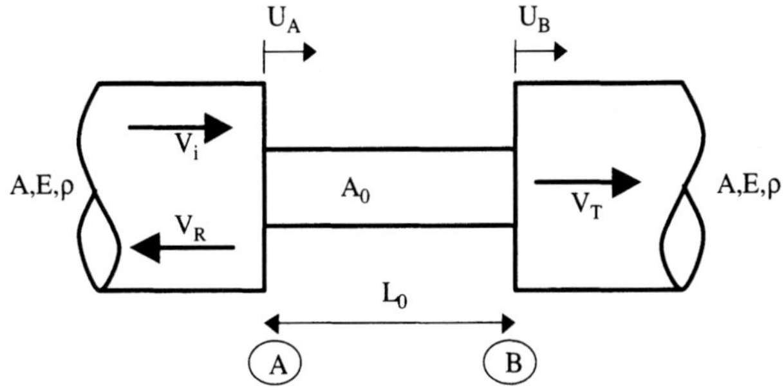

Figure 25: Schematic view of test specimen and stress pulses

The nominal strain in the specimen can be expressed as

$$e_s = \frac{u_A - u_B}{L_0} \tag{2.127}$$

From this expression the strain rate becomes

$$\dot{e}_s = \frac{\dot{u}_A - \dot{u}_B}{L_0} = \frac{(v_i - v_R) - v_T}{L_0}$$

or

$$\dot{e}_s = \frac{c}{L_0}(e_i - e_R - e_T) \tag{2.128}$$

when using

$$v_i = c \cdot e_i, v_R = c \cdot e_T, v_T = c \cdot e_T, c = \sqrt{\frac{E}{\rho}}$$

The forces acting on the specimen are

$$P_A(t) = EA(e_i + e_R) \tag{2.129}$$

and

$$P_B(t) = EAe_T \tag{2.130}$$

The average stress in the specimen can now be obtained

$$\sigma_s = \frac{P_A + P_B}{2A_0} = \frac{EA}{2A_0}(e_i + e_R + e_T) \tag{2.131}$$

Assuming $P_A = P_B$ gives that $e_T = e_i + e_R$. Thus

$$\sigma_s = \frac{EA}{A_0} e_T \tag{2.132}$$

$$e_s = -\frac{2c}{L_0}\int_0^t e_R dt \tag{2.133}$$

$$\dot{e}_s = -\frac{2c}{L_0} e_R \tag{2.134}$$

It is worth noting that we do not consider wave propagation in the specimen. Actually, when the wave enters the specimen several reflections will take place and the specimen will obtain a quasi equilibrium after some time. This stage can be revealed by the recorded signals in the strain gauges, which shows that $P_A \approx P_B$. Due to the numerous stress-wave reflections the initial part of the stress-strain curve does not give the actual material behaviour. When plastification takes place the real responses are often monitored.

2.7.4. Constitutive model and calibration

The high strain-rate plastic deformation of materials is often described by constitutive equations that link stress with strain, strain-rate and temperature. From a uniaxial test we can express the stress as (see Figure 26)

$$\sigma = f(\varepsilon_p, \dot{\varepsilon}_p, T) = \sigma_0 + R + \sigma_v \tag{2.135}$$

A commonly used constitutive equation is that proposed by Johnson and Cook

$$\sigma = (A + B\varepsilon_p{}^n)(1 + C\ln\dot{\varepsilon}^*)(1 - T^{*m}) \tag{2.136}$$

where

$$\dot{\varepsilon}^* = \dot{\varepsilon}_p / \dot{\varepsilon}_{0p} \quad \text{and} \quad T^* = (T - T_r) / (T_m - T_r)$$

Here, T_r is a reference temperature, T_m is the melting temperature while $\dot{\varepsilon}_{0p}$ is the reference strain rate. It can be shown that σ_0, R and σ_v in Eq. (2.135) can be expressed as

$$\sigma_0(T) = A(1 - T^{*m}) \tag{2.137}$$

$$R(\varepsilon_p, T) = B\varepsilon_p^n (1 - T^{*m}) \tag{2.138}$$

$$\sigma_v\left(\varepsilon_p,\dot{\varepsilon}_p,T\right)=C\ln\dot{\varepsilon}^*\left(A+B\varepsilon_p^n\right)\left(1-T^{*m}\right) \tag{2.139}$$

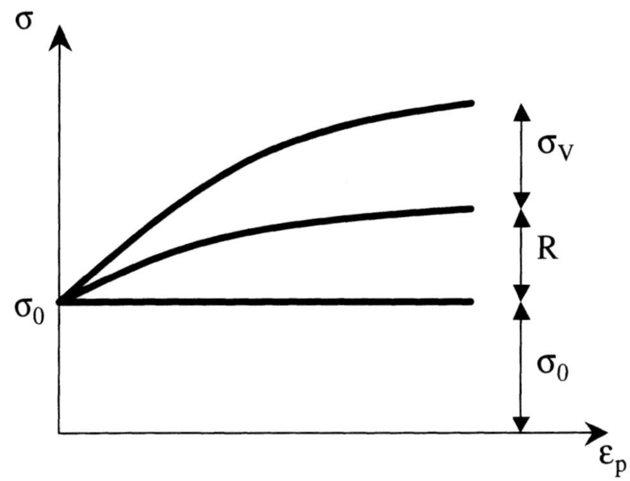

Figure 26: Unaxial stress-strain curve

The different parameters in the constitutive equation can be determined from three separate uniaxial tensile tests:

1. Quasi-static tests at room temperature to determine the constants A, B and n
2. Tests at elevated rates of strain at room temperature to determine the constant C
3. Quasi-static tests at elevated temperatures to determine m.

2.7.5. Strain-rate sensitivity of aluminium

We know from the literature that aluminium alloys for structural applications can be considered almost strain-rate insensitive with respect to the flow stress. However, tests have shown that increasing the strain rate may increase the ductility, Lindholm et al. (1971). For some alloys however, negative strain-rate sensitivity has been found when increasing the strain rate. This effect may cause the localization of the strains during high-rate loading and thus cause failure, Lademo (1999).

3. ENERGY ABSORBING COMPONENTS

3.1. Definitions

A terminology has been developed to describe and compare the energy absorption performance of collapsing and crushing structural parts, see Figure 27.

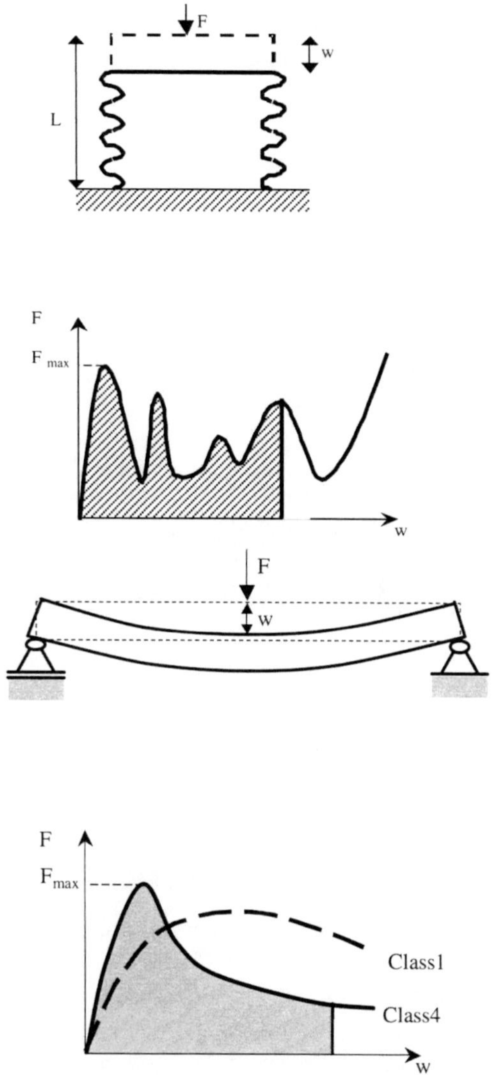

- Absorbed energy: $U(w) = \int_0^w F dw$
- Average force: $F_{avg}(w) = \frac{1}{w}\int_0^w F dw = \frac{U(w)}{w}, w > 0$
- Crush force efficiency: $A_E = \frac{(w)}{F_{max} w} = \frac{F_{avg}(w)}{F_{max}}$

Figure 27: Terminology applied for impact loaded structures

An absorber with a rectangular-shaped force-displacement curve (rigid-plastic) has a crush force efficiency, A_E, of 100 %. Lower values indicate an oscillating force-displacement curve. For axially loaded members such as aluminium extrusions and foams some additional terms can be defined.

The total efficiency is defined as

$$T_E = \frac{U(w)}{F_{\max}(w)L} = \frac{F_{avg}(w)w}{F_{\max}(w)L} = A_E D_C \tag{3.1}$$

where $D_C = w/L$ is defined as the relative deformation (deformation capacity) of the absorber.

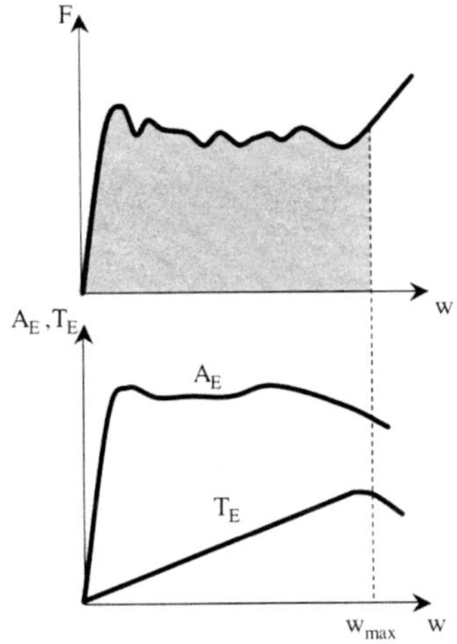

Figure 28: Definition of total efficiency

As seen in Figure 28 a given energy absorber may be defined to show optimum properties when T_E has a maximum value [Hanssen et al. (2000a)]. The maximum value of T_E occurs at deformation, which defines the maximum value of $D_C = S_E = w_{\max}/L$. Here, S_E is denoted as the stroke efficiency of the absorber. When dealing with energy absorbers in the automotive industry the term mass specific energy absorption E_M is often used which relates the absorbed energy to the mass of the absorber. Often only the crushed mass of the absorber is taken into account. The specific energy E_M may then be written as

$$E_M = \frac{U(w_{\max})}{M(absorber\ or\ destroyed\ mass)} \tag{3.2}$$

For some reasons it might be of interest to relate the absorbed energy to the outer volume of the absorber structure to determine the volumetric energy dissipation density, E_V. Thus

$$E_V = \frac{U(w_{\max})}{V(absorber)} \tag{3.3}$$

3.2. Impact loading of beams

3.2.1. Introduction

When a beam is subjected to an impact a very complex process takes place involving the transfer of kinetic energy from the striker to the target. As will be shown in the following, the energy absorption takes place in two phases, namely the transient phase and the global mode phase. In the transient phase elastic and plastic waves will travel away from the impact point, and the energy absorption in this phase is a function of the mass ratio between striker and absorber. Furthermore, the response of the beam will be a function of the material properties, the slenderness of the cross- sectional parts, the support conditions and the contact area between striker and beam. This means that the classification of the cross section and the corresponding rotational capacity is a key issue with respect to energy absorption. A good design principle when dealing with aluminium structures subjected to impact loading conditions is to make a proper selection of alloy and temper. An increased 0.2 % proof stress is preferable, but will normally lead to reduced ductility. This means that a temper T6 alloy is not preferable, so non-heat-treated alloys such as temper T4 or over-aged tempers are often used to avoid localization and thereby increase ductility and the rotational capacity. In the following the behaviour of rigid perfect-plastic beams subjected to impact loading will be shown as well as the behaviour of aluminium beams with a non-linear stress-strain curve.

3.2.2. Rigid perfect-plastic material

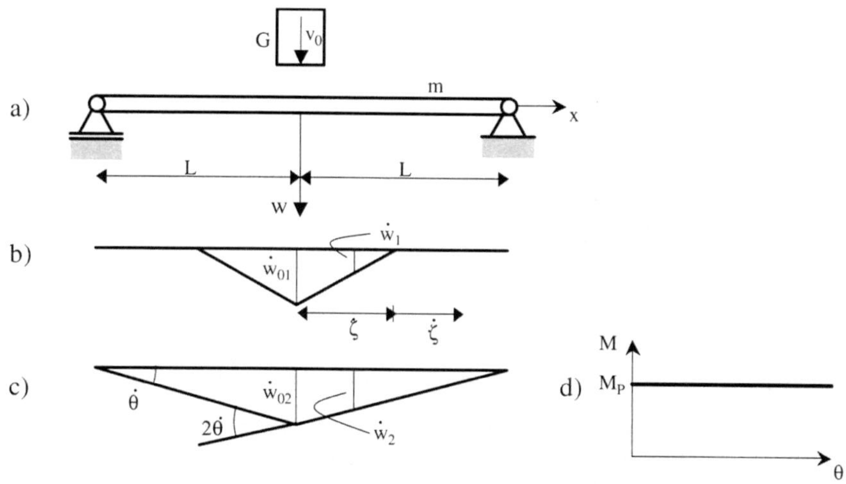

Figure 29: Impact of a simply supported beam

Figure 29a shows a simply supported beam struck at the mid-span by a mass G travelling with an initial velocity v_0. The beam has a mass per unit length equal to m. A plastic hinge develops under the impact point at $t=0$ and two plastic hinges propagate the disturbance away from the mid-span towards the support, Figure 29b. When the two moving plastic hinges reach the support, only one hinge remains at the mid-span, Figure 29c. The striker and beam will come to rest when all the initial kinetic energy of the striking mass $Gv_0^2/2$ is dissipated plastically. The transverse velocity field for the first phase of motion, Figure 29b, can be written as

$$\dot{w}_1 = \dot{w}_{01}\left(1-\frac{x}{\xi}\right), \quad 0<x<\xi \tag{3.4}$$

and

$$\dot{w}_1 = 0\,, \quad \xi<x<L\,, \tag{3.5}$$

while the second phase shown in Figure 29c is given as

$$\dot{w}_2 = (L-x)\dot{\theta}\,, \quad 0<x<L \tag{3.6}$$

It is possible by vertical and moment equilibrium to show [Jones (1989)] that the final permanent transverse displacement w_f can be expressed as

$$w_f = \frac{Gv_0^2}{4M_p}(L-x) \tag{3.7}$$

when $G/mL >> 1$, i.e. a heavy striker. Thus, the kinetic energy of the striking mass is absorbed in the second phase of motion, i.e. the first phase of motion plays an unimportant role. For aluminium beams subjected to impact loading in the automotive industry the mass ratio $G/mL >> 1$. Thus the first phase of motion can be neglected.

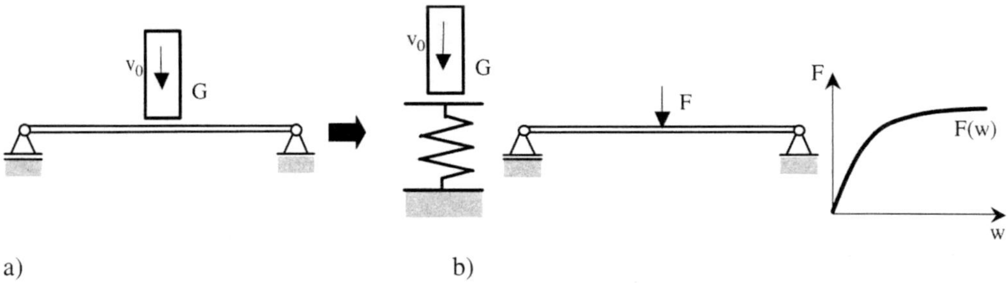

Figure 30: Transformation of a beam into a single degree-of-freedom system

Neglecting the first phase of motion, the beam can be transformed into a single-degree-of-freedom system, see Figure 30a, with a spring stiffness obtained from the quasi-static load-deformation curve, see Figure 30b. The following energy balance will then give the maximum deformation of the beam

$$\frac{1}{2}Gv_0^2 = \int_0^{w_f} F(w)dw \tag{3.8}$$

3.2.3. Strain hardening material

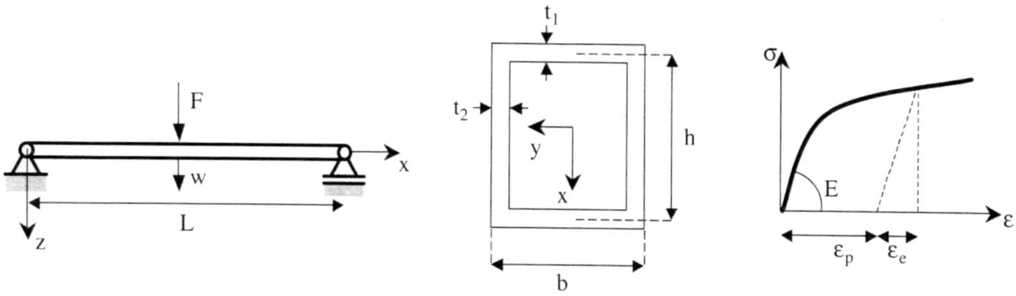

Figure 31: Aluminium beam subjected to a concentrated load

Figure 31 shows a simply supported beam subjected to a concentrated load F at mid-span. The beam has a rectangular hollow cross section with a non-linear stress-strain curve where the relationship between the stress and the plastic strain can be represented by a power law, see Eq. (2.8). It is assumed that the cross section belongs to class 1, i.e. no local buckling will take place. The force-displacement curve will be calculated by assuming that the elastic and plastic deformations are additive, i.e.

$$w_t = w_e + w_p \tag{3.9}$$

Furthermore, the plastic strain ε_p is assumed to have a linear variation over the cross section

$$\varepsilon_p = z\chi_p \tag{3.10}$$

where χ_p is the plastic curvature. The stress and moment in a cross section of the beam are expressed as

$$\sigma = K(\varepsilon_p)^m \tag{3.11}$$

and

$$M = \int_A \sigma z dA \tag{3.12}$$

The variation of Eq. (3.12) yields

$$\delta M = \int_A \delta\sigma z dA \tag{3.13}$$

Introducing the expressions for ε_p and σ into Eq.(3.12) gives

$$= 2\int_0^{\frac{h}{2}} \chi_p^m \quad ^{m+1}(2t_2)d \quad + 2 \quad \chi_p^m (\frac{h}{2})^{m+1} \quad t_1$$

or

$$M = Kh\left(\frac{h}{2}\right)^m \left[\frac{ht_2}{m+2} + t_1\right]\chi_p^m \tag{3.14}$$

and

$$\chi_p = \frac{2}{h}\left[\frac{F}{2Kh\left(\frac{ht_2}{m+2} + bt_1\right)}\right]^{\frac{1}{m}} x^{\frac{1}{m}} \tag{3.15}$$

The plastic deformation is now calculated using the principle of virtual forces

$$w_p \delta F = \int_V \varepsilon_p \delta\sigma dV = 2\int_0^{L/2}\int_A \varepsilon_p \delta\sigma dA dx \tag{3.16}$$

After substitution of Eq. (3.10) and Eq. (3.13), Eq. (3.16) becomes

$$w_p \delta F = 2\int_0^{L/2} \chi_p \left[\int_A \delta\sigma z dA\right] dx = 2\int_0^{L/2} \chi_p \delta M dx \tag{3.17}$$

where $\delta M = \delta F \cdot x/2$. The total displacement can now be written as

$$w_t = \frac{FL^3}{48E} + \frac{mL^2}{2h(2m+1)}\left[\frac{FL}{4Kh\left(\frac{ht_2}{m+2} + bt_1\right)}\right]^{\frac{1}{m}} \tag{3.18}$$

The first term in Eq. (3.18) is the elastic deformation at mid-span while the second term is obtained from Eq. (3.17). The F w_t curve can be generated from Eq. (3.18). The accuracy of the simplified calculation has been checked against tests performed by Welo (1990) and good predictions of the F-w_t curve have been found.

Even if a class 1 section has been chosen, the total deformation can be limited by material tensile instability in the tensile flange and / or local buckling of the compression flange. By using Eq. (3.10) and Eq. (3.15) the requirement to avoid diffuse necking at mid span, see Section 2.1.3, can be written as

$$\varepsilon_p = \chi_p \frac{h}{2} = \left[\frac{FL}{4Kh\left(\frac{ht_2}{m+2} + bt_1\right)}\right]^{\frac{1}{m}} \le \varepsilon_{pu} = m = \frac{1}{n} \tag{3.19}$$

For an alloy in temper T4, $n \approx 10$ and $\varepsilon_{pu} = m = 10\ \%$. The corresponding values for temper T6 are $n \approx 25$ and $\varepsilon_{pu} = m = 4\ \%$. These values are actually in close agreement with the ductility demands in EC9 (AnnexG) [ENV (1999)], i.e. $\varepsilon_{pu} \leq 5\ \%$ and 10 % for brittle and ductile alloys respectively.

Local buckling can be checked by comparing the maximum stress in the compression flange with the buckling stress proposed by Stowell [Iyengar (1988)] for inelastic materials

$$\sigma = K\left(\varepsilon_p\right)^m < \frac{\pi^2 E_s}{9}\left(\frac{t_1}{b}\right)^2\left[2+2\sqrt{\frac{1}{4}+\frac{3}{4}\frac{E_t}{E_s}}\right] \tag{3.20}$$

Here, E_t and E_s are the tangent and sectant moduli of the material respectively. It is assumed that the compression flange is hinged to the webs of the profile in Figure 31.

The proposed expression for the total displacement, Eq. (3.18), can also be used for other cross sections, see Figure 32.

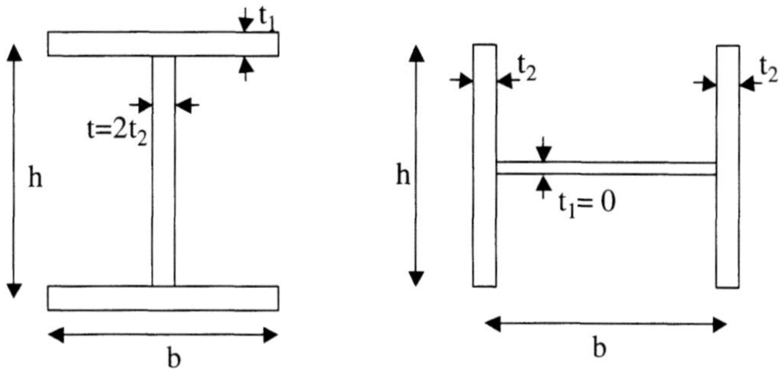

Figure 32: Cross sections

3.2.4. Effect of end restraints

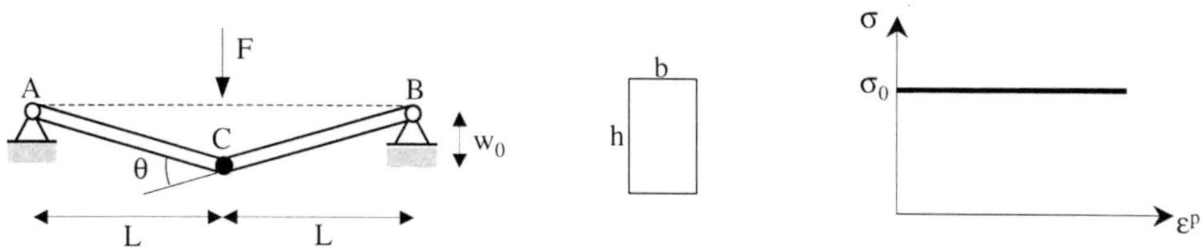

Figure 33: Simply supported beam with horizontally fixed ends

Figure 33 shows a simply supported beam that is axially restrained and is subjected to a concentrated load at mid-span. The beam has a rectangular cross section and the material is assumed to be rigid plastic. All deformations in the beam take place in a plastic hinge at mid-span. Applying the virtual displacement δw_0 from an equilibrium state yields

$$F\delta w_0 = N\delta e + M\delta\theta \tag{3.21}$$

Elongation of each member

$$e_{AC} = e_{CB} = \sqrt{L^2 + w_0^2} - L = L\left(\sqrt{1+\left(\frac{w_0}{L}\right)^2} - 1\right) \tag{3.22}$$

From series expansion it shows that

$$e_{AC} = e_{CB} = 1 + \frac{1}{2}\left(\frac{w_0}{L}\right)^2 - 1 = \frac{1}{2}\left(\frac{w_0}{L}\right)^2$$

or

$$e = 2e_{AC} = \left(\frac{w_0}{L}\right)^2 \tag{3.23}$$

The variation becomes

$$\delta e = 2\frac{w_0}{L}\delta w_0 \tag{3.24}$$

The total rotation is

$$\theta = \frac{2w_0}{L} \tag{3.25}$$

and the variation

$$\delta\theta = \frac{2}{L}\delta w_0 \tag{3.26}$$

The combination of Eq.(3.21), Eq. (3.24) and Eq. (3.26) gives

$$F = \frac{2w_0}{L}N + \frac{2}{L}M \tag{3.27}$$

The ultimate load under pure bending can be expressed as

$$F_0 = \frac{2M_p}{L} \tag{3.28}$$

Bending and membrane capacity of the cross section can be expressed as

$$M_p = \sigma_0 bh^2 / 4 \tag{3.29}$$

and

$$N_p = \sigma_0 bh \tag{3.30}$$

The yield criterion for the cross section

$$f = \frac{M}{M_p} + \left(\frac{N}{N_p}\right)^2 - 1 = 0 \tag{3.31}$$

Eq. (3.27) and Eq. (3.31) give

$$F = F_0\left[1 - \left(\frac{N}{N_p}\right)^2\right] + 2N\frac{w_0}{L} \tag{3.32}$$

The relationship between w_0 and N can now be found by using the flow rule

$$\dot{e} = \dot{\lambda}\frac{\partial f}{\partial N} = \dot{\lambda}2\frac{N}{N_p^{\ 2}} \tag{3.33}$$

$$\dot{\theta} = \dot{\lambda}\frac{\partial f}{\partial M} = \dot{\lambda}\frac{1}{M_p} \tag{3.34}$$

By taking the time derivative of e and θ given in Eq. (3.23) and Eq. (3.26) respectively, the ratio $\dot{e}/\dot{\theta}$ can now be written as

$$\frac{\dot{e}}{\dot{\theta}} = w_0 = 2\frac{NM_p}{N_p^{\ 2}} \tag{3.35}$$

Introducing the expressions for N_p and M_p into Eq. (3.35) gives

$$\frac{N}{N_p} = \frac{2w_0}{h} \tag{3.36}$$

This expression states that when $w_0 = h/2$ the membrane force N equals N_p, i.e. no bending resistance. Introducing Eq. (3.36) into Eq. (3.32) yields

$$\frac{F}{F_0} = 1 + 4\left(\frac{w_0}{h}\right)^2, \quad w_0 < \frac{h}{2} \tag{3.37}$$

and

$$\frac{F}{F_0} = 4\frac{w_0}{h}, \quad w_0 > \frac{h}{2} \tag{3.38}$$

As shown in Figure 34 an increase in the energy absorption will take place when the beam is axially restrained compared to the case with pure bending.

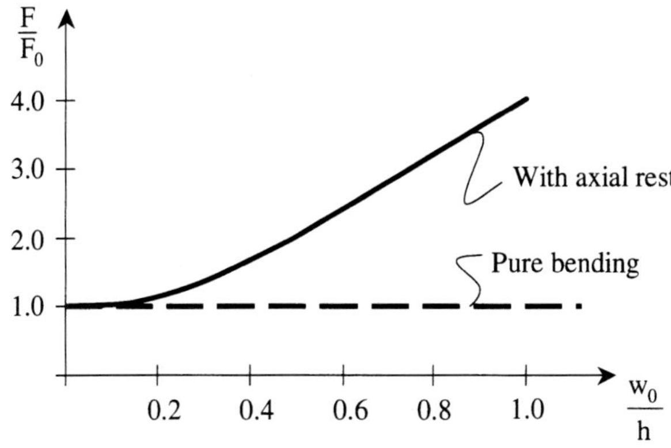

Figure 34: Load-deflection curve

3.3. Dynamic progressive buckling of tubes

3.3.1. Introduction

Ignoring global buckling, a short thin-walled member (circular or square) subjected to a static axial load may have force-displacement characteristics similar to those shown in Figure 35. Figure 36 shows photos, which corresponds to the load-shortening curve.

It is evident that the tube exhibits an unstable behaviour after reaching the first peak load. This load is defined as the capacity of the cross section and is an important parameter when designing thin-walled columns subjected to axial loading for instance. As further axial deformation is applied, the load-carrying mechanism is changed from membrane compression to bending (folding). A repeated pattern of peaks is then observed in the load-displacement curve where each peak is associated with the development of a new fold. This phenomenon is known as progressive buckling.

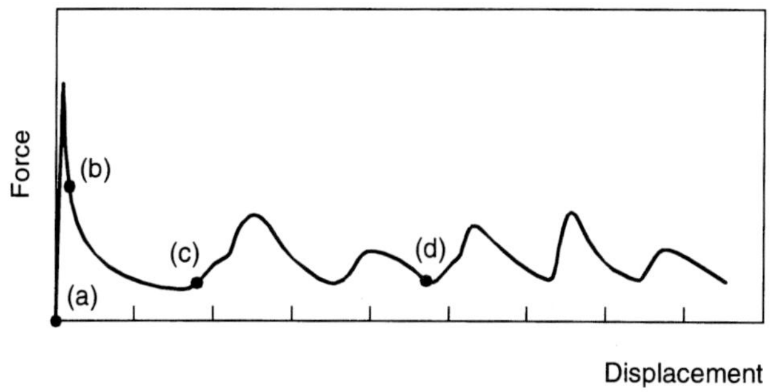

Figure 35: Typical force-displacement curve. Points (a), (b), (c) and (d) correspond to the sequence of photos in Figure 36

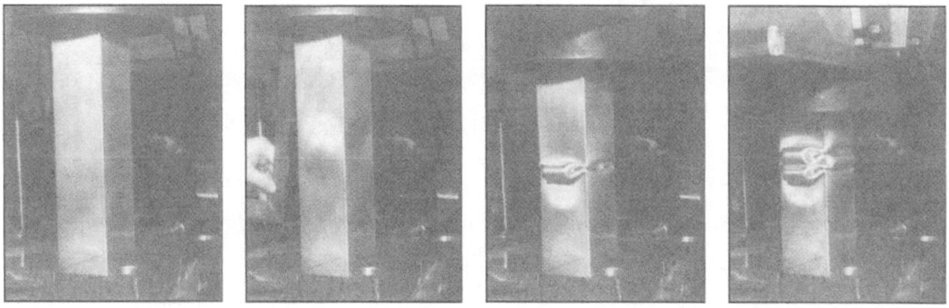

a) b) c) d)

Figure 36: Sequence of photos during axial deformation

When a tube is used as an energy absorber the fluctuations in the load are often ignored and the mean load F_{avg} is often used as a design parameter. This means that an ideal energy absorber is defined with A_E=100 %. Thus, a constant resistance is assumed, and the absorber offers a constant deceleration throughout the entire stroke. Be aware that an actual absorber may have A_E<<100 % due to a high initial peak load and the subsequent oscillating force.

An axially loaded thin-walled aluminium member will be sensitive to the impact velocity due to inertia effects. This will be further discussed in Section 3.3.4.

3.3.2. Crushing of circular tubes

3.3.2.1. Introduction

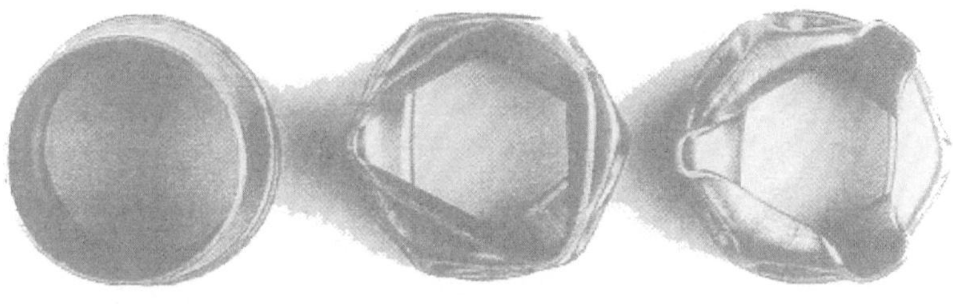

a) b)

Figure 37: Deformation modes of circular tubes. a) Axisymmetric/concertina, b) Non-axisymmetric/diamond [Jones (1989)]

When subjected to an axial force, a thin-walled circular tube of mean radius R and thickness t may develop either axisymmetric buckles, Figure 37a, or a non-axisymmetric (diamond) pattern as shown in Figure 37b (or any combination of these). Various theoretical methods predict that thicker tubes with R/t<40-45 deform axisymmetrically, while thinner tubes with larger R/t-ratios buckle into a diamond mode. However, Hanssen et al. (2000b) have shown that aluminium tubes in alloy AA6060 buckle in a diamond mode for R/t-ratios in the range 17-28. It was believed that the mode of deformation was sensitive to the support condition and hardening of the material.

3.3.2.2. Axisymmetric crushing

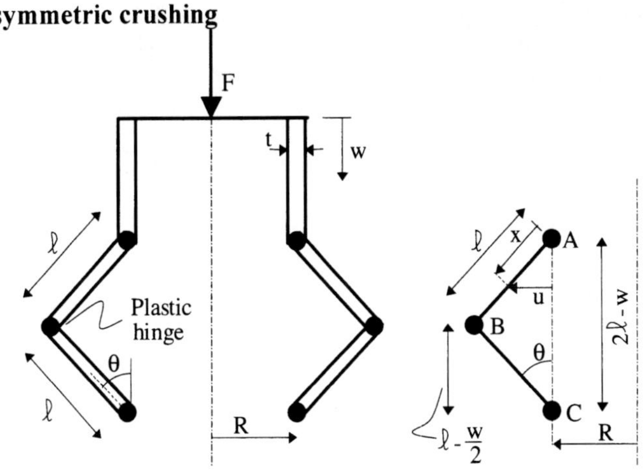

Figure 38: Idealized deformation pattern

An approximate theoretical analysis for the axial crushing of circular tubes was obtained by Alexander (1960). He assumed that the tube was made from a rigid, perfect-plastic material and used the simplified axisymmetric deformation pattern shown in Figure 38. This is an idealization of the actual behaviour, since it is evident from tests that the wrinkles are curved rather than straight.

The principle of virtual work will now be used to calculate the average crushing force, F_{avg}. The kinematic relations shown to the right in Figure 38 give

$$\cos\theta = \frac{l - w/2}{l}$$

or

$$w = 2l(1-\cos\theta) \tag{3.39}$$

and

$$u = x\sin\theta \tag{3.40}$$

The variation of w and u become

$$\delta w = 2l\sin\theta \cdot \delta\theta \tag{3.41}$$

and

$$\delta u = x\cos\theta \cdot \delta\theta \tag{3.42}$$

Assuming small strains ε_θ in the circumferential direction and introducing Eq. (3.40) yields

$$\varepsilon_\theta = \frac{2\pi(R+u) - 2\pi R}{2\pi R} = \frac{u}{R} = \frac{x\sin\theta}{R} \tag{3.43}$$

The variation becomes

$$\delta\varepsilon_\theta = \frac{x}{R}\cos\theta \cdot \delta\theta \tag{3.44}$$

External virtual work reads

$$\delta W_e = F\delta w = F2l\sin\theta \cdot \delta\theta \tag{3.45}$$

When assuming that the wrinkles move outwards as shown in Figure 38, the internal virtual work becomes

$$\delta W_i = \text{Hinge (A+C)} + \text{Hinge B} + \text{Stretching of material between hinges}$$

or

$$\delta W_i = 2\cdot 2\pi RM_p \cdot \delta\theta + 2\pi(R + l\sin\theta)M_p \cdot 2\delta\theta + \int_V \sigma_0 \cdot \delta\varepsilon_\theta \cdot dV \tag{3.46}$$

Note that the integration of the last term on the right-hand side is performed over the initial volume.

$$\delta W_i = \left[4\pi RM_p + 4\pi M_p(R + l\sin\theta)\right]\cdot\delta\theta + 2\int_0^l \sigma_0 \frac{x}{R}\cos\theta \cdot \delta\theta \cdot 2\pi Rt \cdot dx \tag{3.47}$$

or

$$\delta W_i = \left[4\pi RM_p + 4\pi M_p(R + l\sin\theta) + 2l^2\sigma_0 t\pi\cos\theta\right]\cdot\delta\theta \tag{3.48}$$

An energy balance requires

$$\delta W_e = \delta W_i$$

or

$$F = \frac{4\pi RM_p + 4\pi M_p(R + l\sin\theta) + 2l^2\sigma_0 t\pi\cos\theta}{2l\sin\theta} \tag{3.49}$$

The average force is defined as

$$F_{avg}2l = U_1 = \int_0^{2l} F dw = \int_0^{\frac{\pi}{2}} F2l\sin\theta d\theta \tag{3.50}$$

The plastic collapse moment M_p for the cross section (per unit circumferential length) is defined as

$$M_p = \frac{2\sigma_0}{\sqrt{3}}\frac{t^2}{4} \tag{3.51}$$

when assuming plan strain conditions in the hinges, see Section 2.4. The average force can now be written as

$$F_{avg} 2l = 4\pi M_p(\pi R + l) + 2\pi\sigma_0 l^2 t \tag{3.52}$$

or

$$F_{avg} = \pi t \sigma_0 \left[t \frac{\left(\pi \frac{R}{l} + 1\right)}{\sqrt{3}} + l \right] \tag{3.53}$$

The length $l = AB = AC$ between the hinges is unknown and is obtained by minimizing the average crushing force, or $dF_{avg} / dl = 0$. This gives

$$l = \left(\pi Rt / \sqrt{3}\right)^{1/2} = 1.90\left(\frac{Rt}{2}\right)^{1/2} \tag{3.54}$$

Substitution of Eq. (3.54) into Eq. (3.53) gives

$$F_{avg} = \left(4(3)^{\frac{1}{4}} \pi^{\frac{3}{2}} \left(R/t\right)^{\frac{1}{2}} + 2\pi\right) M_p$$

or

$$\frac{F_{avg}}{M_p} = 29.31\left(R/t\right)^{\frac{1}{2}} + 6.28 \tag{3.55}$$

When the wrinkles form internally, F_{avg} can be written as

$$\frac{F_{avg}}{M_p} = 29.31\left(R/t\right)^{\frac{1}{2}} - 6.28 \tag{3.56}$$

An average of the two equations gives

$$\frac{F_{avg}}{M_p} = 29.31\left(R/t\right)^{\frac{1}{2}} \tag{3.57}$$

In the previous calculation small strains are assumed in the circumferential direction. Assuming large strains, ε_θ can be expressed as

$$\varepsilon_\theta = \frac{u}{R} + \frac{1}{2}\left(\frac{u}{R}\right)^2 \tag{3.58}$$

Variation yields

$$\delta\varepsilon_\theta = \frac{\delta u}{R}\left[1+\frac{u}{R}\right] \tag{3.59}$$

Introducing Eq. (3.40) into Eq. (3.59) gives

$$\delta\varepsilon_\theta = \frac{\cos\theta}{R}\left[x+\frac{x^2}{R}\sin\theta\right]\delta\theta \tag{3.60}$$

The term of the internal virtual work related to circumferential stretching, see Eq. (3.46), can now be expressed as

$$\int_V \sigma_0 \delta\varepsilon_\theta d \quad = 2\int_0^\ell \sigma_0 \frac{\cos\theta}{}\left[x+\frac{x^2}{}\sin\theta\right]2\pi \quad t\cdot dx\cdot\delta\theta$$

or

$$\int_V \sigma_0 \delta\varepsilon_\theta dV = 2\pi\sigma_0 t\cos\theta l^2\left(1+\frac{2}{3}\frac{l}{R}\sin\theta\right)\delta\theta \tag{3.61}$$

Equating external to internal virtual work yields

$$F = \frac{4\pi RM_p + 4\pi M_p\left(R+l\sin\theta\right)+2\pi l^2\sigma_0 t\cos\theta\left(1+\frac{2}{3}\frac{l}{R}\sin\theta\right)}{2l\sin\theta} \tag{3.62}$$

Performing the same procedure as before, see Eq. (3.50), the average force can be written as

$$\frac{F_{avg}}{M_p} = 29.4\left(\frac{R}{t}\right)^{1/2} + 11.9 \tag{3.63}$$

where

$$l = 1.76\left(Rt/2\right)^{1/2} \tag{3.64}$$

A rather idealized collapse mechanism of the profile is assumed in the previous calculation of the average force F_{avg}. However, to approximate more closely with the actual deformed shape of a tube, an effective crushing distance L_e is introduced.

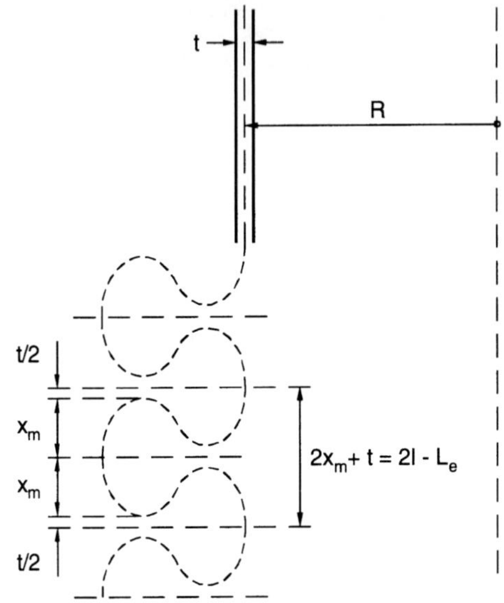

Figure 39: Effective crushing distance [after Jones (1989)]

From Figure 39, $L_e = 2l - 2x_m - t$, where $x_m = 0.28l/2$ (experimentally determined). This gives that

$$L_e = 1.72l - t \tag{3.65}$$

The combination of Eq. (3.64) and Eq. (3.65) yields

$$\frac{L_e}{2l} = 0.86 - 0.568\left(\frac{t}{2R}\right)^{1/2} \tag{3.66}$$

The total plastic energy U_l absorbed during the development of one lobe was divided by 2ℓ in order to obtain F_{avg}, see Eq. (3.50). An alternative estimate of the mean crushing load, which takes a more realistic account of the mode of deformation is given by U_1 / L_e or

$$\frac{\bar{F}_{avg}}{M_p} = \frac{29.40\left(\frac{R}{t}\right)^{1/2} + 11.9}{0.86 - 0.568\left(\frac{t}{2R}\right)^{1/2}} \tag{3.67}$$

3.3.2.3. Diamond deformation mode

For a diamond deformation mode Wierzbicki (1983) has derived the following approximate expression for the average load

$$\frac{F_{avg}}{M_p} = 62.88\left(\frac{2R}{t}\right)^{1/2} \tag{3.68}$$

Introducing $L_e / 2l = 0.73$ gives

$$\frac{\bar{F}_{avg}}{M_p} = 86.14\left(\frac{2R}{t}\right)^{1/3} \tag{3.69}$$

or

$$\bar{F}_{avg} = K_D \sigma_0 (2R)^{1/3} t^{5/3} \tag{3.70}$$

where $K_D = 24.87$.

3.3.2.4. Structural effectiveness and solidity ratio

In order to assist the presentation of experimental data and theoretical predictions two dimensionless ratios known as structural effectiveness and solidity ratio are introduced. Structural effectiveness reads

$$\eta_c = \frac{\bar{F}_{avg}}{A\sigma_0} \tag{3.71}$$

Here $\bar{F}_{avg}$ is the mean crushing force and A is the cross-sectional area of the thin-walled cross section. For a circular tube $A = 2\pi Rt$ and

$$\eta_c = \frac{\bar{F}_{avg}}{2\pi Rt\sigma_0} \tag{3.72}$$

The solidity ratio is defined as

$$\phi = \frac{A}{A_c} \tag{3.73}$$

or

$$\phi = \frac{2\pi Rt}{\pi R^2} = \frac{2t}{R} \tag{3.74}$$

when introducing $A_c = \pi R^2$, i.e. cross sectional area enclosed by the cross section. Using η_c and ϕ the mean crushing force for concertina and diamond mode can be expressed as follows

- Concertina, Eq. (3.67)

$$\eta_c = \frac{\overline{F}_{avg}}{2\pi R t \sigma_0} = \frac{3.36\left(1+0.29\phi^{1/2}\right)}{3.03\phi^{-1/2}-1} \tag{3.75}$$

- Diamond, Eq. (3.70)

$$\eta_c = 3.14\phi^{2/3} \tag{3.76}$$

A comparison shows that a diamond deformation mode is much more efficient in energy absorption than a concertina mode for the same ϕ.

3.3.2.5. Comparison with experimental data

Hanssen et al. (2000b) carried out an extensive experimental study on the behaviour of circular aluminium tubes in alloy AA6060 subjected to axial loading. The tubes had R/t ratio in the range 17-28 and obtained a diamond mode of deformation, see Figure 40. As shown in the figure, most of the specimens had a six-corner mode, but the four-corner mode also occurred.

The derivation of the expression for the mean load is based on a rigid perfect-plastic material. However, when using aluminium, strain hardening must be considered and the yield stress σ_0 has to be replaced by an equivalent flow stress $\overline{\sigma}_0$. Hanssen et al. (2000b) introduced an equivalent flow stress $\overline{\sigma}_0$ which was taken as the average between the 0.2 % and the ultimate stress, i.e.

$$\overline{\sigma}_0 = \frac{\sigma_{0.2}+\sigma_u}{2} \tag{3.77}$$

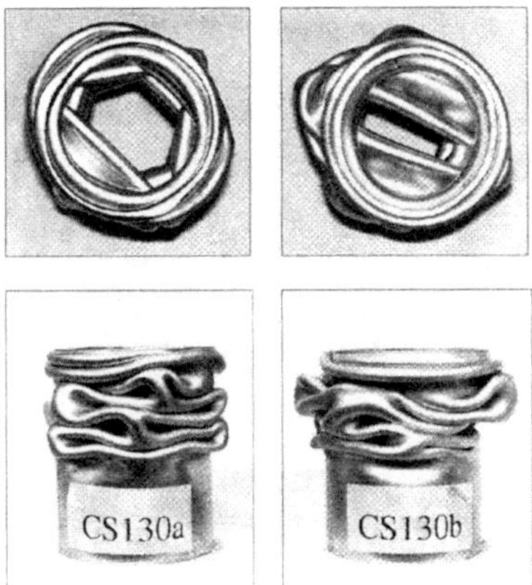

Figure 40: Deformation modes from top view and front view

The expression (Eq. (3.70))

$$\overline{F}_{avg} = K_D \overline{\sigma}_0 \left(2R\right)^{1/3} t^{5/3} \tag{3.78}$$

has been compared with the experimental database by adjusting the constant K_D. This is shown in Table 1 where K_D is given as a function of the deformation capacity D_c. The R^2 value (the squared Pearson value) defines the amount of scatter in the obtained data compared to the suggested theory.

$$\bar{F}_{avg} = K_D \bar{\sigma}_0 (2R)^{1/3} t^{5/3}$$

D_c (%)	20	30	40	50	60	S_E
K_D	16.82	16.45	16.80	17.03	17.45	17.91
R^2	0.974	0.984	0.992	0.987	0.981	0.982

Table 1: [Hanssen et al. (2000b)

As seen, a good approximation that is valid for all relative deformations can be found by setting K_D equal to 17. Be aware that this value is of course a function of the definition of the equivalent flow stress. Wierzbicki (1983) proposed a value of K_D equal to 24.87 for a rigid perfect-plastic material. Figure 41 shows a correlation of the average force at D_c=60 % between the predicted and experimental values for all tests carried out and excellent agreement was found.

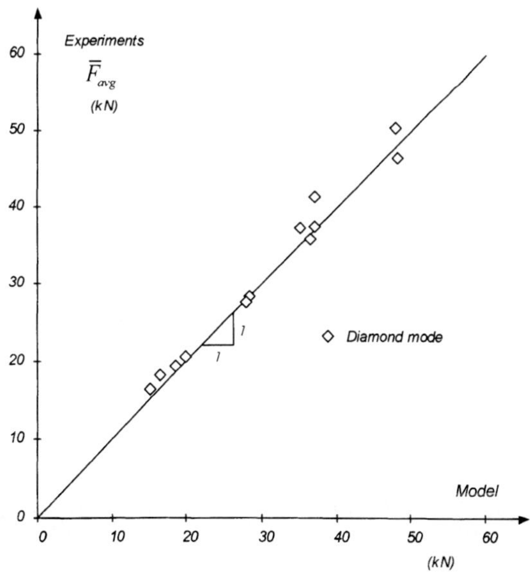

Figure 41: Correlation of average force, non-filled extrusions at D_c=60 %

3.3.3. Crushing of square tubes
3.3.3.1. Introduction

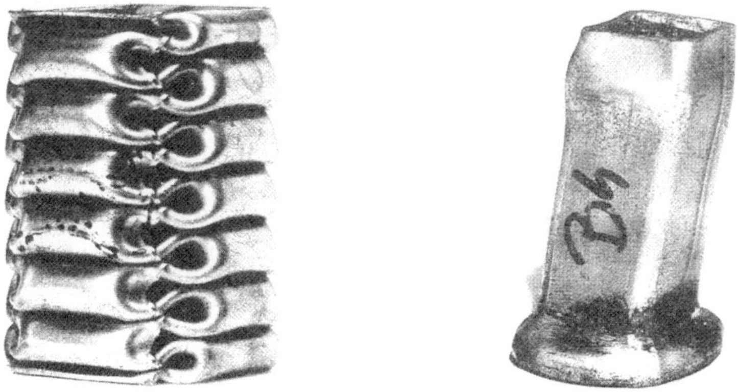

a) Symmetric b) Extensional

Figure 42: Deformation modes [Jones (1989)]

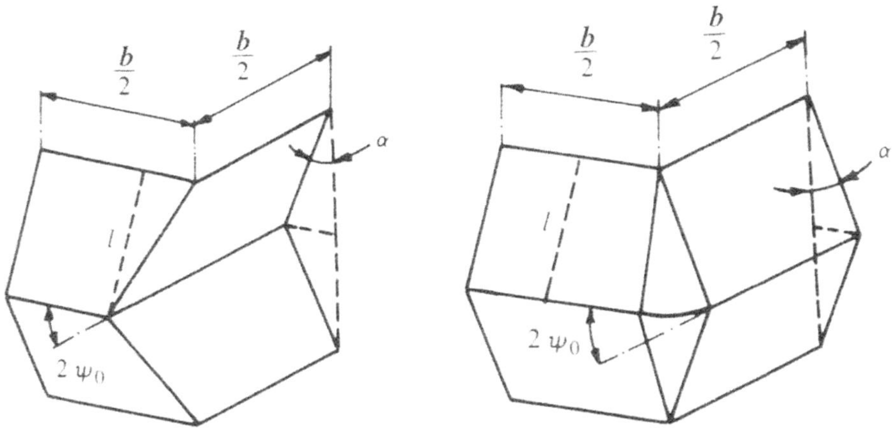

Figure 43: Basic collapse elements [Jones (1989)]

The theoretical analysis for the static crushing of square thin-walled tubes follows the same general procedure as for circular tubes. Wierzbicki and Abramowicz (1983) have identified two basic collapse elements: Type I and Type II shown in Figure 43. These theoretically predict four different collapse modes. Two of the collapse modes are shown in Figure 42, i.e. the symmetric mode and an extensional mode. The symmetric collapse mode is idealized by four Type I elements, while the extensional mode can be described by four Type II elements. The asymmetric mixed modes A and B have a combination of the two types of basic collapse elements [Abramowicz and Jones (1986)]. The mean crushing force is found by equating the external work of the axial crushing force to the internal energy required to form a complete lobe.

3.3.3.2. Symmetric deformation mode

Based on tests performed by Langseth and Hopperstad (1996), the theoretical model proposed by Abramowicz and Jones (1986) for symmetric crushing is found to be adequate to predict the behaviour of square aluminium extrusions subjected to axial loading. Using the basic collapse element Type I, equating external to internal work and minimizing gives the mean crushing force

$$\frac{F_{avg}}{M_p} = 38.12\left(\frac{b}{t}\right)^{1/3} \tag{3.79}$$

and

$$\frac{l}{t} = 0.99\left(\frac{b}{t}\right)^{2/3} \tag{3.80}$$

2ℓ is here defined as the initial distance between plastic hinges at top and bottom of basic folding element, see Figure 43. Tests have shown that the lobes do not flatten completely during crushing and as described for the circular tubes, an effective crushing distance is introduced as

$$\frac{L_e}{2l} = 0.73 \tag{3.81}$$

The mean crushing force can now be written as

$$\frac{\overline{F}_{avg}}{M_p} = 52.22\left(\frac{b}{t}\right)^{1/3} \tag{3.82}$$

or

$$\overline{F}_{avg} = K_D \sigma_0 t^2 \left(\frac{b}{t}\right)^{1/3}, \; K_D = 13.06 \tag{3.83}$$

For circular tubes M_p was calculated using the von Mises yield criterion, while in the present case the Tresca yield condition is used. Thus,

$$M_p = \sigma_0 t^2 / 4$$

Using the structural effectiveness, η_s, and the solidity ratio, ϕ, $\overline{F}_{avg}$ can be expressed as

$$\eta_s = 1.3\phi^{2/3} \tag{3.84}$$

where

$$\eta_s = \frac{\bar{F}_{avg}}{4bt\sigma_0} \tag{3.85}$$

and

$$\phi = \frac{A}{A_c} = \frac{4bt}{b^2} = 4\frac{t}{b} \tag{3.86}$$

when $A_c = b^2$ and $A = 4bt$.

3.3.3.3. Comparison with circular tubes

A comparison can be made between the crushing behaviour of circular and square tubes having the same solidity ratio $\phi = A / A_c$. Thus, the corresponding theoretical values of the structural effectiveness η can be compared very easily. This is shown in Figure 44 where a circular tube with a symmetric mode of deformation, Eq. (3.75), is compared with the symmetric deformation mode for a square tube, Eq. (3.84).

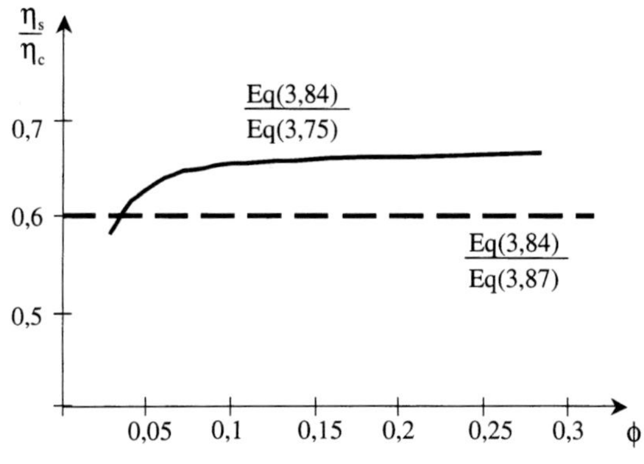

Figure 44: Comparison of structural effectiveness

The full line in Figure 44 shows that the ratio η_s / η_c is an increasing function with respect to ϕ. The superscripts *s* and *c* denote square and circular tubes respectively. However, for most practical applications it can be assumed that a square tube has a value of η_s which is about two-thirds of the corresponding circular tube results.

Hansen et al. (2000b) found a diamond mode of deformation when compressing circular aluminium tubes with a solidity ratio ϕ in the range of 0.050-0.125. Using a value of K_D=17 as found experimentally, Eq. (3.76) can be expressed as

$$\eta_c = 2.15\phi^{2/3} \tag{3.87}$$

The ratio of Eqs. (3.84) and (3.87) is shown as the dotted line in Figure 44, and is equal to 0.6.

The equality in Eqs. (3.74) and (3.86) demands that $b / R = 2$ when the circular and square tubes have the same wall thickness and solidity ratio. In this case the ratio between the crushing forces can be expressed as

$$\frac{F^s_{avg}}{F^c_{avg}} = \frac{\eta_S}{\eta_C}\frac{4}{\pi} \tag{3.88}$$

Introducing $\eta_s / \eta_c = 0.6$ into Eq. (3.88) gives a situation where an aluminium circular tube with a diamond mode of deformation will absorb 24 % more energy than a corresponding square tube when the wall thickness and solidity ratio are equal. It will be shown in the following section that Eq. (3.83) predicts the mean crushing force of square aluminium tubes quite well when, as for circular tubes, it introduces an equivalent flow stress $\sigma_0 = \bar{\sigma}_0$ as given by Eq. (3.77).

3.3.3.4. Comparison with experimental data

Langseth and Hopperstad (1996) carried out an extensive experimental investigation on the behaviour of square aluminium extrusions subjected to axial loading. The extrusions were in alloy AA6060 and had a width of 80 mm and length of 310 mm. The primary variables were the wall thickness of the extrusions (t=1.8, 2.0, 2.5 mm) and the temper (T4, T4* and T6). The temper T4* was obtained by heat treatment of the temper T4 specimens at 190 °C for 1 h. The extrusions where clamped at the lower end and free at the top before applying the load through a rigid steel block fixed to the cross-head of a testing machine.

As shown in Figure 45, a progressive symmetric deformation mode was observed and this was a mode that was independent of the wall thickness and temper. However, the number of lobes formed during deformation seemed to be a function of the hardening properties of the material alone. When the specimens were fully compressed, approximately six lobes were found for temper T4, between six and seven for temper T4* and seven for temper T6. The white patches with specimen number in the figure represent the clamped part of the specimen in the test rig.

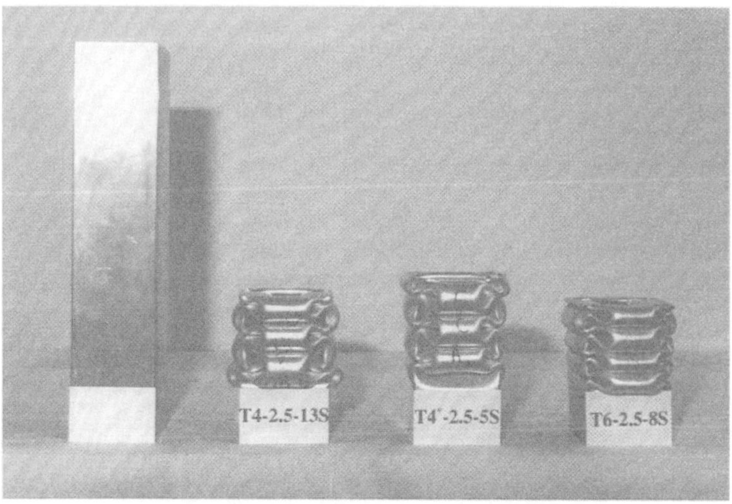

Figure 45: Deformation modes of static tests

For temper T6, the average crushing distance to form a complete lobe was measured to be 32 mm, while the corresponding value for temper T4 was 37 mm. The deformation pattern for temper T4* was somewhere between tempers T4 and T6. Using the model proposed by Abramowicz and Jones (1986)

$$\frac{L_e}{2l} = 0.73 \quad \text{with} \quad \frac{l}{t} = 0.99\left(\frac{b}{t}\right)^{\frac{2}{3}} \tag{3.89}$$

for a rigid perfect-plastic material, the effective crushing distance, L_e, has been predicted to range between 32-36 mm with the highest value occurring for the 2.5 mm thick tubes.

Based on visual inspection of the specimens after testing, incipient fracture was observed in the corners of some of the square tubes of temper T4* and T6, while no such behaviour was found for temper T4. However, the incipient fracture seemed to have no influence on the global response. Failure in the temper T6 specimens was expected due to the low hardening modulus and failure strain obtained from a tensile test. However, during axial loading of the tubes, the dominant collapse mechanism is bending which is a quite different mode of deformation with respect to failure than pure tension. Thus, the failure strain found from a pure tensile test cannot be used to determine failure in a bending mode of deformation.

Figure 46 shows the ratio between the average crushing force according to Eq. (3.83) and the test data as a function of the *b/t* ratio. An equivalent flow stress was defined as for the circular tubes to take strain hardening effects into account, i.e. $\sigma_0 = \bar{\sigma}_0$ as given by Eq. (3.77). Figure 46 also shows that the predicted mean loads and the corresponding test data are within ± 10 %.

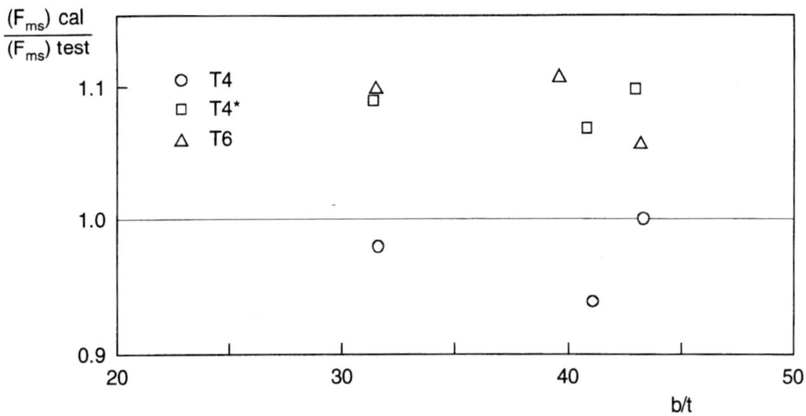

Figure 46: Ratio between calculated and experimental mean crushing force

3.3.4. Dynamic crushing – inertia effects

Limited data are available in the literature regarding the inertia effects set up in an aluminium tube during axial dynamic loading. Langseth and Hopperstad (1996) carried out a study on this issue on square tubes that were initially straight (*LxB*= 80x80 mm), i.e. tubes without any mechanically imposed trigger. In their tests the mass of the projectile was kept constant at 56 kg and the primary variables were the impact velocity (8-20m/s), the thickness of the extrusions (1.8, 2.0, 2.5 mm) and temper (T4 and T6). The dominant deformation mode was symmetric combined with asymmetric and extensional modes. Figure 47 shows some representative modes of deformation for the temper T4 specimens with a wall thickness of 2.5 mm.

The dynamic tests show a significant increase in the dynamic mean load level compared with corresponding quasi-static tests. This effect was mainly related to the first part of the impact, see Figure 48, where the dynamic force level was significantly higher than the corresponding static one. As the aluminium alloy was recognized strain-rate insensitive, see Section 2.7, the observed difference was related to inertia effects set up in the side-walls when at the instant of impact the attempt was made to avoid lateral movement of the side-walls to initiate the folding process. Thus, the initial compression phase is delayed and increased energy absorption will take place in this phase of the impact. It was found that the mean load ratio between a dynamic and a static test was a decaying function with respect to the axial displacement, see Figure 49, and an increasing function with respect to the impact velocity. In addition, a few dynamic tests were carried out to study the effect of an initial geometrical imperfection obtained by static pre-buckling prior to dynamic testing. As seen in Figure 49, an almost constant mean load ratio with respect to the axial displacement was found and the mean load ratio was reduced compared to an initially straight tube.

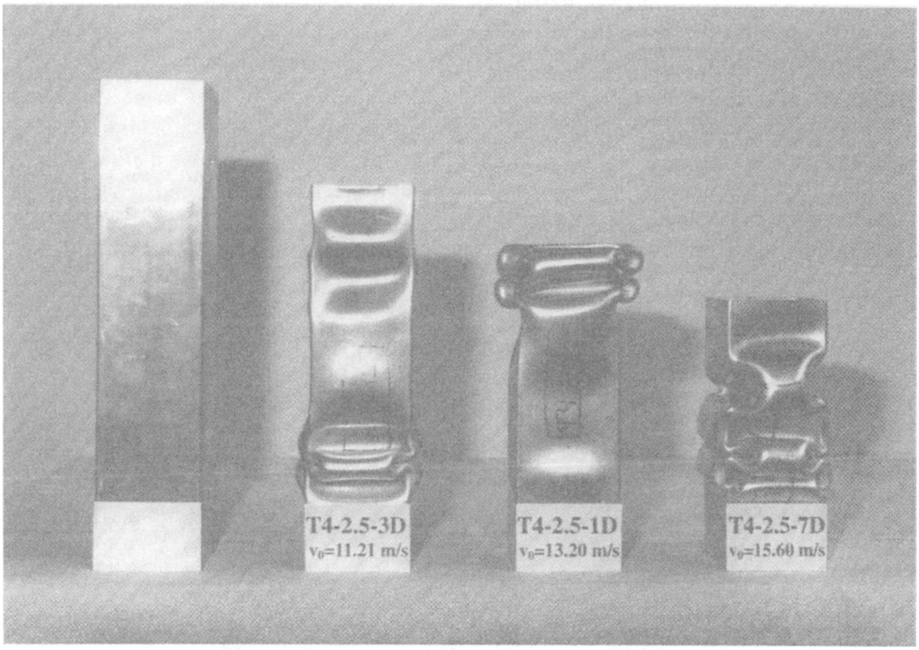

Figure 47: Deformation modes for temper T4

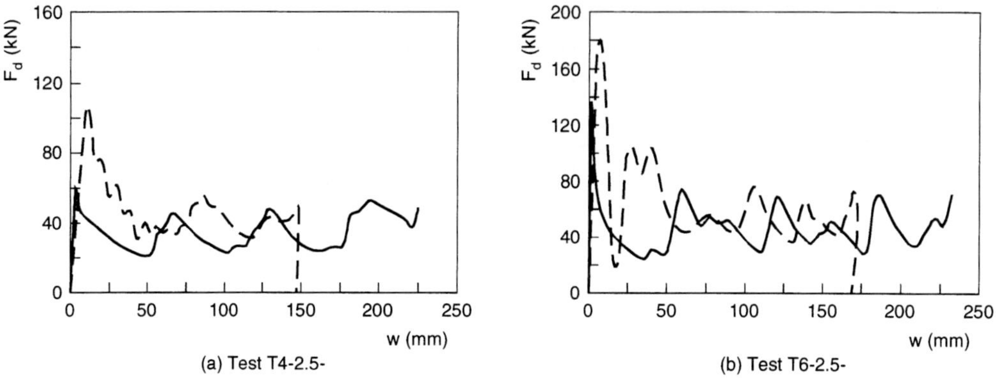

Figure 48: Comparison between static and dynamic force displacement curves; (---) = dynamic, (—) = static

As indicated in Section 3.3.3.4, the static mean load can be predicted with good accuracy using Eq. (3.83) by introducing a modified flow stress taken as the mean value between the 0.2 %-proof stress and the ultimate stress. However, calculating the behaviour during dynamic loading is much more complicated owing to the fact that inertia forces have to be taken into account. A conservative estimate regarding energy absorption would be to use the static value, but the actual load transferred to the rest of the structure could have a much larger value due to inertia effects. One way to control the dynamic amplification is to introduce initial geometrical imperfections. This is a common method used to initiate the desired deformation modes and control the initial buckling load. As shown in Figure 49, the mean load ratio between a dynamic and a static test for an initially straight specimen was a decaying function with respect to the axial displacement, while an almost constant value has been found when introducing initial geometrical imperfections. Another method adopted by the aluminium industry to reduce the buckling load is to introduce a local soft zone in the absorber by means of heat treatment.

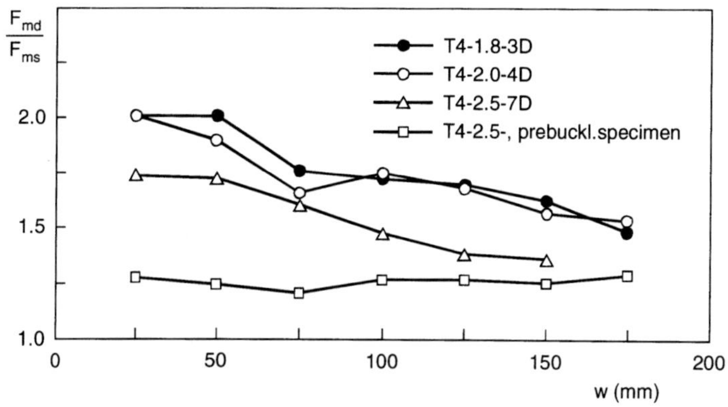

Figure 49: Ratio between dynamic and static mean loads versus displacement. Temper T4

3.4. Aluminium foam

3.4.1. Foam structure

Metal foams have a 3-dimensional cellular structure, see Figure 50. The material making up the structure of the foam is referred to as the base material. Foams are made of different base materials and the most common are polymeric (organic) foams and metal foams. Even wood can be considered a cellular structure. So far, aluminium foams appear to be the most popular among the group of metal foams, but steel, titanium, zinc and magnesium foams also exist. Foams can further be divided into two groups: those having closed cells and those made up with an open cell structure. A direct result of the cellular foam structure is the low apparent density of the material.

There is a wide variety of patented production methods for metal foams. The following section will be based on the work by Hanssen (2000) carried out on aluminium foam produced by Hydro Aluminium AS (the same patented continuous casting process which was sold to the Canadian company CYMAT in 2001). The foam emerging from this process is a closed cell structure with an apparent density in the range from 0.1 to 0.5 g/cm^3.

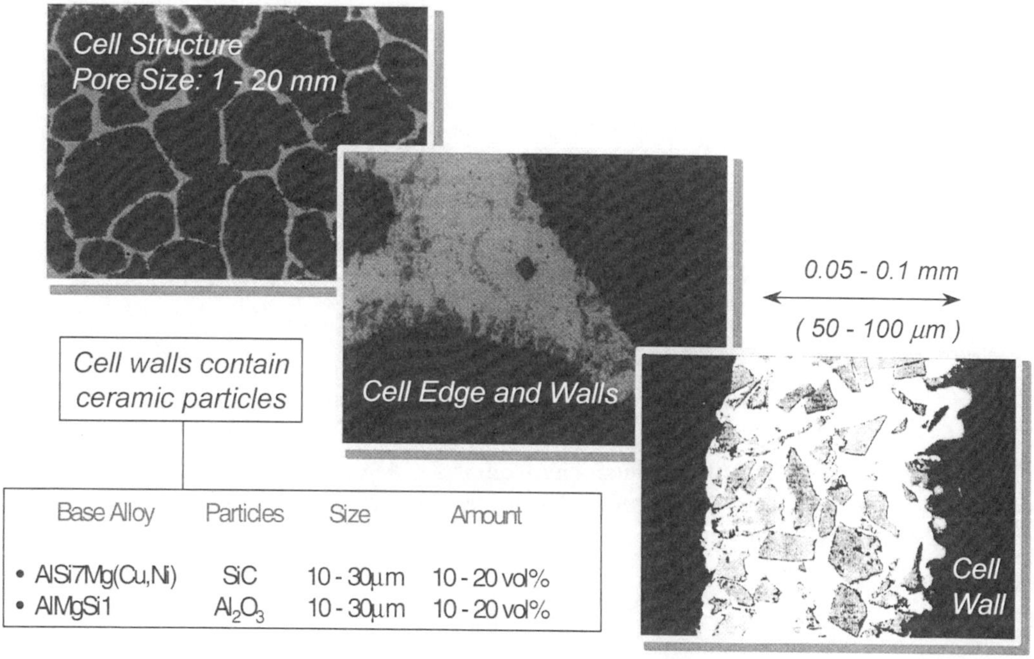

Figure 50: A close-up of aluminium foam

3.4.2. Material properties in uniaxial stress

3.4.2.1. Static behaviour

Aluminium foams are presently being regarded with interest concerning the controlled absorption of kinetic energy, i.e. for crashworthiness applications. An explanation to this can be sought in Figure 51, where generalized compressive as well as tensile properties of a cube of aluminium foam are depicted. The compressive behaviour is characterized by an initial elastic region followed by what is commonly referred to as the plateau region, identified by a more or less constant stress level for strains up to 50-80 %. The average stress level of this region is referred to as the plateau stress σ_f. Further compression of the foam will lead to compaction, quantified by the densification strain ε_D, Figure 51. In addition to the low weight of the foam, it is the constant stress levels in the plateau region that promotes the use of the foam for energy absorption purposes. Hence, kinetic energy can be absorbed at specified force levels, ensuring the integrity of valuable goods as well as the people in the passenger cell of a colliding car.

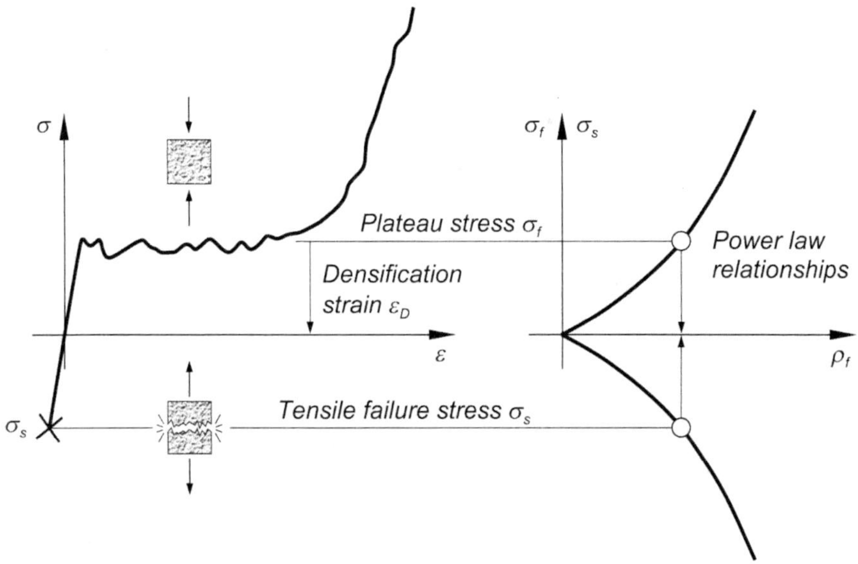

Figure 51: Generalized compressive and tensile properties of aluminium foam

Another characteristic of the aluminium foam is the dependency of mechanical properties on the density of the foam. This is referred to as power law relationships, Figure 51, and for the plateau stress σ_f is written as

$$\sigma_f = C_p \left[\frac{\rho_f}{\rho_{f0}} \right]^n \tag{3.90}$$

where ρ_f is the foam density and ρ_{f0} the density of the base material (2.7 g/cm3). C_p and n are material constants found by model calibration with uniaxial compression tests of foam cubes. For the continuously cast aluminium foam produced by Hydro Aluminium AS, typical values of C_p and n are 450 MPa and 2.0 respectively. The corresponding density range of this foam is from 0.15 to 0.50 g/cm^3.

In tension, the behaviour is quite different from that in compression. The initial elastic region is followed by brittle failure, characterized by the tensile failure stress σ_s. The tensile capacity σ_s of the Hydro foam is approximately equal to the plateau stress σ_f, i.e. they both have roughly the same foam density dependency.

Dividing Eq. (3.90) by the foam density ρ_f yields the mass specific energy absorption E_M (destroyed mass)

$$E_M = \frac{\sigma_f}{\rho_f} = \frac{C_p}{\rho_{f0}}\left[\frac{\rho_f}{\rho_{f0}}\right]^{n-1} \tag{3.91}$$

An important consequence of this expression is that the mass specific energy absorption of the foam increases with the foam density as long as $n > 1$. Hence, solely from this point of view, high foam densities are preferable when minimum weight designs are considered.

3.4.2.2. Dynamic behaviour

For crashworthiness applications it is relevant to assess the influence of dynamic loading conditions on the mechanical properties. Limited data are available, but there are a few independent investigations [Hanssen (2000)], [Tan and Reid (2000)], [Deshpande and Fleck (2000)]. Figure 52 presents tests carried out on aluminium foam cylinders in uniaxial stress using a drop hammer for impact velocities from 5 to 15 m/s. As seen, the effect of loading velocity compared to the quasi- static case is low. Tan and Reid (2000) showed that the maximum dynamic enhancement of the plateau stress of aluminium foam could be approximated by a shock wave theory, giving

$$\sigma_f^D = \sigma_f + \frac{\rho_f}{\varepsilon_D} v^2$$

where σ_f^D is the dynamic plateau stress and v is the loading velocity.

3.4.3. Foam in combination with extrusions

3.4.3.1. Static and dynamic behaviour

The sections in Figure 53 are of a statically crushed non-filled extrusion and a similar foam-filled extrusion. As observed, the number of lobes in the sidewalls of the extrusions is increased due to the foam filling. In general, the number of lobes is an increasing function with respect to the foam density. The foam filler causes the initial buckling of the sidewalls to take place on an elastic foundation, thus reducing the buckling length and increasing the number of lobes created. For circular extrusions, an effect of increasing foam density may also be to change the deformation mode, Figure 54.

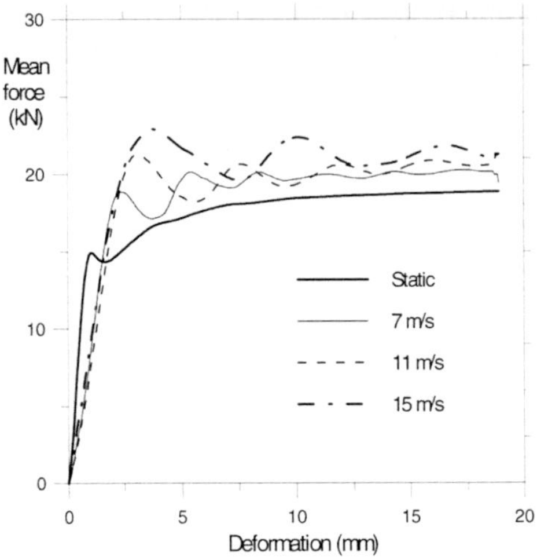

Figure 52: Effect of loading velocity on mean crushing force of foam [Hanssen (2000)]

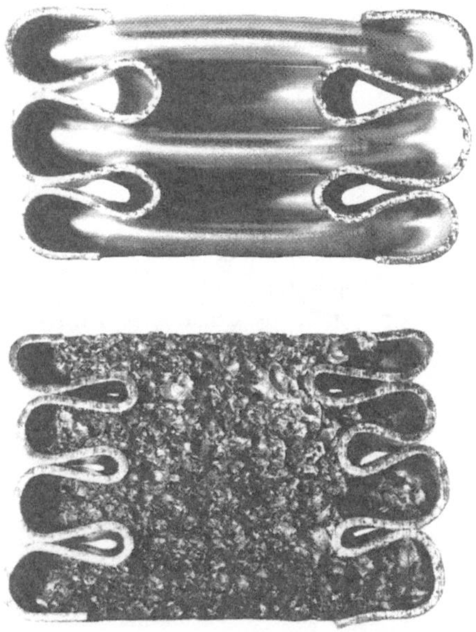

Figure 53: Non-filled vs. foam-filled square extrusion, quasi-static loading

The dynamic loading condition does not significantly alter the apparent deformation behaviour of the square and circular extrusions, although there are some differences. However, a difference between static and dynamic loading conditions can be observed by studying the force vs. deformation curves of tested foam-filled extrusions. Figures 55, 56 and 57 compare the static with the corresponding dynamic force vs. deformation behaviour of foam alone (Figure 55), a non-filled extrusion (Figure 56) and finally a foam-filled extrusion (Figure 57). A simple means of determining the effect of the loading condition is the ratio between dynamic and static energy absorption. This ratio is referred to as the "energy ratio" in Figures 55 to 57 and is plotted as a function of axial deformation. As was observed by Hanssen (2000) and also illustrated in Figure 55, a loading velocity of 15 m/s appears to have no significant effect on the force-deformation behaviour of aluminium foam alone. The contrary is observed when studying the crushing of non-filled extrusions, Figure 56. Here, the force levels are increased significantly. Experiments on the extrusion material discovered no significant strain rate sensitivity; hence the increase in force is likely to be due to transverse inertia forces arising in the extrusion walls during crushing (see also Section 3.3.4). It was observed that the energy ratio generally was a decreasing function of deformation. Since the strength of the extrusion material can become grater without increasing the material density, an increase in material strength will lead to reducing the relative importance of the dynamic inertia effects (reduced energy ratio). The same effect is observed when filling the extrusion with the rate insensitive foam, Figure 57. The increase in the total force level compared with the non-filled extrusion diminishes the relative importance of the inertia effects of the extrusion sidewalls, leading to a lower energy ratio.

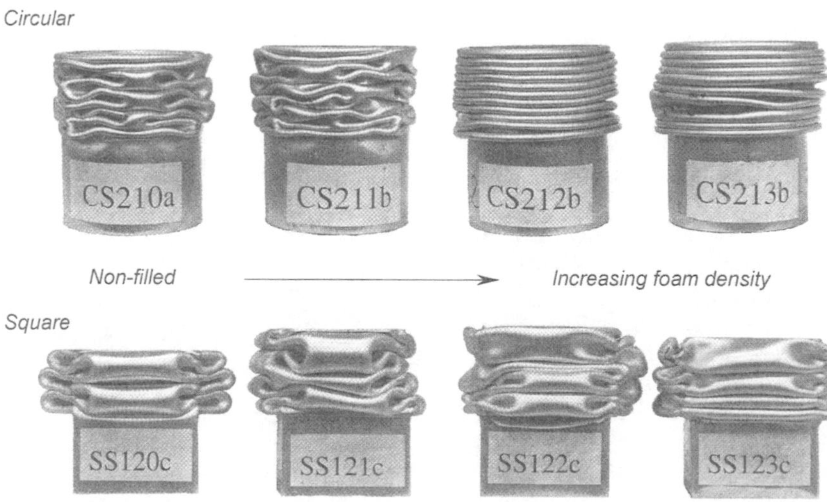

Figure 54: Axial crushing of non-filled and foam-filled circular and square extrusions, quasi-static loading

Figure 55: Static and dynamic force deformation curves of circular foam specimens

Specimen diameter:	79 mm
Foam density:	0.38 g/cm^3
Impact velocity:	15 m/s

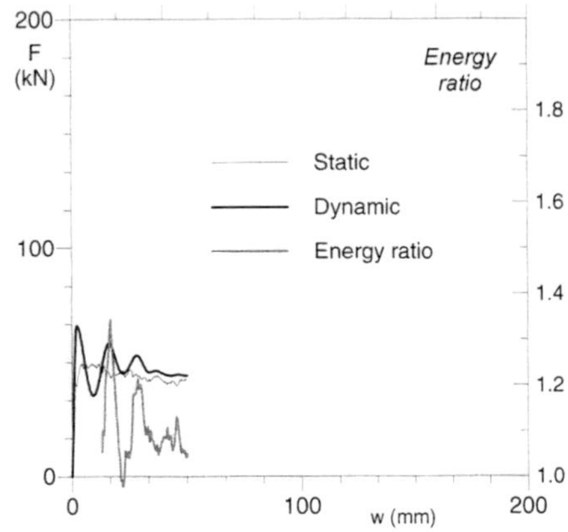

Figure 56: Static and dynamic force deformation curves of square non-filled extrusions

Outer width, b	80 mm
Wall thickness, t	1.95 mm
Foam density, ρ_f	0 g/cm^3
Characteristic stress, $\bar{\sigma}_0$	151 MPa
Impact velocity, v_0	14 m/s

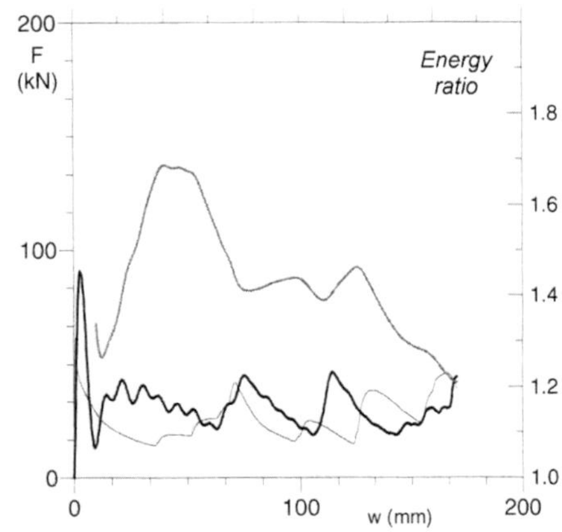

Figure 57: Static and dynamic force deformation curves of square foam-filled extrusions

Outer width, b	80 mm
Wall thickness, t	1.95 mm
Foam density, ρ_f	0.34 g/cm^3
Characteristic stress, $\bar{\sigma}_0$	151 MPa
Impact velocity, v_0	23 m/s

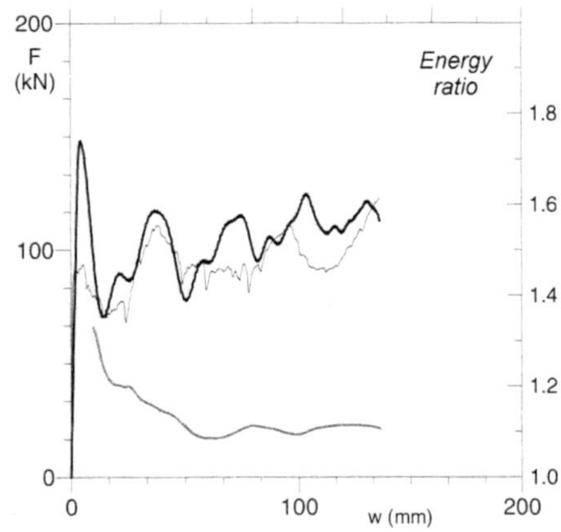

3.4.3.2. Design methods

The following design formula has been found to represent the average crush force F_{avg} of both square and circular foam-filled extrusions with satisfactory accuracy

$$F_{avg} = \bar{F}_{avg} + \sigma_f A_f + C_1 \sqrt{\bar{\sigma}_0 \sigma_f} A \tag{3.92}$$

Three terms are present in the above equation. The first is simply the average crush force of the corresponding non-filled extrusion, whereas the second term constitutes the uniaxial resistance of the foam filler. As observed experimentally, the average crush force of foam-filled extrusions always exceeds that of the sum of these two terms, see Figure 58. This increase in capacity is referred to as an interaction effect and is represented by the third term in Eq. (3.92). The parameters involved are:

$\bar{F}_{avg}$	average crush force of corresponding non-filled extrusion
σ_f	foam plateau stress, see Figure 51
A_f	foam core cross-sectional area
C_1	cross-section dependent dimensionless constant, (Table 2)
$\bar{\sigma}_0$	characteristic stress of extrusion material, see Figure 63
A	cross-sectional area of extrusion
$\sigma_f / \bar{\sigma}_0$	relative foam stress

In order to make the design formula complete, an expression for the average crush force of non-filled extrusions has been included, see Eqs. (3.84) and (3.87).

$$\bar{F}_{avg} = C_0 \phi^{2/3} \bar{\sigma}_0 A \tag{3.93}$$

This equation is represented by the following parameters:

C_0	cross-section dependent dimensionless constant (Table 2)
ϕ	A / A_f: solidity ratio

Eq. (3.93) assumes that there is a symmetric deformation pattern for the square extrusions and diamond modes for the circular ones.

By substituting Eq. (3.93) into Eq. (3.92), the normalized average crush force of foam-filled extrusions can be written as

$$\frac{F_{avg}}{\bar{\sigma}_0 A} = C_0 \phi^{2/3} + \frac{1}{\phi} \frac{\sigma_f}{\bar{\sigma}_0} + C_1 \sqrt{\frac{\sigma_f}{\bar{\sigma}_0}} \tag{3.94}$$

The maximum force level F_{max} occurring during crushing can be approximated by the following expression

$$\frac{F_{max}}{\bar{\sigma}_0 A} = \frac{1}{A_{E0}} C_0 \phi^{2/3} + \frac{1}{A_{Ef}} \frac{1}{\phi} \frac{\sigma_f}{\bar{\sigma}_0} + C_2 \sqrt{\frac{\sigma_f}{\bar{\sigma}_0}} \tag{3.95}$$

where three new parameters have been introduced:

A_{E0}	ratio between average and maximum force, non-filled extrusion (Table 2)
A_{Ef}	ratio between average and maximum force, foam alone (Table 2)
C_2	cross-section dependent dimensionless constant (Table 2)

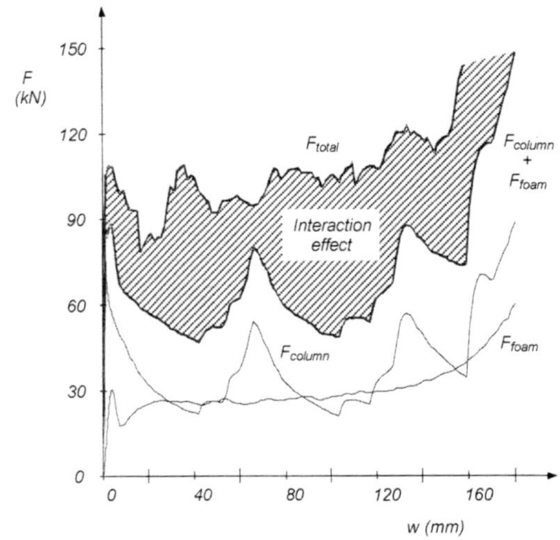

Figure 58: Illustration of interaction effect

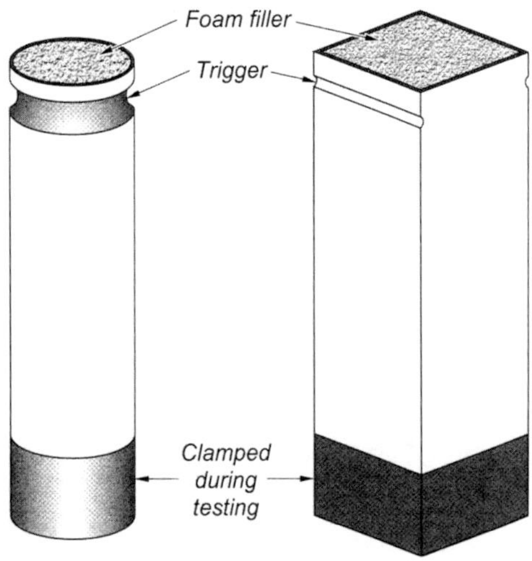

Figure 59: Triggers used

A_{E0} and A_{Ef} are referred to as crush force efficiencies, and they can be determined from compression tests on non-filled extrusions and foam alone respectively. If the foam-filled extrusion has a trigger in order to reduce the peak force and initiate folding, see Figure 59, tests on the corresponding triggered non-filled extrusion will yield a value of A_{E0} for use with Eq. (3.95). The maximum force of non-filled extrusions is usually related to the initial peak force, even if the extrusions are triggered. Table 2 gives the values of A_{E0} if the initial peak force is neglected, being the equivalent of designing a "perfect" trigger. Table 2 also gives the minimum value of A_{E0} found for the triggered non-filled components. If tests on the triggered non-filled extrusions are not available for the determination of A_{E0}, it should be possible to utilize the results in Table 2 as a rough guideline for estimation of the maximum force. The crush force efficiency of the foam filler A_{Ef} should be set to 0.85.

	Square	Circular
A_f	$(b-2t)^2$	$\pi(d-2t)^2/4$
A	$4t(b-t)$	$\pi t(d-t)$
ϕ	$4t(b-t)/(b-2t)^2$	$4t(d-t)/(d-2t)^2$
C_0	1.30	2.15
K_D	13.06	17
C_1	1.4	0.9
C_2	0.5	0.9
C_3	1.7	17
A_{E0}	0.40 – 0.65*	0.55 – 0.75*
A_{Ef}	0.85	0.85
m	0.8	2
S_E^0	0.76	0.76

*) initial peak force neglected

Table 2: Parameters needed for properties of foam-filled tubes

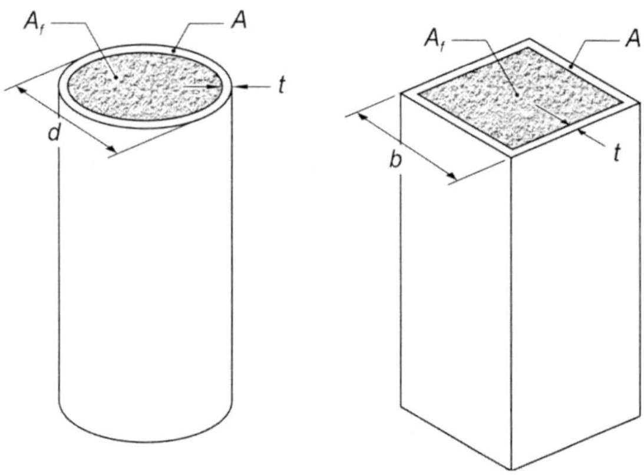

A summary of all the parameters involved in Eqs. (3.94) and (3.95) is given in Table 2. Here b is the outer width of the square cross section, d is the outer diameter of the circular extrusion, whereas t represents the wall thickness. It must be noted that the dimensionless constants C_1 and C_2 were found to increase slightly with deformation, Figure 62. The values given in Table 2 correspond to an approximate deformation of 50 %.

The corresponding correlation plots for Eqs. (3.94) and (3.95) vs. experiments are shown in Figures 60 and 61. The diamond and square symbols in Figure 60 represent tests from different series. For most tests, the accuracy of the proposed design formula for the average force is within an error margin of ±10 %. The maximum force level is predicted with somewhat less accuracy. The parameter range for the tests was given by $0<\phi<0.125$ and $0<\sigma_f/\bar{\sigma}_0<0.10$ (however, the upper limit for $\sigma_f/\bar{\sigma}_0$ is infinite for single foam blocks).

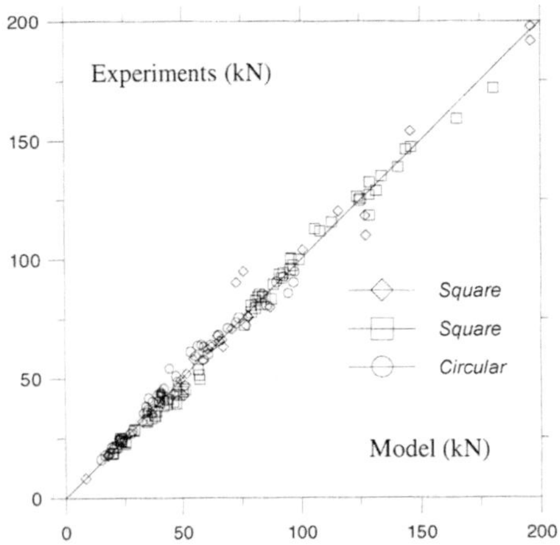

Figure 60: Correlation, average force

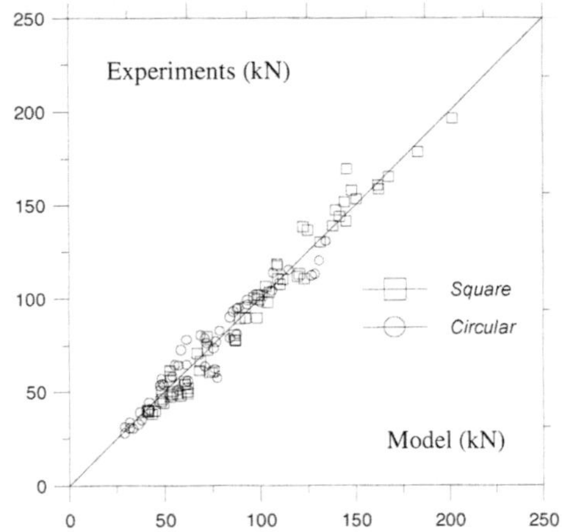

Figure 61: Correlation, maximum force

In Table 2, the dimensionless constant K_D for determination of the average crushing forces of non-filled extrusions is based upon the definition of the characteristic extrusion material stress $\bar{\sigma}_0$ as given in Figure 63. Here $\bar{\sigma}_0$ is defined as the average of the yield stress σ_{02} (0.2 % plastic strain) and ultimate stress σ_u. Using this definition, the values of K_D in Table 2 were determined from correlation with experiments on non-filled aluminium extrusions.

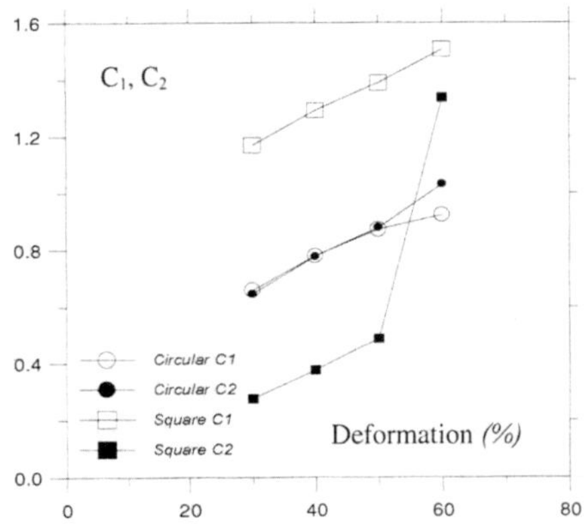

Figure 62: Development of C_1 and C_2

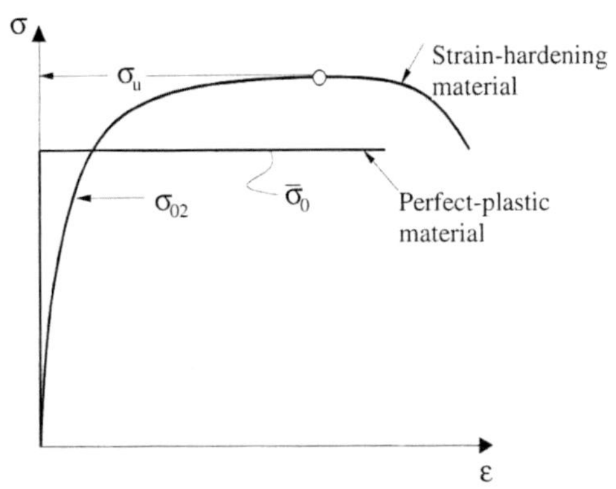

Figure 63: Definition of $\bar{\sigma}_0 = 1/2(\sigma_{0.2} + \sigma_u)$

The final design parameter of interest is the effective crushing length of the components, $w_{max} = S_E \times L$. Here S_E is referred to as the stroke efficiency, whereas L is the total length of the component. For a square non-filled extrusion, S_E is approximately 73 %, meaning that the component can be crushed three-quarters of its original length before bottoming up occurs (leading to rapidly increasing force levels). When filling foam inside tubes, the effective crushing length is generally reduced. This has implications for practical design cases, since the total capacity for energy absorption is given by $E = F_{avg} \times S_E \times L$.

The stroke efficiency S_E is given by

$$S_E = \frac{\left(C_0 \phi^{2/3} + C_1 \sqrt{\frac{\sigma_f}{\bar{\sigma}_0}} \right) \cdot S_E^F + \frac{1}{\phi} \frac{\sigma_f}{\bar{\sigma}_0} \cdot \varepsilon_D}{C_0 \phi^{2/3} + \frac{1}{\phi} \frac{\sigma_f}{\bar{\sigma}_0} + C_1 \sqrt{\frac{\sigma_f}{\bar{\sigma}_0}}} \tag{3.96}$$

where

$$S_E^F = S_E^0 \left(1 - C_3 (\rho_f / \rho_{f0})^m \right) \quad \text{and} \quad \varepsilon_D = 1 - 1.5(\rho_f / \rho_{f0}) \tag{3.97}$$

Here ε_D denotes the densification strain of the foam filler. The constants S_E^0, C_3 and m are given in Table 2 for both square and circular extrusions. Figure 64 shows the correlation between this model and experiments.

Finally, the total mass of the foam-filled tube is

$$M = (\rho_0 A + \rho_f A_f) \cdot L \tag{3.98}$$

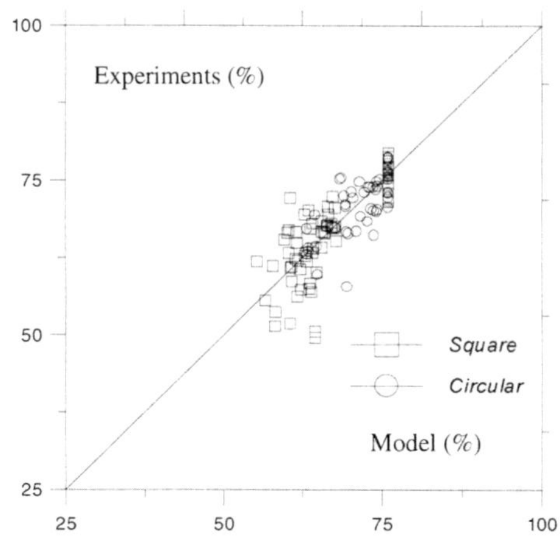

Figure 64: Correlation, model for stroke efficiency

3.4.3.3. Dynamic loading effect on average force level

The increase in force levels due to dynamic loading conditions was described in Section 3.4.3.1. For square foam-filled extrusions Hanssen et al. (2000a) assumed that the average force levels of dynamically loaded specimens could be described by Eq. (3.94) above.

$$F_{avg} = \overline{F}_{avg} + \sigma_f A_f + C_1 \sqrt{\overline{\sigma}_0 \sigma_f} A \tag{3.99}$$

The increase in force levels of the foam-filled component was related to the average crush force of the non-filled extrusion in the following way

$$\overline{F}_{avg} = C_0 \phi^{2/3} \overline{\sigma}_0 A \left[1 + C_4 \left(\frac{1}{\phi} \frac{\rho_0}{\overline{\sigma}_0} v_0^2 \right)^{1/2} \right] \tag{3.100}$$

Here, the following parameters have been introduced:

ρ_0 density of extrusion material

v_0 initial impact velocity

C_4 factor describing the dynamic effect

The C_4 parameter was found to be a decreasing function of the relative deformation measure w/l where w and l are defined as

w deformation

l half the relative plastic lobe length of a non-filled tube, $l = 0.99 b^{2/3} t^{1/3}$

Hence, the complete formula giving the average crush force for square foam-filled extrusions subjected to dynamic loading conditions is

$$\frac{F_{avg}}{\overline{\sigma}_0 A} = C_0 \phi^{2/3} \left[1 + C_4 \left(\frac{1}{\phi} \frac{\rho_0}{\overline{\sigma}_0} v_0^2 \right)^{1/2} \right] + \frac{1}{\phi} \frac{\sigma_f}{\overline{\sigma}_0} + C_1 \sqrt{\frac{\sigma_f}{\overline{\sigma}_0}} \tag{3.101}$$

The C_4 factor as a function of w/l is given in Figure 65.

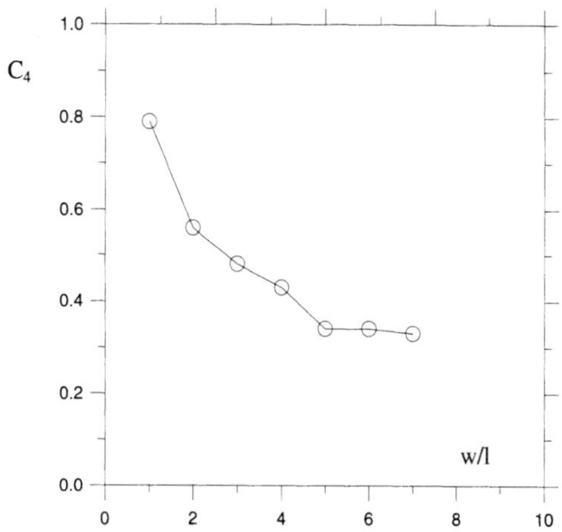

Figure 65: Development of dynamic amplification factor C_4 (square extrusions)

3.4.3.4. Optimization -general definition of optimization problem for a crash box

Designing a structure like a car is a complex task since often conflicting needs have to be taken into account, e.g. esthetical appearance, comfort, noise, stiffness, vibration, safety and available compartment space. Different technical solutions have to be introduced that often result in limitations of space for each component in the total structure. These restrictions will be referred to as geometrical restrictions. Table 3 illustrates these restrictions applied to the foam-filled column. Here L, b and t are the length, outer width (diameter) and wall thickness of the column respectively.

Moreover, the material selection is an equally important task. Some materials may have disadvantages like toxic combustibility, costly recycling properties or high production costs. Even if the type of material has been determined, say limited to a group of aluminium alloys, within this group there can be found materials with minimum and maximum strength. For the aluminium foam-filled column, a proper foam density ρ_f has to be selected in order to obtain optimum performance. Since all foam densities varying from 0.1 to 0.5 g/cm^3 can be specified, this will be the material restriction for the selection of the optimum foam density.

For design formulas to be easily applicable, simple obtainable model parameters have to represent the material behaviour of both the column wall and the foam. As illustrated in Table 3, the characteristic stress $\bar{\sigma}_0$ is defined for the column material as the average value of column wall yield stress (σ_{02}) and ultimate stress (σ_u). This may be regarded as a transformation of a general material showing hardening effects into a perfect plastic material. A transformation like this is needed since existing theoretical formulas for the average crush force of non-filled columns is based upon a perfect plastic material behaviour.

A characteristic of the aluminium foam is the plateau level of constant stress σ_f, herein evaluated as average stress at 50 % deformation. The power law is widely accepted for foam materials, see Eq. (3.90), linking the plateau stress σ_f with the foam density ρ_f. Summing up, the characteristic stress $\bar{\sigma}_0$ and the foam plateau stress σ_f are applied here to represent the material properties of column wall material and foam.

As seen from the listing of Table 3, criteria like that of global stability during loading must also be taken into consideration. However, restrictions like these are generally dependent upon both the geometrical and material properties selected. A simplified restriction that is only dependent upon geometrical properties is formulated for the foam-filled column, Table 3. As seen, the maximum allowed *L/b* ratio is set equal to 3 based upon experimental observations. Moreover, this *L/b* ratio corresponds well with crash box designs found in existing vehicles.

Finally, the component or structure has to be designed in agreement with a given performance. For the energy absorber this is measured in terms of a minimum required amount of energy absorbed ($E > E^*$) coupled with force levels that have to stay below a fixed maximum allowable level ($F_{max} < F^*_{max}$). These restrictions will be referred to as target restrictions. The maximum force restriction ensures that the minimum elastic capacity of the rest of the structure (front rail) is not exceeded and thus not functionally damaged, Table 3. On the other hand, the minimum energy absorption restriction ensures that the kinetic energy of the car is absorbed.

It is possible to optimize a structure with regard to several parameters, such as minimum mass, minimum manufacturing costs or recycling costs. Whereas optimization with regard to one parameter is straightforward in principle, optimization with regard to several parameters must be based upon a set of compromises between the different parameters. In the following, only the search for the component of minimum mass will be emphasized.

3.4.3.5. Foam filling of an extrusion with constant outer dimensions

One application of the design formulas presented above is to investigate the optimum-weight combination of foam-filler density ρ_f, extrusion wall thickness t, extrusion width/diameter (b or d) and extrusion strength $\bar{\sigma}_0$. That is, what combination of these parameters will give the component of minimum mass satisfying the given requirements for the necessary energy absorption at restricted force levels. A simplified approach is adopted here that is focused on describing the effects rather than quantifying them.

Geometrical restrictions $t_{\min} < t < t_{\max}$ $b_{\min} < b < b_{\max}$ $L_{\min} < L < L_{\max}$ ***Global stability*** $\frac{L}{b} \leq 3$	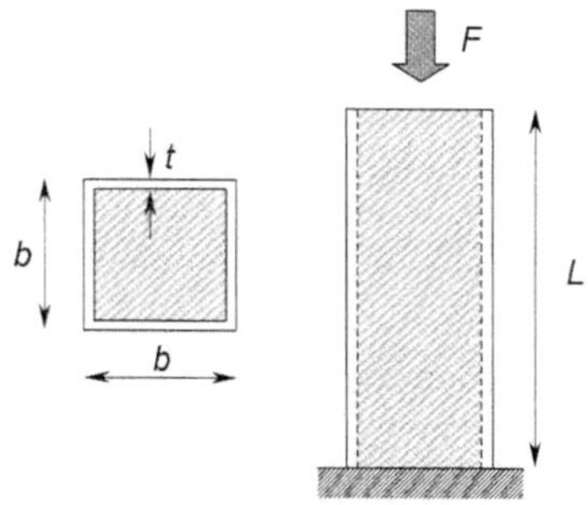
Material restrictions ***Column wall:*** $\bar{\sigma}_0^{\min} < \bar{\sigma}_0 < \bar{\sigma}_0^{\max}$ $\bar{\sigma}_0 = \frac{1}{2}(\sigma_{0.2} + \sigma_u)$ ***Foam filler:*** $\rho_f^{min} < \rho_f < \rho_f^{max}$	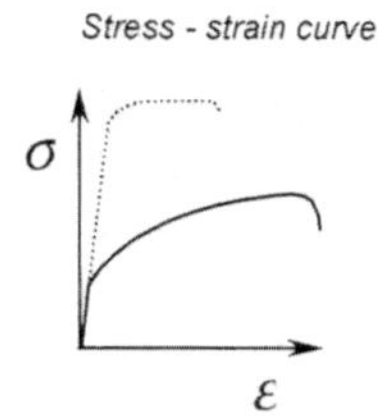 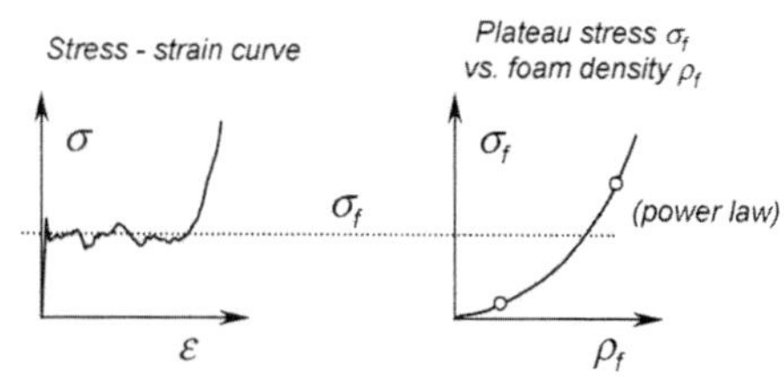
Target restrictions $F_{\max} < F_{\max}^*$ $E > E^*$	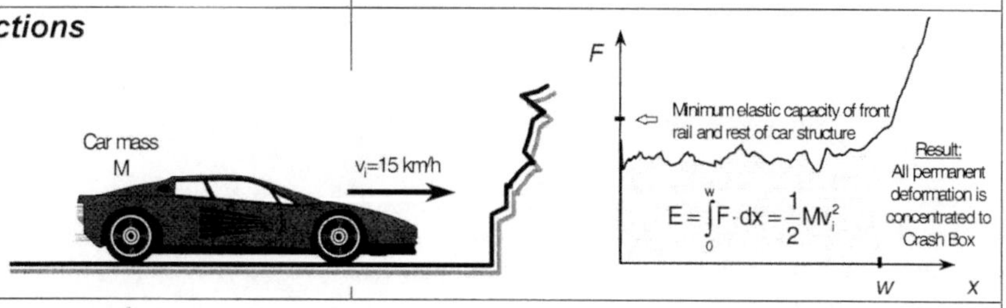
Optimization parameter M (mass of crash box)	

Table 3: Design restrictions, optimizing a foam-filled column for energy absorption

A first approach in gaining understanding of the optimization problem is to keep some variables fixed and carry out a graphical interpretation of the reduced problem. In this section information on the effective stroke length of the absorbers will be neglected, and an optimization carried out on "cross-section level" will be performed. Moreover, the outer dimension (width/diameter) of the columns will be fixed.

Figure 66 compares the mass specific energy absorption *E/M* (average force divided by mass per unit of length) of square and circular extrusions as function of the foam-filler density. It is assumed that the outer dimensions (*b* and *d*) and the extrusion material strength $\bar{\sigma}_0$ are kept constant during the comparison. In addition, both cross sections have the same mass per unit of length ($M_{circ} = M_{square}$) and this value is taken to be independent upon the density of the foam filler. Hence, for a given foam density, the comparison is carried out under the assumption that the solid cross-section area A and the cross section of the foam filler A_f are equal for both square and circular components. Assuming thin-walled sections, it can be shown that this is achieved when both the diameter and wall thickness of the circular extrusion is increased by a factor of $2/\sqrt{\pi}$ compared to the width and wall thickness of the corresponding square cross section. Equalling A and A_f for circular and square sections implies that the comparison should be carried out at equal solidity ratios $\phi = A/A_f$. Now, since the foam density is increased while keeping the outer dimensions fixed, the wall thickness (equivalent to A or ϕ) has to be reduced when moving from left to right in Figure 66. Hence, the left-hand side of the graph compares non-filled circular vs. square extrusions, whereas the right-hand side compares a circular vs. a square foam block.

Two important observations can be made from Figure 66. First,

> *Circular extrusions are preferable to square extrusions when regarding the mass specific energy absorption. However, this difference decreases as the density of the foam-filler increases.*

Moreover it is seen that both curves have a minimum point, hence:

> *Assuming a square or circular column with constant outer width/diameter and constant material properties, all combinations of foam-filler density and extrusion wall thickness giving a constant mass per unit of length, will result in lower energy absorption per unit of mass than that exhibited by foam or extrusion alone.*

Whether it is the foam alone or the non-filled extrusion that exhibits the highest energy, absorption per unit of mass is dependent upon the foam strength compared to the material strength of the column. An important consequence of this observation is that an effort for improving an existing energy absorbing device by a reduction of wall thickness and the introduction of foam filler are meaningless since lower mass specific energy absorption is likely to be the result (unless the new wall thickness is very small and the foam filler has a high specific energy absorption).

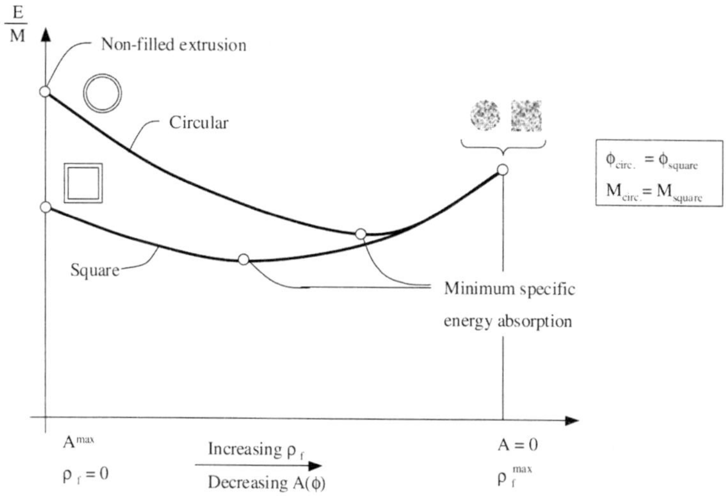

Figure 66: Comparison of cross sections as functions of foam filler density

3.4.3.6. Foam filling of an extrusion with different outer dimensions

The above comparison was carried out at constant outer cross-section dimensions. Figure 67 explores what happens if two cross sections with different outer dimensions are compared. Again, the mass per unit of length M and the extrusion material strength $\bar{\sigma}_0$ are equal and constant for both the large (A) and small (B) cross section. Also, the comparisons are still on "cross-section level" operating with energy absorbed per unit length. The graphs show average and maximum force as function of foam-filler density ρ_f. Assume that the large non-filled cross section (A) is a traditional solution for an energy absorber. The characteristics are given by a maximum allowable force level as well as a minimum requirement to the average force (energy absorption).

Again, as explained above in connection with Figure 66, the introduction of foam filler to this cross section leads to lower average force and thus lower energy absorption per unit of mass. This is seen by the decrease in average force level as function of foam-filler density. On the other hand, the smaller, non-filled cross section (B) has much higher force levels than cross-section type A. This is due the higher value of the wall thickness, which is necessary if the two cross sections are to have the same mass. However, it is seen that the force levels also decrease here when reducing the wall thickness and simultaneously introducing foam filler. Hence, at a given foam-filler density, the maximum force level of cross section B will correspond to the maximum force level of the traditionally designed non-filled extrusion of type A. Now, a comparison of the average force levels shows that the smaller cross section will have a higher average force in the valid density range than the non-filled cross section A. Since components A and B have the same mass, the following statement can be made

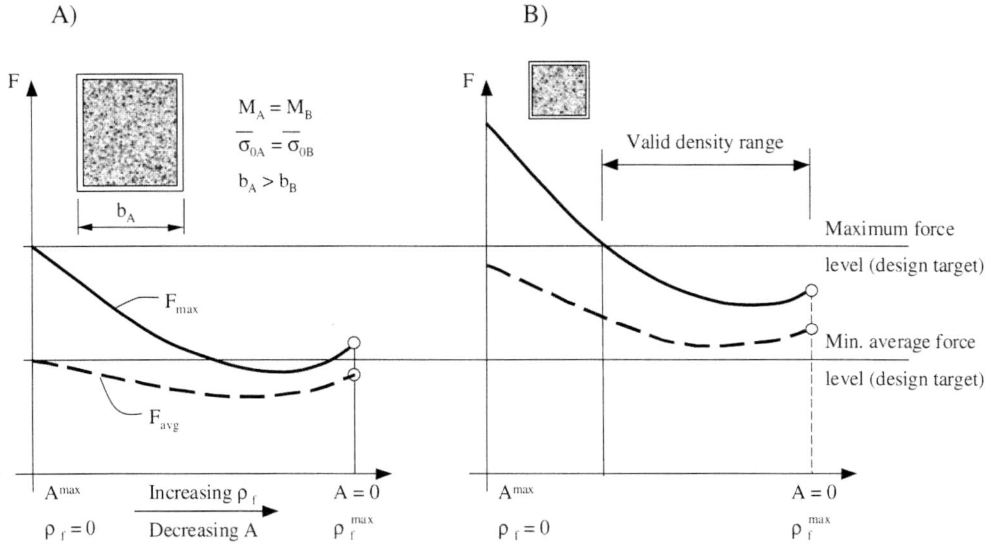

Figure 67: Comparison of crash boxes with different outer dimensions (width)

In order to obtain similar or better energy absorption performance per unit of weight, the outer geometry of the foam-filled extrusions as well as the wall thickness have to be reduced compared to traditionally designed non-filled extrusions having the same material characteristics.

A reason for this has been given previously in Eq. (3.91) which shows that the mass specific energy absorption increases with the foam density. This leads to high foam densities and small cross sections with low values of the wall thickness being preferable. However, for practical design cases this observation must also be regarded with respect to 1) global buckling (reduction of cross section increases column slenderness), 2) the maximum available foam density from the manufacturer and 3) the minimum practical wall thickness of the extrusion. It is important to keep in mind that the above statement is only valid if a design requirement is introduced to the maximum allowable force levels.

3.4.3.7. The complete design case, numerical results

The general problem given by the set of Eqs. (3.94), (3.95), (3.96), (3.97) and (3.98) and corresponding problem constraints of Table 3, can easily be solved by applying non-linear optimization codes found as standard options in most commercially available worksheets and mathematically oriented software. Table 4 shows a design case where the target restrictions have been a maximum force level of 70 kN coupled with a minimum energy absorption of 7 kJ. The other restrictions that exist together with the resulting optimum material and geometry parameters constituting the final solution are also shown. In addition, the optimum non-filled column together with the optimum foam-alone solution that satisfies the restrictions are given for comparison. The foam-alone solution has only been given for reference since a surrounding column should appear to sustain the structural integrity during normal load conditions. The material properties of the foam have been represented by the power law, Eq. (3.90), using C_p = 540 and n = 2.2.

In order to obtain a fair amount of design conservatism regarding the behaviour of the foam filled columns, values of the crush force efficiency above 85 % have not been allowed to occur.

Several characteristics of the general solution, exemplified in Table 4, are important to be aware of when designing these components. The above section focused on the influence of the column material strength on the design of least mass absorbers. As seen from Table 4, the highest column strength (250 MPa) is selected for the non-filled column, whereas the optimum strength for the foam-filled column is 239 MPa. Obviously, other parameters of the constrained problem have put restrictions on the maximum column strength for the foam-filled component. In this case, the minimum value of the column wall t_{min} restricts the column strength to a lower value than maximum possible. Thus, since a high column material strength can be selected without increase of column mass, the optimum solution always tends to select the highest column strength coupled with the lowest column wall thickness possible. Deviations from selecting the maximum column strength are due to geometrical restrictions on the minimum value of the wall thickness.

Regarding the foam alone, the maximum foam density has been selected as the optimum one. In the general optimization problem, this is no surprise as the power law, Eq. (3.91) when n>1, implies that energy absorbed per unit of mass increases as the foam density increases. Hence for optimization, the maximum possible foam density is selected, and the cross section adjusted correspondingly to keep the maximum force below the maximum allowable level. This is also reflected in the optimum solution for the foam filled column, a high foam density (0.41 g/cm^3) is chosen. However, the minimum requirement on the column wall thickness constrains the maximum allowable foam density (0.50 g/cm^3) from being selected. The other factor, which may limit the maximum possible foam density, is global instability. If a small outer cross-sectional width is chosen, the necessary effective crush distance needed in order to fully absorb all energy may be limited by a global stability criterion.

As pointed out above, the cross section of foam-filled columns has to be reduced in order to compare with the performance of non-filled columns. This is also seen from the general solution of Table 4 where the cross-section reduction is from 91 mm to 62 mm. However, 62 mm is higher than the minimum cross-section width specified to be 50 mm. The reason for not selecting this value is due to the minimum restriction of column wall thickness coupled with maximum restriction on foam density.

The weight savings achieved by introducing foam to the column is 8 % for this specific case. However, the reductions of component length and volume are 32 % and 68 % respectively. The latter are significant reductions important in today's concepts of small, compact cars. The first reason for these improvements can be found from the efficiency numbers in Table 4. As seen, the stroke efficiency S_E is somewhat reduced both for foam-filled columns and foam alone compared to the non-filled column, but this negative effect is balanced by a large increase of the crush force efficiency, A_E. As a result, the total efficiency T_E is increased compared to the non-filled column. The second reason for the weight savings is given by the value of the mass specific energy absorption of the foam. Foam with better strength-to-weight ratios than that assumed in this case will increase the weight savings potential of the final components even more.

Restrictions			
Type	*Variable*	*Min. value*	*Max. value*
	F^*_{max}		*70 kN*
Target-	E^*	*7 kJ*	
	b	*50 mm*	*100 mm*
Geometrical-	t	*1 mm*	*3 mm*
	L		*400 mm*
Global stability	L/b		*3*
	$\bar{\sigma}_0$	*100 MPa*	*250 MPa*
Material-	ρ_f	*0.10 kg/dm^3*	*0.50 kg/dm^3*
Optimized results			
Optimum values	*Non-filled*	*Foam filled*	*Foam alone*
b *(mm)*	*91*	*62*	*67*
t *(mm)*	*1.6*	*1.0*	*0*
L *(mm)*	*274*	*186*	*163*
$\bar{\sigma}_0$ *(MPa)*	*250*	*239*	*0*
ρ_f *(MPa)*	*0*	*0.41*	*0.50*
Mass (kg)	*0.438*	*0.403*	*0.365*
A_E *(%)*	*48*	*85*	*85*
S_E *(%)*	*76*	*63*	*72*
T_E *(%)*	*37*	*54*	*61*

Table 4: Example of problem restrictions and solution of the general optimization problem

It is evident from the above discussion that, within a geometric and target restricted boundary value problem, there is a foam quality that breaks even both in performance and mass with a given material quality of the column walls. Knowledge of this foam quality is of great importance for cost analyses in connection with the energy absorber system.

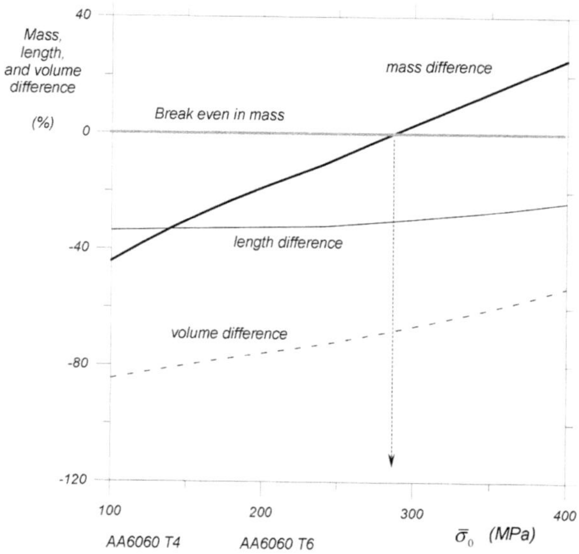

Figure 68: Mass-, length- and volume difference. Foam-filled column compared to non-filled column

Figure 68 shows the mass-, length- and volume reduction of the components as function of characteristic column material strength $\bar{\sigma}_0$. The problem constraints given in Table 4 have been assumed to be valid. As seen, the difference in mass between foam-filled and non-filled columns of equal performance decrease as the column strength increases. At a characteristic column stress of approximately 300 MPa there is a break-even point in performance. Above this stress level, the non-filled column is superior to the foam-filled ones. However, in the range below the break-even point, significant weight savings can be achieved by introducing foam filler to the columns. As mentioned above, Figure 68 also shows that large savings in length and volume can be achieved by utilizing foam filler.

Figure 69 shows the typical development in strength-to-weight ratio of aluminium foam produced by Hydro Aluminium during the last few years. Further improvements, as indicated in this figure, are already a reality. The corresponding effect of foam quality on the possible mass reduction of foam-filled columns is shown in Figure 70. As seen, improvements in foam quality make foam-filled columns preferable for higher column material strengths. For assumed future foam qualities, mass reductions will be possible for column material strengths above 400 MPa. (However, high strength aluminium alloys are produced on the cost of ductility. For axial crushing of tubes the occurrence of large strains will cause rupture if the ductility is not satisfactory. It may then be that the overall optimum is found for alloys with lower strengths.) Finally it must be noted that these conclusions are based upon the problem constraints of Table 4.

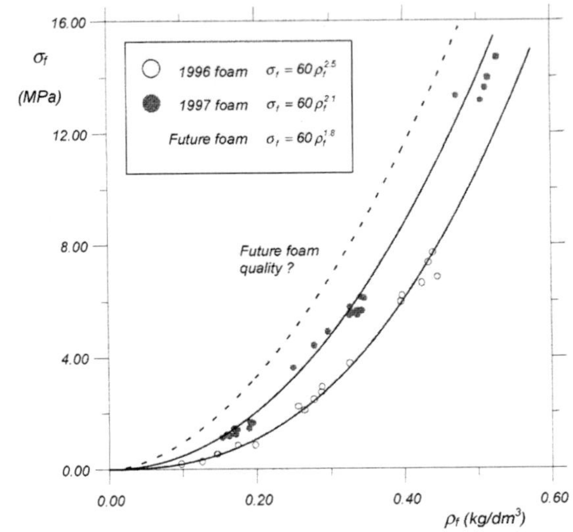

Figure 69: Developments in foam quality

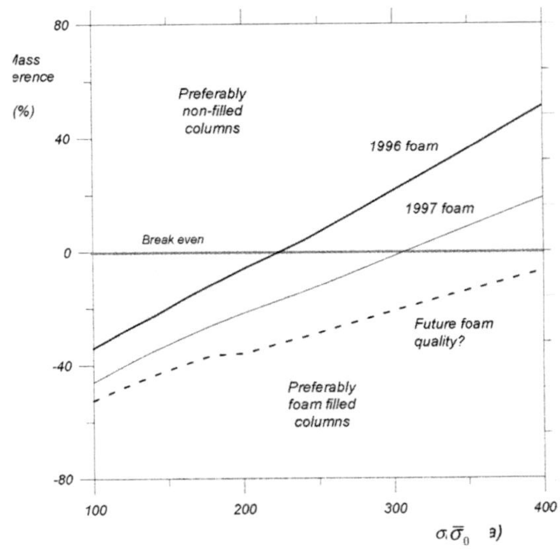

Figure 70: Effect of foam-filler quality on the mass difference between foam-filled vs. non-filled columns

Acknowledgements

The authors would like to thank Øystein Malm, Siv. ing. and Bernt Natvig, Siv. ing. who prepared the drawings and typed the draft of the manuscript.

References

Abramowicz, W. and Jones, N. (1986). Dynamic progressive buckling of circular and square tubes." *International Journal of Impact Engineering*. 4, 234-270.

Alexander, J.M. (1960). An approximate analysis of the collapse of thin cylindrical shells under axial loading. *Quarterly Journal of Mechanics and Applied Mathematics*, 13, 10-15.

Børvik, T. (2001). Ballistic Penetration and Perforation of Steel Plates. Dr.ing. thesis, Department of Structural Engineering, Norwegian University of Science and Technology, Trondheim.

Deshpande, V.S., and Fleck N.A. (2000): High strain rate compressive behaviour of aluminium alloy foams. *Int. J. Impact Engineering.*, Vol. 24 pp. 277-298

ENV 1999-1-1 CENITC 250ISC9 Eurocode 9. Design of aluminium structures. Part 1.1: General rules.

Hanssen, A.G. (2000). Structural crashworthiness of aluminium foam-based components. Doctoral thesis 2000:76, Department of Structural Engineering, Norwegian University of Science and Technology, Trondheim. ISBN 82-7984-102-4.

Hanssen, A.G., Langseth, M. and Hopperstad, O.S. (2000a). Static and dynamic crushing of square aluminium extrusions with aluminium foam filler. *Int. J. of Impact Eng.* 24 (2000) 347-383.

Hanssen, A.G., Langseth, M. and Hopperstad, O.S. (2000b). Static and dynamic crushing of circular aluminium extrusions with aluminium foam filler. *Int. J. Impact Eng.* 24, 475-507.

Hill, R. (1952). On Discontinuous Plastic States, with Special Reference to Localised Necking in Thin Sheets. *J. Mech. Phys. Solids*, 1, pp. 19-30.

Hill, R. (1989). The mathematical theory of plasticity. *Clarendon Press. Oxford.*

Hosford and Caddell (1993). Metal forming, Mechanics and Metallurgy. *PTR Prentice Hall, Inc.*

Iyengar, N. G. R. (1988). Structural Stability of Columns and Plates. *Wiley, New York.*

Jones, N. (1989). Structural Impact. *Cambridge University Press, Cambridge.*

Lademo, O. (1999). Engineering Models of Elastoplasticity and Fracture for Aluminium Alloys. Dr.ing. thesis, Department of Structural Engineering, The Norwegian University of Science and Technology, Trondheim.

Langseth, M. and Hopperstad, O.S. (1996). Static and dynamic axial crushing of square thin walled aluminium extrusions. *Int. J. Impact Engineering,* Vol. 18, Nos 7-8, pp 949-968.

Langseth, M. and Hopperstad, O.S. (1997). Local buckling of square thin-walled aluminium extrusions. *Thin-walled structures*: Vol. 27, No. 1, pp. 117-126.

Lindholm, K.S., Bessey, R.L. and Smith, G.V. (1971). Effect of Strain Rate on Yield Strength, Tensile Strength and Elongation of Three Aluminium Alloys. *Journal of Materials*, Vol. 6, No. 1.

Marciniak, Z., Wreczinsky, K., Pokora, T., (1973). Influence of the Plastic Properties of a Material on the Forming Limit Diagram for Sheet Metals in Tension. *Int. J. Mech. Sci.*, 15, pp. 789-805.

Matusiak, M. (2000). Strength and Ductility of Welded Structures in Aluminum Alloys. Dr. ing. thesis, Department of Structural Engineering, The Norwegian University of Science and Technology, Trondheim.

Opheim, B.S. (1996). Bending of Thin Walled Aluminium Extrusions. Dr.ing. thesis, Department of Structural Engineering, The Norwegian University of Science and Technology, Trondheim.

Swift. H.W. (1952). Plastic Instability Under Plane Stress. *Journal of the Mechanics and Physics of Solids*, 1, pp. 1-18.

Tan, P.J. and Reid, S.R. (2000). Dynamic characterisation of "Hydro" Aluminium foams subjected to high strain rate compression. UMIST report ME/AM10.00/NTNU2

Welo, T (1990). Inelastic deformation capacity of flexurally-loaded aluminium alloy structures. Dr.ing. thesis, Department of Structural Engineering, The Norwegian University of Science and Technology, Trondheim.

Wierzbicki, T. (1983). Optimum design of integrated front panel against crash. Report for Ford Motor Company, Vehicle Dept. 15 July.

Wierzbicki, T. and Abramowicz, W. (1983). On the crushing mechanics of thin-walled structures. *Journal of Applied Mechanics,* 50, 727-34.